Energy Unit Conversion Factors

		J	kWh	Btu
1 Joule (J)	equals	1	2.78×10^{-7}	9.49×10^{-4}
1 kilowatt hour (kWh)	equals	3.60×10^{6}	1	3413
1 calorie (cal)	equals	4.184	1.16×10^{-6}	3.97×10^{-3}
1 British thermal unit (Btu)	equals	1055	2.93×10^{-4}	1
1 foot-pound (ft·lb)	equals	1.36	3.78×10^{-7}	1.29×10^{-3}
1 electron-volt (eV)	equals	1.60×10^{-19}	4.45×10^{-26}	1.52×10^{-22}

Energy Equivalents

	J	kWh	Btu
Crude petroleum (42 gallon barrel)	6.12×10^{9}	1700	5.80×10^{6}
Bituminous coal (1 ton[a])	2.81×10^{10}	7800	2.66×10^{7}
Natural gas (1000 cubic feet[b])	1.09×10^{9}	303	1.035×10^{6}
Gasoline (1 gallon[c])	1.32×10^{8}	36.6	1.25×10^{5}
Uranium-235 (1 gram)	8.28×10^{10}	2.30×10^{4}	7.84×10^{7}
Deuterium (1 gram)	2.38×10^{11}	6.60×10^{4}	2.25×10^{8}

[a]1 ton = 2000 lb = 0.907 tonne.
[b]At STP.
[c]The U.S. gallon is used in this text. The Imperial gallon used in Canada and Great Britain equals 1.201 U.S. gallons

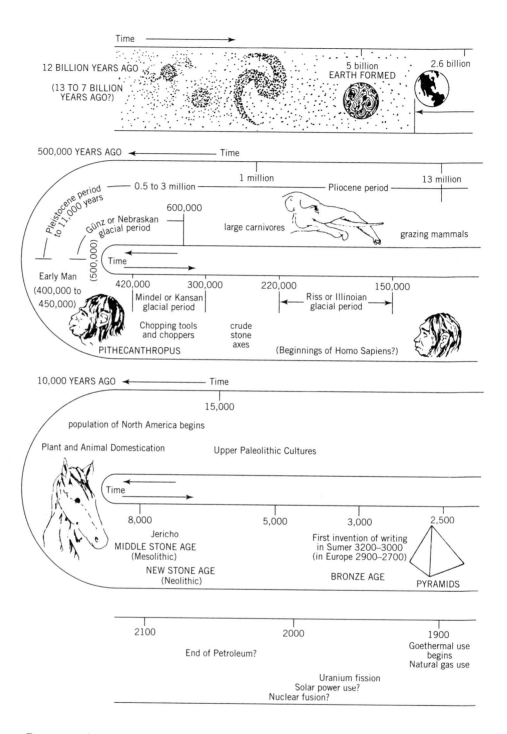

Representation of the vast difference in time between our rapid use of fossil fuels and their early formation. Adapted from Wilson and Jones, *Energy, Ecology, and the Environment*, copyright © 1974, Academic Press, New York. Reprinted by permission.

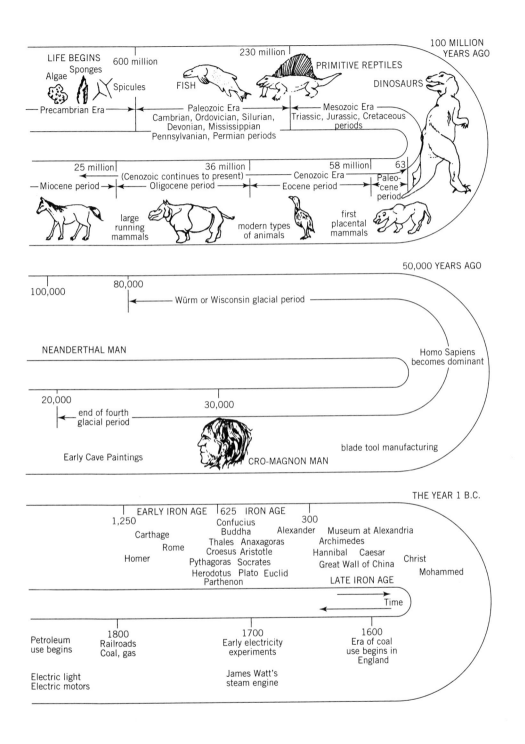

Energy and the Environment

Energy and the Environment

Robert A. Ristinen
University of Colorado–Boulder

Jack J. Kraushaar
University of Colorado–Boulder

John Wiley & Sons, Inc.

Acquisitions Editor	*Stuart Johnson*
Senior Production Editor	*Elizabeth Swain*
Marketing Manager	*Amanda Wygal*
Senior Photo Editor	*Lisa Gee*
Senior Designer	*Madelyn Lesure*
Senior Illustration Editor	*Anna Melhorn*
Cover Photo	*NASA/Photo Researchers, Inc.*

This book was typeset in 10/12 Times Ten by Matrix and printed and bound by Malloy. The cover was printed by Phoenix Color Corp.

This book is printed on acid-free paper. ∞

Copyright © 2006 by John Wiley & Sons, Inc. All rights reserved.

No part of this publication may be reproduced, stored in a retrieval system or transmitted in any form or by any means electronic, mechanical, photocopying, recording, scanning or otherwise, except as permitted under Sections 107 or 108 of the 1976 United States Copyright Act, without either the prior written permission of the Publisher, or authorization through payment of the appropriate per-copy fee to the Copyright Clearance Center, Inc. 222 Rosewood Drive, Danvers, MA 01923, website *www.copyright.com*. Requests to the Publisher for permission should be addressed to the Permissions Department, John Wiley & Sons, Inc., 111 River Street, Hoboken, NJ 07030-5774, (201)748-6011, fax (201) 748-6008, website *http://www.wiley.com/go/permissions*.

To order books or for customer service please, call 1-800-CALL WILEY (225-5945).

Library of Congress Cataloging in Publication Data
Ristinen, Robert A.
   Energy and the environment / Robert A. Ristinen, Jack J. Kraushaar.—2nd ed.
    p. cm.
   ISBN-13: 978-0-471-73989-0 (pbk.)
   ISBN-10: 0-471-73989-8 (pbk.)
   1. Power resources—United States. 2. Pollution—United States. I. Kraushaar, Jack J.
  II. Title.
TJ163.25.U6R57 2006
333.79—dc22
                                      2005055150

Printed in the United States of America
10 9

CONTENTS

CHAPTER 1 *Energy Fundamentals, Energy Use in an Industrial Society* **1**

    1.1   Introduction *1*
    1.2   Why Do We Use So Much Energy? *4*
    1.3   Energy Basics *7*
        1.3.1   General *7*
        1.3.2   Forms of Energy *9*
        1.3.3   Power *11*
    1.4   Units of Energy *12*
        1.4.1   The Joule *13*
        1.4.2   The British Thermal Unit *13*
        1.4.3   The Calorie *13*
        1.4.4   The Foot-Pound *13*
        1.4.5   The Electron-Volt *14*
    1.5   Scientific Notation *14*
    1.6   Energy Consumption in the United States *16*
    1.7   The Principle of Energy Conservation *22*
    1.8   Transformation of Energy from One Form to Another *23*
    1.9   Renewable and Nonrenewable Energy Sources *25*
        1.9.1   Nonrenewable Energy Sources *25*
        1.9.2   Renewable Energy Sources *25*
Key Terms *27*
Suggested Reading and References *27*
Questions and Problems *28*
Multiple Choice Questions *29*

CHAPTER 2 *The Fossil Fuels* **31**

    2.1   Introduction *31*
    2.2   Petroleum *32*
    2.3   History of the Production of Petroleum in the United States *33*
    2.4   Petroleum Resources of the United States *35*
    2.5   World Production of Petroleum *39*
    2.6   The Cost of Gasoline in the United States *39*
    2.7   Petroleum Refining *40*
    2.8   Natural Gas *43*
    2.9   The History of Use of Natural Gas *45*
    2.10 The Natural Gas Resource Base in the United States *47*
    2.11 The Natural Gas Resource Base for the World *48*
    2.12 The Formation of Coal *50*

2.13 Coal Resources and Consumption 50
2.14 Shale Oil 54
2.15 Tar Sands 56
2.16 Summary 57
Key Terms 58
Suggested Reading and References 58
Questions and Problems 59
Multiple Choice Questions 60

CHAPTER 3 *Heat Engines* 63

3.1 The Mechanical Equivalent of Heat 63
3.2 The Energy Content of Fuels 64
3.3 The Thermodynamics of Heat Engines 65
3.4 Generation of Electricity 67
3.5 Electric Power Transmission 71
3.6 Practical Heat Engines 72
   3.6.1 Steam Engines 73
   3.6.2 Gasoline Engines 74
   3.6.3 Diesel Engines 76
   3.6.4 Gas Turbines 78
3.7 Heat Pumps 79
3.8 Cogeneration 82
Key Terms 84
Suggested Reading and References 85
Questions and Problems 85
Multiple Choice Questions 86

CHAPTER 4 *Renewable Energy Sources I: Solar Energy* 90

4.1 Introduction 90
4.2 Energy from the Sun 92
4.3 A Flat-Plate Collector System 97
4.4 Passive Solar 101
4.5 Solar Thermal Electric Power Generation 105
   4.5.1 Power Towers 108
   4.5.2 Parabolic Dishes and Troughs 110
4.6 The Direct Conversion of Solar Energy to Electrical Energy 111
4.7 Solar Cooling 119
Key Terms 120
Suggested Reading and References 120
Questions and Problems 121
Multiple Choice Questions 121

CHAPTER 5 *Renewable Energy Sources II: Alternatives* 124

5.1 Introduction 124
5.2 Hydropower 125

5.3 Wind Power 132
5.4 Ocean Thermal Energy Conversion 138
5.5 Biomass as an Energy Feedstock 145
5.6 Biomass: Municipal Solid Waste 151
5.7 Biomass-Derived Liquid and Gaseous Fuels 153
5.8 Geothermal Energy 157
5.9 Tidal Energy 162
5.10 Wave Energy 164
5.11 Summary 165
Key Terms 166
Suggested Reading and References 166
Questions and Problems 167
Multiple Choice Questions 168

CHAPTER 6 *The Promise and Problems of Nuclear Energy* 171

6.1 Introduction 171
6.2 A Short History of Nuclear Energy 172
6.3 Radioactivity 175
6.4 Nuclear Reactors 177
6.5 The Boiling Water Reactor 179
6.6 Fuel Cycle 181
6.7 Uranium Resources 183
6.8 Environmental and Safety Aspects of Nuclear Energy 185
6.9 The Chernobyl Disaster 189
6.10 Nuclear Weapons 190
6.11 The Storage of High-Level Radioactive Waste 192
6.12 The Cost of Nuclear Power 195
6.13 Nuclear Fusion as An Energy Source 196
6.14 Controlled Thermonuclear Reactions 198
6.15 A Fusion Reactor 200
Key Terms 204
Suggested Reading and References 204
Questions and Problems 205
Multiple Choice Questions 207

CHAPTER 7 *Energy Conservation* 210

7.1 A Penny Saved Is a Penny Earned 210
7.2 Space Heating 212
    7.2.1 Thermal Insulation 213
    7.2.2 Air Infiltration 219
    7.2.3 Furnaces, Stoves, and Fireplaces 221
    7.2.4 Solar and Other Sources of Heat Energy 223
    7.2.5 Standards for Home Heating 225

                7.3   *Water Heaters, Home Appliances, and Lighting*   226
                        7.3.1   Water Heating   226
                        7.3.2   Appliances   227
                        7.3.3   Lighting   231
                        7.3.4   The Energy-Conserving House   233
                7.4   *Energy Conservation in Industry and Agriculture*   235
                        7.4.1   Housekeeping   235
                        7.4.2   Waste Heat Recovery and Cogeneration   235
                        7.4.3   Process Changes   235
                        7.4.4   Recycling   236
                        7.4.5   New Developments   237
                        7.4.6   Help from Public Utilities   238
        Key Terms   239
        Suggested Reading and References   239
        Questions and Problems   240
        Multiple Choice Questions   241

CHAPTER 8   *Transportation*   244

                8.1   *Introduction*   244
                8.2   *Power and Energy Requirements*   247
                8.3   *Electric Batteries, Flywheels, Hybrids, Hydrogen, Alcohol*   254
                        8.3.1   Electric Vehicles   256
                        8.3.2   Flywheel-powered Vehicles   258
                        8.3.3   Hybrid Vehicles   262
                        8.3.4   Hydrogen, Fuel Cells   264
                        8.3.5   Alcohol as a Transportation Fuel   268
                8.4   *Traffic Safety*   271
                        8.4.1   The Record   271
                        8.4.2   Collisions   274
                        8.4.3   Nontechnical Traffic Safety Measures   277
                8.5   *Mass Transportation*   279
        Key Terms   282
        Suggested Reading and References   282
        Questions and Problems   283
        Multiple Choice Questions   283

CHAPTER 9   *Air Pollution*   287

                9.1   *Spaceship Earth*   287
                9.2   *The Earth's Atmosphere*   288
                9.3   *Thermal Inversions*   290
                9.4   *Carbon Monoxide*   294
                9.5   *The Oxides of Nitrogen*   300
                9.6   *Hydrocarbon Emissions and Photochemical Smog*   301

9.7 Reduction of Vehicle Emissions 304
9.8 Sulfur Dioxide in the Atmosphere 307
9.9 Particulates as Pollutants 310
9.10 Acid Rain 315
9.11 Summary of Air Pollution in 2005 319
Key Terms 321
Suggested Reading and References 321
Questions and Problems 322
Multiple Choice Questions 323

CHAPTER 10 *Global Effects* 326

10.1 Introduction 326
10.2 Ozone Depletion in the Stratosphere 327
10.3 The Greenhouse Effect and World Climate Changes 332
Key Terms 345
Suggested Reading and References 345
Questions and Problems 346
Multiple Choice Questions 347

APPENDIX 349

A.1 Linear Plots, Semilogarithmic Plots, and Exponential Growth 349
A.2 Fahrenheit, Celsius, and Kelvin Temperature Scales 353

ANSWERS TO SELECTED END-OF-CHAPTER PROBLEMS 355

INDEX 357

PREFACE

In the year 1973 the term "energy" became common in households throughout the United States. At that time, an energy crisis suddenly fell upon the country and for some time it was not unusual for motorists to spend hours waiting in line to obtain gasoline at a filling station. Customers were sometimes limited to a 5-gallon purchase. The speed limit on all highways throughout the nation was reduced to 55 miles per hour, and it stayed that way for 15 years. Decorative lighting was markedly reduced during the holiday season as an energy-saving measure.

The experience of 1973 gave immediate significance to energy for a wide audience. Much has happened since that energy crisis. Gasoline is now widely available, as is electrical energy. However the underlying problems remain.

The problems of energy are complex and go far beyond questions of the immediate availability of motor fuel. These issues affect the entire world and the problems are becoming more severe with the passage of time as our fossil fuel reserves diminish and the citizens of developing countries aspire to share more fully in the use of the world's energy resources. Beyond the supply side of the energy problem, we see the effects of emissions on local and global scales.

The topics of energy and the environment are obviously crucial to all of us, and effective policies at all levels of government depend on an informed citizenry. To address this need, courses dealing with the subject are being offered at colleges and universities in the United States and elsewhere. *Energy and the Environment* was created from experience the authors have had in teaching such courses starting more than 30 years ago at the University of Colorado in Boulder. An earlier text first published more than 20 years ago was intended for a two-semester course; it was more quantitative and technical than the present text, published in its first edition 7 years ago.

The closing date for compilation of information and statistics for the first edition of *Energy and the Environment* was not long after the middle of the 1990s. The field is changing rapidly, and the step from one decade to the next can be a long one. We continually see new developments in every aspect, from fossil fuels to alternatives, from hydrogen to hybrids. Our major environmental problem, global warming, has become an issue of broad concern, and there are now serious efforts to seek means of mitigation. In this second edition we have included recent statistical information on fossil fuel reserves and consumption as well as new data on stratospheric ozone and global warming.

Energy and the Environment deals with the core subjects of energy and the environment. With respect to energy, we have tried to cover the basic concepts, resources, applications, and problems of current interest. With respect to the environment, we have included most of the major concerns; unfortunately, because of space limitations, we have had to omit some areas such as water resources and pollution. When the problems covered in this book are examined together, it is seen that many, but not all, of our environmental problems have their origin in our quest for abundant and inexpensive energy.

One major change that has come about between our preparing the first and second editions of this text is the emergence of the internet as a primary information source. The many internet sites now addressing this field are often broad-ranging and current. With cautious judgment on the part of the reader these sites can provide abundant, authoritative, and up-to-date information. We have made frequent use of several websites, notably those of the United States Energy Information Administration. The World Energy Council, The Oak Ridge National Laboratory, and numerous others, which we have tried to cite when information from those sites is used. In addition, at the end of each chapter we have added new references to some that have been carried over from the first edition.

Energy and the Environment is intended for students having little or no background in science or mathematics. Some elementary calculations are included in the subject matter, but these calculations do not involve mathematics beyond introductory algebra, and this is introduced along with the material under discussion.

In this updated second edition we recognize that many of the troubling issues that have been so apparent for more than 30 years continue on their course. We continue to face declining domestic production of petroleum and natural gas, increasing dependence on imports, the approach to the peak of world petroleum production, a growing threat of nuclear weapons proliferation, no politically acceptable means of nuclear waste disposal, changing global climate, burgeoning human population, and so forth. However, encouraging developments are appearing in the areas of renewable energy, energy conservation, and energy-efficient transportation as well as in several other areas.

To extend a comment put forth by Aldo Leopold many years ago, it is our hope that this text will help to bring its readers beyond thinking that "heat comes from the furnace, food comes from the store, water comes from the faucet, gasoline comes from the filling station, truth comes from the experts".

Acknowledgements

Many people were of great help as we sought to obtain and organize material for the first edition of this text. In particular we thank Dr. Robert Cohen for providing material on ocean energy, especially the subject of OTEC. The Colorado State Department of Health provided valuable information on air quality and pollution control measures. The National Center for Atmospheric Research made available useful information on global warming. In addition, Stacy Davis of Oak Ridge National Laboratory was helpful in providing data on transportation energy matters. Thomas Boden of Oak Ridge National Laboratory helped us obtain data on atmospheric carbon dioxide. We are grateful for this assistance.

Throughout our years of teaching about energy and the environment, and in the preparation of this text, our colleague, Professor A. A. Bartlett of the Department of Physics at the University of Colorado in Boulder, has been a constant source of information and perspective on this subject. We thank him for his continuing support and encouragement.

Several reviewers evaluated the initial proposal for the first edition of this text. Those who helped at this early stage were: Professor Gregory Greer of Normandale Community College, Professor Jack A. Kaeck of Chicago State University, Professor John R. Kalafut of the University of Scranton, Professor V. Paul Kenney of the University of Notre Dame, and Professor Roger E. Mills of the University of Louisville. The manuscript for the first edition was reviewed at a late, almost final, stage of its development by Professor Phillip E. Best of the University of Connecticut-Storrs, Professor Ljubisa R. Radovic of Pennsylvania State University, and Professor Don D. Reeder of the University of Wisconsin-Madison.

The assistance provided on the first edition by the staff at Wiley was invaluable. In particular, our interactions with Stuart Johnson, Cynthia Rhoads, Catherine Donovan, Kim Katchatourian, Dawn Stanley, and Sandra Dumas were important.

In addition to the many people who provided help for the first edition, our preparation of this second edition has been assisted greatly by many others. In particular, we wish to thank Dr. J. Herring of the National Center for Atmospheric Research and Dr. D. Murphy of the National Oceanic and Atmospheric Administration for technical information on global warming. We also thank Professor Kim Griest of the University of California, San Diego, for his helpful comments on several areas, Dr. Matthew Kohler for going over the entire text, John Katers, of the the University of Wisconsin-Green Bay, who reviewed the accuracy of the manuscript, and Richard Monson of the University of Colorado Facilities Management Department who provided information on the CU cogeneration system. The staff at Wiley, including Stuart Johnson, Krista Jarmas, Aly Rentrop, Elizabeth Swain, Lisa Gee, Anna Melhorn, and copyeditor Connie Parks have made our task go as well as reasonably possible.

Robert A. Ristinen
Jack J. Kraushaar

CHAPTER 1

Energy Fundamentals, Energy Use in an Industrial Society

(*Source*: Media Bakery)

1.1 Introduction

Energy enters our everyday lives in many different ways. The energy in the food we eat maintains our body temperature and lets us walk, talk, lift things, and toss frisbees. The use of energy in food has been essential for the existence of all humankind and animals throughout our evolution on this planet. In some developing countries the supplying of food for energy and nutrition is a difficult task that requires most of the waking hours of the population. Food acquisition is just as essential in the more developed countries, but because of the greater mechanization of agricultural production, the effort of only a relatively small number of persons is devoted to obtaining food. This leaves most of the rest of us free to pursue other activities throughout our lives.

Energy in forms other than food is also essential for the functioning of a technical society. For example, in the United States, many times more energy in the form of engine fuel goes into the agricultural enterprise than is obtained in the useful food Calorie content of the food produced. Prodigious amounts of energy are also used to power automobiles, heat homes, manufacture products, generate electricity, and perform various other tasks. In order for our society to function in its present patterns, vast amounts of coal, natural gas, and oil are extracted from the earth and burned to provide this energy. To a lesser extent we also derive energy from hydroelectric plants, nuclear reactors, electric wind generators, and geothermal plants, and, of course, we all benefit enormously from the energy obtained directly from the sun.

The fossil fuels: coal, natural gas, and oil, supply about 86% of the energy used in the United States. These resources evolved hundreds of millions of years ago as plant and animal matter decomposed and was converted under conditions of high temperature and pressure under the earth's surface into the hydrocarbon compounds that we now call fossil fuels. Since the beginning of the machine age, industrial societies have become increasingly dependent on fossil fuels. A hundred and fifty years ago, the muscular effort of humans and animals played an important role in the American economy, and firewood supplied most of the heat energy. Now only a small fraction of our energy comes from firewood and we rely much less on the physical effort of people and animals. The process by which we have moved to our present dependence on coal, oil, and natural gas is illustrated in Figure 1.1, where the energy consumed in the United States each year from various sources is shown in terms of quadrillion British thermal units (QBtu) for the years 1850 to 2003. The definition of QBtu will be given in Section 1.6.

Should we be concerned that so much of our energy is now coming from fossil fuels? Here are two of many factors that should cause concern.

First, the fossil fuel resource is limited in amount. The fossil fuels were produced by solar energy hundreds of millions of years ago, and when they are gone, there will be no more. It is true that the fuels are still being formed, but at an entirely negligible rate compared to the rate at which we are consuming them. We first began consuming the fossil fuels at an appreciable rate only about 150 years ago. How long will they last? On a global scale we will still have some coal for a few centuries, but natural gas and oil will be in short supply in only a few decades. In the United States, the situation is worse than the global average because we are depleting our resources at a faster rate than in other fossil fuel-rich areas around the globe. Figure 1.2 shows the narrow blip of our fossil fuel use set against a time scale of thousands of years. As you consider the brief duration of this blip, remember that we have living trees thousands of years old, a much longer time than what will be spanned by the entire era of fossil fuel consumption. It is clear from this figure that we live in an extraordinary time in the many billion year history of the earth. The entire stock of fossil fuels available for our use has been held in storage under the earth's surface for more than a hundred million years, and now it is being completely exploited in only a few centuries.

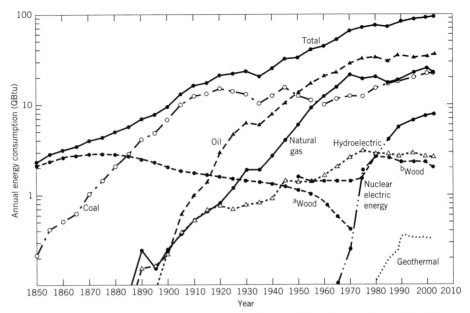

Figure 1.1 Various forms of energy consumed in the United States since 1850. This type of graph is called a semilogarithmic plot, an explanation of the scales is given in the Appendix. *Sources: Historical Statistics of the United States, Colonial Times to 1970*, U.S. Department of Commerce. Bureau of the Census, 1975; U.S. Energy Information Administration, *Annual Energy Review*, 2003. (a) The wood data set from 1850 to 1970 is from the first source. (b) The wood data set from 1950 to 2003 is from the second source; it includes wood, black liquor (a byproduct of the wood-based paper production process), and wood waste.

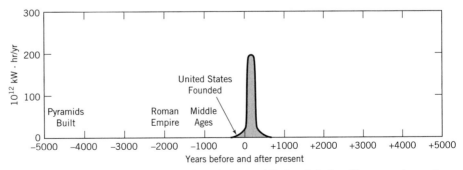

Figure 1.2 The complete exploitation of the world's fossil fuels will span only a relatively brief time in the 10,000 year period shown centered around the present. (*Source*: Reprinted with permission from M. K. Hubbert, *Resources and Man*, Washington, D.C., National Academy of Sciences, 1969. Historical events added.)

Second, unintended environmental consequences result from the extensive scale of our use of the fossil fuels for everything from heating our homes to powering our automobiles. When we burn coal, natural gas, or oil to obtain energy, gaseous compounds are formed and dumped into the atmosphere. This is causing problems we are just beginning to face. For many years it was felt that the emitted gases were not significant, given the vastness of the earth's atmosphere. But now with increasing world population, and industrialization, this is no longer true. The atmospheric pollution is producing health problems and even death, and it is now becoming recognized that carbon dioxide emissions are threatening to produce climate changes over the entire globe.

Can we find solutions to these problems of resource depletion and environmental pollution? Clearly the answers are not simple or the solutions would have been put into effect by now. The subject is complex and involves some understanding of topics such as patterns of resource depletion, the workings of heat engines, solar cells, wind generators, nuclear reactors, and a myriad of other specialized subjects. We do not have to become experts on each of these individual topics to be sufficiently well informed as voting citizens to influence a rational decision-making process. Our goal is to gain understanding concerning the essential points.

1.2 Why Do We Use So Much Energy?

A partial answer to this question is simple—we don't use our energy resources as efficiently as we could. The standard of living we enjoy in the United States could be maintained with an expenditure of far less energy per person than at present. This side of solving the energy problem will be explored later under the heading of Energy Conservation. There is a large discrepancy between the rate of energy use by a typical citizen of an industrialized society and the typical citizen of a developing country, and it is accompanied by a notable difference in what we perceive as the standard of living. This is illustrated in Figure 1.3, where we see the per capita Gross Domestic Product (GDP) and the per capita energy use for several countries of the world. Although not indicated on this figure, several developing countries have very low rankings by either measure, and they would be located within the small quarter-circle shown at the extreme lower left corner of the figure.

There is no essential relationship between GDP per capita and the standard of living, but both are often related to the use of energy. A citizen of a developing country might use the energy equivalent of less than one barrel of oil per year, compared to an annual energy equivalent of 20 to 60 barrels per capita for the most industrialized countries. The nonindustrialized countries derive a large fraction of their necessary energy from the muscular effort of people and

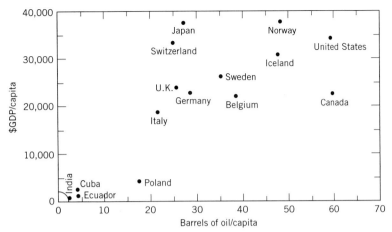

Figure 1.3 The Gross Domestic Product (GDP) per capita in U.S. dollars is compared to the total energy consumed per capita in equivalent barrels of oil for several countries. The small quarter-circle at the lower left corner is discussed in the text. (*Source: United Nations Statistical Yearbook*; data January 2003.)

animals. There is an interesting quotation from an early physics textbook written by J. Dorman Steele in 1878:

The combustion of a single pound of coal, supposing it to take place in a minute, is equivalent to the work of three hundred horses; and the force set free in the burning of 300 pounds of coal is equivalent to the work of an able-bodied man for a lifetime.

This observation, while a bit off the mark in exact technical detail, is essentially correct, and it sets the stage and justifies the enormous effort that has gone into our learning to exploit the fossil fuels—energy reserves held in waiting for hundreds of millions of years—until we have learned to use them with high efficiency to ease human labor. Whether we refer to tons of coal or barrels of oil, it is indeed the fossil fuels that have had the major effect. Without fossil fuels we surely would have made progress toward labor-saving technology based on waterpower, firewood, windpower, and perhaps even nuclear power, but we would not have gone nearly so far in developing the energy-intensive society in which we now live.

We may take the average power available to a person to be a measure of the productive output of a society. As seen in Figure 1.4, in the United States in 1850, about 0.38 horsepower per person was available, of which 0.26 horsepower was provided by work animals. We now have a few hundred times that from other sources. Most of the difference is due to our use of fossil fuels to make the wheels go around.

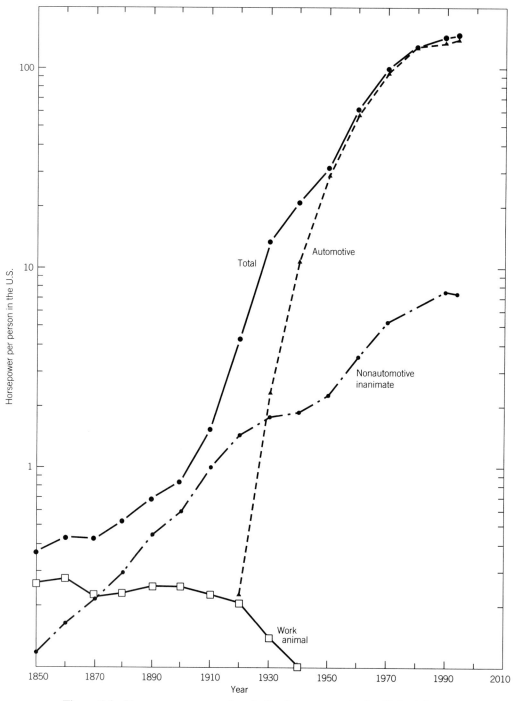

Figure 1.4 Horsepower per capita of all prime movers in the United States since 1850. Only a small fraction of this available horsepower is in use at any given time. (*Source: Historical Statistics of the United States, Colonial Times to 1970; Statistical Abstracts of the United States 2003*. Washington, D.C.: U.S. Department of Commerce, Bureau of the Census.)

Example 1.1

Using generally available information, estimate the dollar value of the equivalent amount of oil which we each use annually.

Solution

Given:

    58 barrel/(yr · person); see Figure 1.7.
    42 gallons/barrel; see Energy Equivalents chart inside front cover.
    Oil is approximately $1.25/gallon; estimated from reported crude oil prices.

$$58 \frac{\text{bbl}}{(\text{yr} \cdot \text{person})} \times 42 \frac{\text{gal}}{\text{bbl}} \times 1.25 \frac{\$}{\text{gal}} = 3045 \frac{\$}{(\text{yr} \cdot \text{person})}$$

Note that the units of bbl and gal cancel in this calculation. We can extend the answer to obtain the cost per day of this oil.

$$3045 \frac{\$}{(\text{yr} \cdot \text{person})} \times \frac{1 \text{ yr}}{365 \text{ day}} = \mathbf{8.34} \frac{\$}{(\textbf{day} \cdot \textbf{person})}$$

Here the units of yr have canceled.

1.3 Energy Basics

1.3.1 General

Our discussions of energy use and resources can proceed effectively only if we have a common understanding of exactly what energy is, and what forms it can take.

Physicists and engineers define energy as *the capacity to do work*, leaving us then with the need to define *work*. Work is a general term to most of us; it signifies everything from shoveling snow off the driveway to making out an income tax form, studying for an examination, or writing an essay. But we may not think of taking a bicycle ride on a nice Saturday afternoon as being work; it's too pleasant an experience. In order to make work a useful concept for scientific purposes, we must forget about the pleasant and unpleasant aspects and come up with a definition suitable for quantitative analysis. We can achieve this by defining work to be the product of force times the distance through which the force acts. A common example of this definition of work is given by a force pushing an object along a rough surface. The force could be exerted by any agent: human, steam engine, sled dog, or electric motor. In the British system of units, the force is given in pounds (lb) and the distance in feet (ft), so work will then be in units of pound-feet, or more commonly foot-pounds (ft·lb). In the metric system, work has the units of newton-meter (N·m), where the newton is the metric unit of force and the meter is the metric unit of distance. The metric unit of energy, the joule, is defined as 1 J = 1 N·m. The two systems of units (British and metric) are both in common use in the United States, and conversions between them are not difficult. The numerical conversion factors are given inside the covers of this book. It is important to note that the same units are used for energy and work. We will often find that energy and work are equivalent; the units are identical and, in many cases, the work done on an object is equal to the energy gained by the object. A more complete discussion of energy units is given in Section 1.4 of this chapter.

In the example given above, the work, equal to the product of force times distance, comes out to be zero if the pushed object doesn't move through some distance. A person can push against a solid wall all day long, but if the wall doesn't move, no work is done, even though the experience will be tiring to the person doing the pushing. In another case, the work being done also comes out to be zero if an object moves through a distance but with no force being exerted on it in the direction of the motion. A hockey puck sliding freely along a perfectly slippery ice surface represents a situation where no work is being done on either the puck or the ice, and no energy is being expended. Both the force and the distance must have nonzero values if work is to be done.

Here's an example that will help us gain a feeling for magnitudes and units of work and energy: Imagine that you slowly lift a 10 pound sack of sugar upward 1 foot. The force is 10 pounds and the distance is 1 foot, so the work (force times distance) you do on the sack of sugar is 10 ft·lb. The energy to do this work would have come from the food you ate. The work done can also be expressed in metric units. From the chart of conversion factors, we see that 1 ft·lb is the same as 1.36 joules, so the 10 ft·lb is 13.6 joules. Or we could deal in terms of British thermal units (Btu), another unit of energy. From the chart of conversion factors, 1 ft·lb is seen to equal 0.00129 Btu. Thus the 10 ft·lb of energy expended would be the same as 0.0129 Btu.

Example 1.2

A force of 50 pounds pushes a box along a floor a distance of 100 feet. How much work (in ft·lb) has been done? How much energy (in joules) has been expended?

Solution

$$\text{Work} = \text{force} \times \text{distance}$$
$$= 50 \text{ lb} \times 100 \text{ ft} = \textbf{5000 ft·lb}$$

Energy expended = work done

$$= 5000 \text{ ft·lb} \times 1.36 \frac{\text{joule}}{\text{ft·lb}} = \textbf{6800 joules}$$

The conversion factor between ft·lb and joule has been taken from the table of conversion factors.

1.3.2 Forms of Energy

Energy comes in many forms and can in principle be transformed from one form to another without loss. This is consistent with the Principle of Energy Conservation, which we will address later in Section 1.7. Some of the common forms of energy are discussed here.

(a) Chemical Energy Chemical energy is the energy stored in certain chemicals or materials that can be released by chemical reactions, often combustion. The burning of wood, paper, coal, natural gas, or oil releases chemically stored energy in the form of heat energy and, as discussed earlier, most of the energy used in the United States is of this form. We heat our homes, power our automobiles, and turn the generators that provide electricity primarily with chemical energy.

Other examples of chemical energy sources are hydrogen, charged electric batteries, and food in the stomach. Chemical reactions release this energy for our use.

(b) Heat Energy Heat energy is the energy associated with random molecular motions within any medium. The term thermal energy is interchangeable with heat energy. Heat energy is related to the concept of temperature. Increases of heat energy contained in any substance result in a temperature increase and, conversely, a decrease of heat energy produces a decrease of temperature.

(c) Mass Energy Albert Einstein taught us that there is an equivalence between mass and energy. Energy can be converted to mass, and mass can be converted to energy. The famous formula

$$E = mc^2$$

gives the amount of energy, E, represented by a mass, m. This energy is often referred to as the *mass energy*. The symbol c stands for the speed of light.

The most dramatic recent examples of this equivalence are in nuclear weapons and nuclear reactors, but our entire existence is now known to depend on nuclear reactions in the sun. There we have atomic nuclei coming together in a reaction with the resulting products having less mass than what went into the reaction. The mass that is lost in the reaction appears as energy according to the Einstein equation

$$\Delta E = \Delta m c^2,$$

where Δm (read it as *delta m*) is the missing mass, and c is the speed of light. The energy that appears, ΔE, is in joules if Δm is in kilograms and c is in meters per second. Because c is such a very large number, 3×10^8 m/sec, a small loss of mass results in a huge release of energy. At a detailed level, any reaction, of any type, chemical or nuclear, which releases energy does so in association with a loss of mass between the inputs and outputs, according to the Einstein equation.

The idea of mass energy is relatively new in human experience. Einstein put forth the $E = mc^2$ equation in the early 1900s. It was not until the 1920s and 1930s that the nuclear fusion processes in stars were first understood, and in the 1940s that energy release from man-made nuclear fission reactions was first demonstrated.

(d) Kinetic Energy   Kinetic energy is a form of mechanical energy. It has to do with mass in motion. An object of mass m, moving in a straight line with velocity v, has kinetic energy given by

$$KE = \frac{1}{2} mv^2.$$

If the object in question is an automobile, work must be done to bring the auto up to speed, and, conversely, a speeding car must do work in being brought to rest. The work done on the accelerating car is derived from the fuel, the work done by the stopping car will appear mainly as heat energy in the brakes, if the brakes are used to stop the car.

In a similar manner, an object rotating around an axis has kinetic energy associated with the rotation. It is just a matter of all the mass elements which make up the object each having velocity and kinetic energy according to the description given above. These combined kinetic energies make up the kinetic energy of the rotating object. We commonly see rotational kinetic energy in a potter's wheel, a child's top, an automobile flywheel, and so forth. Some day rapidly rotating flywheels may provide the stored energy needed to power a car.

(e) Potential Energy   Potential energy is associated with position in a force field. An obvious example is an object positioned in the gravitational field of the earth. If we hold an object having weight w at a height h above the earth's surface, it will have potential energy

$$PE = w \times h$$

relative to the earth's surface. If we then release the object and let it fall to the earth, it will lose its potential energy but gain kinetic energy in the same amount. Another example would be at a hydroelectric dam where water is effectively, but usually not literally, dropped onto a turbine below. In this example, the water hitting the blades of the turbine has kinetic energy equal to the potential energy it would have had at the top of the reservoir surface. This potential energy is measured relative to the turbine's location. The kinetic energy of the water becomes electric energy as the turbine spins a generator.

(f) Electric Energy   The idea of electric energy is less obvious than the examples of other types given previously. Not surprisingly, electric energy is one of the last types of energy to have been brought into practical use. With electric energy, nothing can be seen, either stationary or in motion, but the effects can be readily apparent. In spite of this difficulty, an understanding of electric en-

ergy is necessary for the functioning of a complex industrial society. It is electric energy that allows us to have telephones, television, lighting, air-conditioning, electric motors, and so forth.

If an electric charge q is taken to a higher electric potential (higher voltage) V, then it is capable of releasing its potential energy, given by $PE = q \times V$, in some other form such as heat or mechanical energy. A battery, such as we have in a flashlight or automobile, is a common device for storing electric energy. The chemicals in a battery have an inherent difference of electric potential. When the battery is charged, electric charges are brought to the higher-potential so that energy is stored as chemical energy for later use as electric energy. Thus a battery works both ways; it can convert electric energy to chemical energy, or chemical energy to electric energy.

Mechanical energy is converted to electric energy in a generator, where conductors are forced to move through a magnetic field to induce a voltage between the ends of the conductor. And, if a voltage is applied to the terminals of a common type of generator, it can function as a motor, thereby converting electrical energy to mechanical energy.

(g) Electromagnetic Radiation The energy radiated by the sun travels to the earth and elsewhere by electromagnetic radiation. That part of the spectrum of electromagnetic energy to which our eyes are sensitive is known as visible light, and a large fraction of the solar energy we receive is in the form of visible light.

The electromagnetic spectrum covers a very wide range of frequency, and visible light is only a small part of the entire spectrum. Electromagnetic radiation is characterized by a wavelength, λ (the Greek letter lambda), and a frequency, f. In a free space, the velocity of light, c, is related to these quantities by the equation $c = f \times \lambda$. The numerical value of c is 3×10^8 meters/second. The electromagnetic spectrum ranges from radio waves ($\lambda = 200$ m) to microwaves ($\lambda = 0.1$ m), to light ($\lambda = 5 \times 10^{-7}$ m), to x-rays ($\lambda = 1 \times 10^{-8}$ m) and beyond.

Various portions of the electromagnetic spectrum are important to the transformation and use of energy on earth. The portion that includes radio waves and microwaves is generated by electronic devices. Light and x-rays have their origin in atomic excitations and radiating electrons. Gamma rays are produced by the decay of excited states of atomic nuclei.

1.3.3 Power

Energy is often expressed in the units of joules or foot-pounds. We may also find it convenient to work in terms of the *rate* of use of energy, as well as with the energy amounts themselves. We will then speak in terms of joules per second. This is analogous to the way in which we commonly discuss our wages in dollars per hour (a *rate*) as well as in dollars (an *amount*). These two quantities are related, but different.

Power is the time rate of using, or delivering, energy:

$$\text{Power} = \frac{\text{energy}}{\text{time}}$$

and

$$\text{Energy} = \text{power} \times \text{time}.$$

In the metric system, for power we use units of watts, where 1 W = 1 J/sec. In the British system, the unit of power is the horsepower, where one horsepower is 550 foot-pounds per second. The rating of an electric power plant should be in the power unit of *watts*, as it can supply electrical *energy* at a certain *rate*.

The units of *kilowatts* (10^3 W), *megawatts* (10^6 W), and *gigawatts* (10^9 W) are also often used. If a power plant operating at a steady power P has run for a time t, then the energy produced is

$$E = P \times t.$$

The common unit for energy in this case of electricity generation is the kilowatt-hour (kWh). Over a given time, such as a day, the amount of electrical energy in kilowatt-hours delivered is given by multiplying the power rating in kilowatts by the number of hours in a day. One kilowatt-hour is 3.6×10^6 joules because one watt is one joule per second and there are 3600 seconds in an hour. When you pay your electric bill, you pay for energy, or the number of kilowatt-hours used.

In the United States, automobile engines and electric motors are often rated in horsepower. Although this may be historically related to the power a horse can deliver, it is more precisely equal to 550 ft·lb per second or, equivalently, to 746 watts. In recent years, it has been common, especially in Europe, to rate engine power in terms of kilowatts rather than in horsepower. In terms of human capabilities, one horsepower is an impressively large unit, as it is equivalent to raising a 55 pound weight a distance of 10 feet every second, and continuing to do this. Even the most powerful human is capable of working at a rate of only a small fraction of a horsepower for any extended time.

1.4 Units of Energy

We have already found that various units are attached to the numbers we use in discussing energy. These units are arranged into consistent systems. The two most common in the treatment of energy are the metric and the British systems. The metric units are also known as the *Système International*, usually abbreviated SI, and this system is becoming standard throughout the world.

Discussing energy without using an orderly system of units would be like trying to discuss American currency without using dollars and cents. It could make a difficult conversation if the units of dollars, pesos, and francs were included all in the same sentence. In dealing with energy, we must frame the discussion in a consistent system of units, using a single system of our choice. The variety of possibilities is often confusing and can lead to errors in even simple computations. Let us start by sorting through some of the common units used to measure energy. Later on we can become concerned with the units for the related concepts of power, work, distance, force, and so forth.

1.4.1 The Joule

The joule is the metric unit of energy. It has its fundamental definition in terms of force and distance. One metric unit of force (the newton) acting through one metric unit of distance (the meter) is equivalent to the expenditure of one joule of energy. The kinetic energy of a tennis ball moving at 14 miles per hour is about one joule. There are constant numerical factors relating the joule to the other energy units: the Btu, the calorie, the foot-pound, and the electron-volt.

1.4.2 The British Thermal Unit

We often encounter the British thermal unit (Btu) in discussions of fuel and insulation. The unit has a simple definition based on the amount of heat energy which must be given to a known amount of water to increase its temperature by a given amount. One Btu is defined to be the amount of heat energy required to raise the temperature of one pound of water by one degree Fahrenheit. Similarly, it is the amount of heat energy given off by one pound of water when it cools by one degree Fahrenheit. As a rough approximation, the burning of a wooden match releases 1 Btu. A Btu is a relatively large amount of energy; it is the same as 1055 joules.

1.4.3 The Calorie

The calorie, like the Btu, is also defined in terms of the heating of water. It is the amount of energy required to raise the temperature of one gram of water by one degree Celsius, or the amount of energy given off when one gram of water cools by one degree Celsius. The calorie is much smaller than the Btu. This should seem reasonable when considering that a gram of water is several hundred times smaller than a pound, and that a Celsius degree is only 1.8 times larger than a Fahrenheit degree. The conversion factor is 252 calories per Btu.

We often speak of food energy in terms of Calories (with a capital C), but here we must be careful. The food Calorie, or Calorie with a capital C, is 1000 times larger than the calorie used in physics or chemistry. This measure of food energy is also known as the kilocalorie. The prefix "kilo" denotes 1000. One Calorie is equal to about 4 Btu.

1.4.4 The Foot-Pound

In the United States it is common to use the ordinary measures of distance (the foot) and force (the pound) along with the equivalence of energy and work to give the foot-pound as a useful unit for energy, as discussed briefly in Section 1.3. A force of one pound acting through a distance of one foot by definition expends one foot-pound of energy, and one foot-pound of work is done. It may be of interest to note that 1 Btu is the same as 778 foot-pounds, or what it would take to lift a one pound weight to a height of 778 feet.

1.4.5 The Electron-Volt

In dealing with problems in electronics, or atomic and nuclear physics, it is convenient to have a very small unit for the extremely small amounts of energy involved. This unit, the electron-volt, abbreviated eV, is related to the idea of moving one electron through an electric potential difference of one volt. This would take an applied external force if the motion is to be against the force exerted on the electron by the electric field. If the motion is in the opposite direction, the electron would gain kinetic energy as the electric field acts on it, giving it a velocity. The electron-volt is so small that it takes 6×10^{18} of them to equal one joule.

1.5 Scientific Notation

Because the numerical quantities of interest to us range from the extremely small to the enormously large, it is hopeless to try to express them by the use of many zeroes before or after the decimal point. For example, in 2003 we consumed about 98×10^{15} Btu of energy in the United States. To write this as 98 followed by 15 zeroes would be awkward and would lead to many mistakes. Similarly, we have about 1.6×10^{-19} joules in an electron-volt. We could write this as a decimal point followed by 18 zeroes and then 16, but there is no reason to do this when we can use the efficient notation 1.6×10^{-19}. This approach is known as powers-of-ten, or scientific, notation.

There are some simple rules to follow when using powers of 10 in multiplication and division. For multiplication, say of 4.30×10^8 by 6.21×10^3, we first multiply in the usual way the 4.30 by the 6.21 and obtain 26.7. In any multiplication, the powers of ten are simply added. For example, $10^8 \times 10^3 = 10^{11}$. We have just added the 8 and the 3 to get the 11. In this case, then,

$$(4.30 \times 10^8) \times (6.21 \times 10^3) = 26.7 \times 10^{11}.$$

If we want to divide 3.50×10^5 by 2.10×10^3, the 2.10 is divided into the 3.50 in the usual way to get 1.67. The power of ten in the denominator is then subtracted from that in the numerator to get $5 - 3 = 2$, or $10^5 \div 10^3 = 10^2$. Putting this together, we get

$$\frac{3.50 \times 10^5}{2.10 \times 10^3} = 1.67 \times 10^2.$$

A special condition is encountered when the power of ten is zero. 10^0 is always equal to 1. In fact, as strange as it may seem at first, *any* number raised to the zero power is equal to 1.

When adding or subtracting numbers with various powers of 10, the numbers must first be converted to the same power of ten and then added or subtracted in the usual way. For example, $(4.8 \times 10^6) + (1.2 \times 10^7)$ should be written $(4.8 \times 10^6) + (12.0 \times 10^6)$, with the result 16.8×10^6. Similarly, $(4.80 \times 10^{-4}) - (3.6 \times 10^{-5})$ should be written

$$(48.0 \times 10^{-5}) - (3.6 \times 10^{-5}) = 44.4 \times 10^{-5}.$$

Example 1.3

The temperature of 15 pounds of water in a tank has been raised by 10 degrees Fahrenheit. How many Btu of heat energy was added to the water? What is this energy in joules?

Solution

For water:

$$\text{Energy (Btu)} = \text{weight (lb)} \times \Delta T \text{ (°F)}$$
$$= 15 \text{ lb} \times 10°F = \mathbf{150 \text{ Btu}}$$

$$\text{Energy (joule)} = 150 \text{ Btu} \times 1055 \text{ joule/Btu}$$
$$= \mathbf{158{,}250 \text{ joules}}$$

Example 1.4

The mass of a pencil is 10 grams. What is the equivalent mass energy in joules?

Solution

$$E \text{ (joules)} = m \text{ (kilograms)} \times c^2 \text{ (meters}^2/\text{second}^2)$$

and

$$m = 10 \text{ g} \times 1 \text{ kg}/1000 \text{ g} = 0.01 \text{ kg}$$
$$c = 3 \times 10^8 \text{ meter/second}$$
$$c^2 = 9 \times 10^{16} \text{ (m/sec)}^2$$

Therefore,

$$E = 0.01 \times 9 \times 10^{16}$$
$$= \mathbf{9 \times 10^{14} \text{ joules}}$$

1.6 Energy Consumption in the United States

We frequently hear various figures put forth for energy consumption rates, and these numbers are sometimes given in terms of so many QBtu/year, where Q is a symbol for quadrillion, or 10^{15}. This number is huge and seems disconnected from our everyday experience. How can we gain a feeling for what the numbers actually mean? To make the situation more real, Table 1.1 shows the origin of the energy consumed in the United States in 2003 in terms of tons of coal, barrels of oil, and so forth. We see that coal, natural gas, and petroleum provided most of the energy, but nuclear and renewables also made significant contributions.

In Table 1.1, the values in the third column for coal, natural gas, petroleum, and other sources are for the heat energy released by the burning of the given fuel. The numbers given for nuclear and renewables are different. For these generating technologies, what is shown is the number of QBtu which would be needed to produce the given number of kilowatt-hours of electric energy if fossil fuels were the energy source. This consideration would apply, for instance, if there were a severe drought and hydroelectricity had to be replaced by electricity from a coal-burning plant. Calculation of this number of QBtu uses a prevailing annual heat rate factor for fossil-fueled electric power plants in the United States. This number corresponds to an efficiency for converting thermal energy to electrical energy of approximately 30 to 35%.

Figure 1.5 shows in more detail the sources and final uses of energy in the United States as compiled by the U.S. Energy Information Administration for 2003. Here and in Table 1.1 renewables include biofuels, conventional hydroelectric power, geothermal energy, solar energy, and wind energy. The many footnotes shown for Figure 1.5 explain some of the discrepancies with Table 1.1

Table 1.1 U.S. Energy Consumption in 2003

Source	Amount	QBtu	Percent	1 QBtu Equiv.
Coal	1.08×10^9 tons	22.6	23.0	47.8×10^6 tons
Natural gas	21.8×10^{12} ft^3	22.5	22.9	0.97×10^{12} ft^3
Petroleum	6.72×10^9 bbl	39.1	39.8	172×10^6 bbl
Nuclear elec.	757×10^9 kWh	7.97	8.1	95×10^9 kWh
Renewables	578×10^9 kWh	6.15	6.3	94×10^9 kWh
Total		98.3	100	

The data in column 3 are from U.S. Energy Information Administration, *Annual Energy Review*, 2003. The numbers in columns 2 and 3 are given in thermal energy for the first three entries and in terms of equivalent thermal energy for the next two. The numbers in column 5 are from the same source. The last entry includes typical conversion efficiencies from thermal to electrical energy.

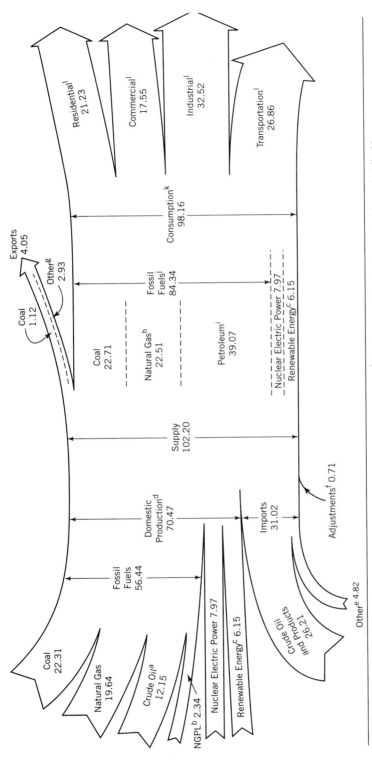

Figure 1.5 Energy flow from source to use in the United States in 2003 in units of QBtu. (*Source*: Washington, D.C.: U.S. Department of Energy, Energy Information Administration, *Annual Energy Review*, 2003.)

such as the total energy being 98.16 QBtu instead of the exact sum of the third column of Table 1.1.

Figure 1.6 gives information similar to that of Figure 1.5, but arranged in a different way to provide more detail. For example, this figure shows that in the United States in 2002, 3.4 QBtu of electrical energy was used by industry and 8.4 QBtu of electric energy went to residential and commercial uses, while 26.3 QBtu was lost in generating and distributing electricity. The reasons why most of this loss is unavoidable are discussed later in the text, especially in Chapter 3. Some of the numbers in Figure 1.6 differ slightly from those of Figure 1.5 because they come from different sources and are not for the same year.

If the data of Table 1.1 are divided by the population of the United States (291 million people), we obtain the average amount of energy that we each used in 2003. These numbers are shown in Table 1.2, which indicates that each of us on average consumed in 2003 about 23 barrels of oil, 3.7 tons of coal, etc. Our average per capita energy consumption is equivalent to the burning of 58 bar-

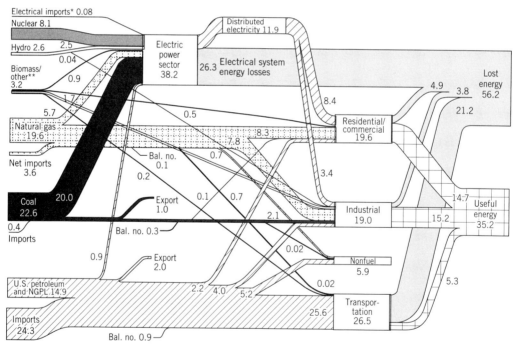

*Net fossil-fuel electrical imports.
**Biomass/other includes wood, waste, alcohol, geothermal, solar, and wind.

Figure 1.6 Energy flow in the United States in 2002 in units of QBtu, arranged to separate out useful energy from lost energy. (*Source*: Lawrence Livermore National Laboratory (2004) and United States Energy Information Administration, Annual Energy Review 2002.)

Table 1.2 U.S. Energy Consumption per Person in 2003

Source	Amount
Coal	3.7 tons
Natural gas	74,900 ft^3
Petroleum	23.2 bbl
Nuclear electric	2600 kWh
Renewables	1986 kWh

The first three entries are for the thermal energy from the fuel. The last two entries are in terms of equivalent thermal energy, as in Table 1.1. The amounts given here are for a population of 291 million.

rels of oil. This is an impressively large amount of energy by any standard. See Figure 1.7.

Only in Canada do people use as much energy per person as we do in the United States; most industrialized countries use only about one-half. Figure 1.3 shows the energy use per capita along with the Gross Domestic Product per capita for several representative countries. The energy use is in terms of equivalent barrels of oil; the GDP is in terms of U.S. dollars.

It could be argued from the data shown in Figure 1.3 that in order to have a large per capita GDP, a relatively large per capita energy consumption is also needed. Although there is indeed such a general trend, there is no consistent correlation. For example, Switzerland consumes less than half the per capita en-

Figure 1.7 Each person in the United States consumes an energy equivalent of 58 barrels of oil burned as fuel each year.

ergy of the United States, but has about the same per capita GDP. The striking element illustrated by Figure 1.3 is the great disparity between the industrialized and the developing nations. A number of developing countries such as Angola, Haiti, and Somalia fall within the little quarter-circle shown at the lower left-hand corner of the figure. The world as a whole used the equivalent of 11.4 barrels of oil per person in 2002. Most of the countries of the world use relatively little energy per capita and have a small per capita GDP compared to the most developed countries.

The world consumption of energy in 2002 was about 410 QBtu. That same year the United States' consumption was about 98 QBtu. With just 4.5%, or one-twentieth, of the world's population, we were using one-quarter of the world's energy. There are several causes for the relatively high rate of energy consumption in the United States. One is related to transportation. The United States is a large and affluent country where a good deal of the personal transportation is by automobile. Although public transportation can be much more energy efficient in regions of high population density such as Europe or Japan, it is not as common in the United States because our population density is far less. It is also true that a large fraction of our population lives in separated houses that are more expensive to heat and cool than are multifamily dwellings. The development of the United States into an industrialized society took place in a time when energy sources appeared to be abundant, even inexhaustible, and it has not been until recently that serious energy conservation steps have been widely undertaken.

How do we use energy in the United States? Table 1.3 shows the breakdown by major categories. Excluding electricity, residential and commercial use involves mostly natural gas and petroleum; the industrial category mainly involves coal, natural gas, and petroleum. The energy for transportation fuel is supplied almost entirely by petroleum with some inroads made recently by natural gas. The energy resources for electric utilities are quite mixed: 50% coal, 20% nuclear, 18% natural gas, 7% hydroelectric, 3% petroleum, and 3% other.

Table 1.3 Percentage U.S. Energy Use in Various Sectors in 2003

Sector	Amount
Electric utilities	39.0%
Transportation	27.3%
Industrial	22.1%
Residential and commercial	11.6%

The last three categories do not include electric energy from the utilities. Data from U.S. Energy Information Administration, *Annual Energy Review*, 2003.

The history of consumption and product of energy in the United States since 1950 is shown in Figure 1.8. In this figure, and elsewhere in this text, energy *production* refers to the mining of coal and the bringing of oil and natural gas to the earth's surface, or to the making of useful energy by nuclear power, hydroelectric power, geothermal power, biomass fuel, solar collectors, and other means. Energy *consumption* occurs when the fossil fuel is burned or when energy is put to use by the consumer.

It is clear from Figure 1.8 that we now are in a persistent pattern of consuming more energy than we are producing in our nation. The difference must be made up from imports. Over this period, energy consumption has generally gone up with time except for a dip after 1980. The decrease in the 1980s reflected a movement toward greater energy conservation following increases in fuel costs and the "energy crisis" that started in 1973. In this period there was a major turn toward cars with greater fuel efficiency, improved standards for building insulation, and various energy conservation measures in industry. Once the easiest and

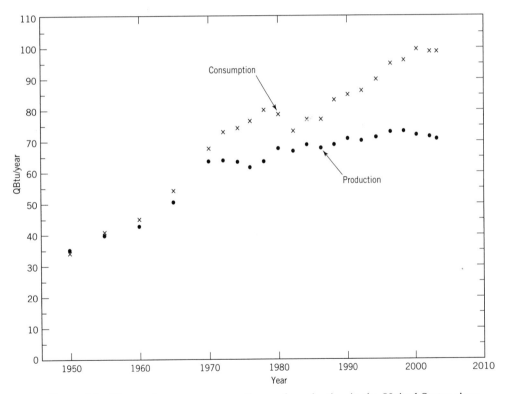

Figure 1.8 The total energy consumption and production in the United States since 1950 in quadrillion British thermal units (QBtu) per year. (*Source*: Washington, D.C.: U.S. Department of Energy, Energy Information Administration, *Annual Energy Review 2003*.)

least expensive energy conservation measures had been implemented, the energy consumption curve began to rise again. During this entire period, the national population increased at about 1% per year; this accounts for some of the increases seen.

In Figure 1.8, we see that until about 1957, the energy consumed was about equal to the energy produced in the United States. In fact, because of coal and petroleum exports, for some years in the 1940s and 1950s, production exceeded consumption. Beginning in 1958, however, our energy consumption has been greater than production for every year, indicating a generally increasing trend in this direction. Our energy production has remained relatively constant since about 1980. Production of oil and natural gas has gone down, and coal production has increased. Our country is now experiencing a serious imbalance of international trade due in part to the cost of importing oil to fill the gap between our rising consumption and declining production.

Energy consumption can be related to the general well-being of a country's populace. The advantages of convenient transportation, abundant food and water, comfortably heated and cooled residences and places of work, ample production of goods, and many other aspects of the good life involve the consumption of energy. So what is the problem with a high level of energy consumption? As we discussed earlier in Section 1.1, the problems arise in two areas. The first has to do with resource depletion. We saw earlier that the majority of our energy supply is derived from fossil fuels. The resources of coal, petroleum, and natural gas are being depleted both in the United States and throughout the rest of the world at alarming rates. For the moment, the United States has sufficient resources of coal and natural gas to meet the needs of the country. We have consumed more natural gas than we have discovered nearly every year for the past 25 years, and we will probably see shortages within a decade or two. Production of petroleum from domestic sources has gone down rather steadily since 1970. In 2003 we produced only 62% of the amount of petroleum produced in 1970, while our consumption went up by 24%. We now import considerably more than half of the petroleum that we use. The balance-of-trade deficit is seriously affected by these purchases, which are likely to increase with time.

The second aspect of the problem is more general in that it relates to the emission into the atmosphere of the by-products of burning any fossil fuel. With petroleum and coal being the worst culprits, effects on the atmosphere both locally and globally are well documented. The rate at which we are burning fossil fuels worldwide has reached a point where restraint must be considered.

1.7 The Principle of Energy Conservation

A well-established law of physics states that the total energy in an isolated region cannot change. By "isolated," it is meant that energy can neither enter nor escape that region. In other words, the total energy in the region is conserved, even though it may be transformed from one form of energy to another. Energy cannot be created or destroyed. This is *The Principle of Energy Conservation*, quite different from the more general ideas about reducing waste of energy.

As an example of The Principle of Energy Conservation, imagine that you have a perfectly insulated box (the isolated region) containing only air and a battery connected to a lightbulb by wires. When the connection is made, a current flows and the chemical energy in the battery becomes electric energy and heats up the filament in the lightbulb. Then electromagnetic energy in the form of light and infrared radiation travels throughout the interior of the box where it becomes heat energy in the air and other materials. After the wires are disconnected, the amount of heat energy that has been added to the lightbulb, the air, the wires, the battery, and the interior of the box will be exactly equal to the chemical energy taken from the battery. The filament in the bulb eventually transfers its heat energy to everything inside of the box and cools from its highest temperature to a new temperature somewhat warmer than its original temperature before the connection was made. The total energy inside the box will be the same before and after the electrical connection is made. It should be noted, however, that the energy originally stored in the battery was in a much more useful form than the heat energy that was created in the lightbulb. Although, according to The Principle of Energy Conservation the total energy remains the same, its usefulness for performing tasks has certainly been diminished.

When the energy was in the form of chemical energy in the battery, it could have been used for powering a motor, lighting a bulb, sounding a horn, and so forth. None of these things can be done with the heat energy finally stored in the box, even though the number of joules is the same. One might say that in this experiment the usefulness of the energy has been degraded, or that it is of a lower form, but nevertheless the energy has been conserved.

Our other common use of the words *energy conservation* is distinct from The Principle of Energy Conservation and applies to the idea of using less energy to perform a given task. An example of energy conservation would be the installing of better insulation in the walls of a house in order to allow one to maintain the interior at a comfortable level on a cold winter day with the use of less fuel in the furnace. Chapter 7 of this book is devoted to discussion of practical ways of achieving energy conservation goals.

At times the two usages of *energy conservation* can be confusing, but usually the nature of the discussion leaves little doubt as to what is meant.

1.8 *Transformation of Energy from One Form to Another*

We can gain an understanding of energy in our everyday lives by examining some of the ways in which transformation of energy from one form to another takes place. Let's start with the sun. Consider first Figure 1.9. We have known for at least the past 70 years that nuclear fusion reactions are taking place on a very large scale deep in the interior of the sun. Before the 1930s it was not understood in detail why the sun is so hot. But now we know that the dominant solar

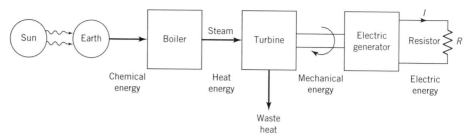

Figure 1.9 Steps in the transformation of the nuclear fusion energy in the sun to the electric energy used in a residence or industry. This example involves a time period of hundreds of millions of years.

fusion reaction is one in which helium is formed following the fusion of hydrogen nuclei. The net effect is to transform nuclear mass into heat energy. This heat energy raises the temperature of the sun's surface to a very high value, high enough to make the surface glow and radiate energy in the form of sunlight. Sunlight is a form of electromagnetic energy; it is radiated outward in all directions and a small part of the total radiated energy is incident on our planet, earth.

The earth has been bathed in this solar radiation since the time of its formation about 5 billion years ago. For the past hundreds of millions of years, plants have captured electromagnetic energy and formed the organic material which makes up the plants through the process of photosynthesis. The ancient plants, and the animals that lived off the plants, were then transformed over the ages into the chemical compounds we call fossil fuels.

We now extract the fossil fuels from beneath the earth's surface to provide fuel for the needs of society. To follow the energy transformation pathway further, let us consider coal that was mined to fuel the boilers of an electric utility plant. The chemical energy in the coal is transformed through the process of burning into heat energy which boils water into high-pressure steam to drive a turbine. The mechanical energy of the rapidly rotating turbine is then transformed into electrical energy by the generator attached to the shaft of the turbine. Then the electrical energy is transmitted over high-voltage transmission lines to the final users.

The electrical energy entering a home is used for light, heat, television, and so forth. For example, suppose that an electric oven is turned on for 45 minutes to cook a frozen pizza. The heat energy in the materials of the oven itself, in the air inside, and in the pizza will soon result in a warming of the air in the kitchen. The warm kitchen air will lose its energy to the rest of the air in the house, and then the heat energy in the house air will eventually find its way to the outside atmosphere. The heat added to the atmosphere will very slightly raise its temperature. The atmospheric heat energy then is radiated off into the coldness of space. The radiated energy does not disappear; it's added to the background energy of the universe.

Thus, in heating up the kitchen oven, we have drawn on nuclear fusion energy from the sun, transforming it from one kind of energy to another many times in the process. Energy is conserved in each transformation.

1.9 Renewable and Nonrenewable Energy Sources

In dealing with energy resources and energy use it is often necessary to distinguish between *renewable* and *nonrenewable* resources. The *nonrenewable resources* are those that could be exhausted within a relatively short time as a result of our exploiting them; *renewable resources* can never be consumed to completion. There is not always complete agreement on the definitions of *renewable* and *nonrenewable*. Some would classify a given category of resource under the heading of renewable, while others, for equally valid reasons, would consider it nonrenewable. We will see this in the case of geothermal energy as an example.

1.9.1 Nonrenewable Energy Sources

Examples of nonrenewable energy sources would be the following: all the fossil fuels (coal, oil, natural gas, shale oil, tar sands, etc.), uranium-235 nuclear fission fuel, deuterium nuclear fusion fuel, and some types of geothermal energy. The time to exhaust a nonrenewable resource depends on the rate of use and size of the resource. We might estimate a time of several centuries to exhaust the entire stock of fossil fuels, several decades for the earth's uranium-235 with a vigorous program of energy from nuclear fission, and a few decades to use up the heat energy at some local geothermal sites. In this sense we could classify geothermal energy as nonrenewable. Although it may be true that fossil fuels are continually being produced far underground from biomass, the time scale for this conversion is hopelessly long for our purposes. It takes perhaps a hundred million years for natural processes to produce useful amounts of petroleum, natural gas, or coal.

It is likely that most forms of nonrenewable energy will get more expensive when they are near exhaustion. It's simply a matter of supply falling short of demand.

1.9.2 Renewable Energy Sources

We have only three sources of renewable energy: solar, geothermal, and tidal.

All energy sources based on the solar energy incident on earth—direct sunlight, wind, hydroelectric power, ocean currents, ocean thermal gradients, and biomass—are renewable. In these cases the time to exhaustion depends on the life of the sun itself, which is certainly far beyond the time period important to any discussion of energy sources for humanity. The rate at which we use solar energy does not affect its lifetime. Whatever solar energy source we put into use

26 Chapter 1 **Energy Fundamentals, Energy Use in an Industrial Society**

will continue to be available, or can be soon renewed. An example would be firewood, which can be grown indefinitely, season after season, as long as water and plant nutrients are available. Another example would be food derived from biomass. This energy can be put to various uses, as shown in Figure 1.10.

Some local geothermal energy sites might be depleted fairly quickly through use, but they will again become available in several centuries as the earth's inner heat sources (which have effective lifetimes of billions of years, because of the half-lives of radioactive nuclear decays) supply heat energy to the geothermal regions near the earth's surface. This is renewal on a fairly long time scale but is still consistent with geothermal energy being considered renewable.

The energy in the oceans' tides has its origin in the gravitational interaction between the earth and the moon and, to a lesser extent, between the earth and the sun. Our harnessing of this tidal energy will not reduce the magnitude of the total gravitational energy appreciably within the foreseeable future, so we classify tidal energy as renewable.

It is a remarkable fact that, in contrast to the case of nonrenewables, renewable energy will tend to become less expensive as the scale of its use in-

Figure 1.10 The world's first sustained flight powered by renewable energy in the form of food for humans. On August 23, 1977, the *Gossamer Condor* flew a distance of 1.3 miles in under seven minutes in a figure-eight pattern around two markers a half mile apart. It was powered by Bryan Allen, who provided the one-third horsepower required for the flight. (*Source*: Courtesy Gossamer Ventures, CA)

creases. To the extent that the supply is unlimited, mass production of the energy extraction technologies will lead to reduced per unit costs. In some limited cases this argument may fail, as when all the prime hydroelectric sites are being utilized, or when we rely too much on marginal yields of firewood. In general though, *the more renewable energy we use, the cheaper it will be*, and there will always be more for succeeding generations.

Because of concerns over resource depletion and environmental damage, the use of renewable energy presents an option that is often preferable to the use of nonrenewables. We often are tempted to associate the use of renewables with "good" and the use of nonrenewables with "bad," but let's not rush to dam the Grand Canyon for renewable hydroelectricity when we can still find other options, even including nonrenewables. In the United States, we are seeing increasing interest in obtaining energy from renewable sources, but the progress is painfully slow. In 2003, only about 6.3%, or 6.15 QBtu of a total of 98 QBtu, came from renewable sources.

Key Terms

Energy efficiency	Hydroelectric
The Principle of Energy Conservation	Energy conservation
Energy consumption	Nuclear fission
Energy production	Nuclear fusion
Fossil fuels	Work
Renewable and nonrenewable energy	Chemical energy
British Thermal Unit (Btu)	Newton (N)
Foot-pound (ft·lb)	Heat energy
Joule (J)	Mass energy
calorie (cal)	Mechanical energy
Calorie (Cal)	Potential energy
Watt (W)	Electric energy
Horsepower	Electromagnetic radiation
Geothermal	Energy transformation

Suggested Reading and References

1. *Basic Petroleum Data Book*, Vol XIV. Washington, D.C.: American Petroleum Institute, January 1994.

2. U.S. Energy Information Administration, *Annual and Monthly Energy Review*. Washington, D.C., 2004.

3. Kraushaar, Jack J., and Ristinen, Robert A. *Energy and Problems of a Technical Society*, second edition. New York: John Wiley, 1993.

4. Priest, Joseph. *Energy, Principles, Problems, Alternatives*, third edition. Reading, MA: Addison-Wesley, 1984.
5. Cassedy, Edward S., and Grossman, Peter Z. *An Introduction to Energy: Resources, Technology, and Society.* Cambridge: Cambridge University Press, 1990.
6. "Energy for Planet Earth." *Scientific American*, **263**, No. 3 (September 1990), pp. 54–163.
7. Aubrecht, Gordon J., *Energy*. Columbus, OH: Merrill Publishing Company, 1989; 2nd Ed., Prentice-Hall, 1995.
8. Parfit, Michael; photographs by Leen, Sarah; After Oil, Powering the Future; *National Geographic* 208 2 (August 2005), pp. 2–31.

Questions and Problems

1. What are the two major problems created by depending on fossil fuels for most of our energy?
2. Give three examples of work that you have done the past week. Use the physics definition of work.
3. If you push a cart along a horizontal surface with a force of ten pounds, and the cart moves ten feet, how much work have you done in ft·lbs? In joules?
4. Since energy is conserved, where has the energy gone in question 3? Where did it come from?
5. If you throw a ball straight up into the air, at some time it will stop rising and return to you. What form of energy did the ball have just after being thrown? What form of energy did it have exactly at the top of its path? What forms of energy did it have just before and just after being caught?
6. What are some possible ways to provide energy for society after all the fossil fuels on earth have been consumed?
7. How many tons of coal would be needed each year to provide for the entire energy needs for the average person in the United States?
8. A bicyclist on a flat road expends energy at the rate of 80 watts. How many calories of energy are expended in five minutes of pedaling?
9. What happens to all the energy radiated by the sun that does not impinge on the earth or other planets?
10. Solar energy is incident on a black parking lot with an intensity of 1000 W/m^2 and 90% of it is absorbed. What is this in Btu/hr per square meter?
11. A windmill produces 1400 watts of electric power that is used to heat water. The efficiency is 100%. How long will it take to raise the temperature of 40 gallons of water by 50°F?
12. Assume that the population of the United States increases by 1%/yr. How many Btu of energy will have to be added to the national annual energy budget this year to maintain the same per capita expenditure? What is this in gallons of petroleum if it all comes from petroleum? In tons of coal?

Multiple Choice Questions

1. The product of $(5 \times 10^5) \times (6 \times 10^6) \times (7 \times 10^7) = $ _____
   a. 18×10^{18}
   b. 18×10^{20}
   c. 18×10^{210}
   d. 18×10^{180}
   e. 210×10^{210}
   f. 2.1×10^{20}
   g. 20×10^{21}
   h. 2.1×10^{212}

2. A car has a mass of 2000 kg and is traveling at a speed of 30 m/sec. What is its kinetic energy in joules?
   a. 4.5×10^3
   b. 1.5×10^4
   c. 3×10^4
   d. 1.5×10^5
   e. 3×10^5
   f. 4.5×10^5
   g. 9×10^5
   h. 4.4×10^6

3. A 5 kg mass is attached to the end of a string 2 meters long. The other end of the string is fixed to a hook to make a simple pendulum. Initially, the mass is held so that the string is horizontal. The mass is then released. At the point when the string is vertical, what is the kinetic energy of the mass?
   a. 10 joules
   b. 49 joules
   c. 98 joules
   d. cannot be determined because we do not know the velocity

4. Per person, the energy consumption in the United States is about 3.3×10^8 Btu/yr. This is equivalent to a power of
   a. 11.0 W
   b. 11.0 W
   c. 11.0 MW
   d. 11.0 kW
   e. 110 kW
   f. 1.10 kW
   g. 11.0 QW
   h. 0.11 kW

5. A typical U.S. citizen consumes ___ times as much energy as does a typical citizen of India.
   a. 6
   b. 300
   c. 2
   d. 25

6. What is the potential energy increase of a 1000 kg auto driven up the 1000 meter elevation gain from Boulder to Nederland, CO?
   a. 1×10^6 joules/sec
   b. 9.8 joules
   c. 1×10^5 Btu
   d. 1×10^6 joules
   e. 9.8×10^5 watts
   f. 9.8×10^6 joules

7. In the United States, the average person eats about 3000 food calories per day. What is the average power of this energy intake?
   a. 14.5 watts
   b. 145 watts
   c. 1450 watts
   d. 14,500 watts

30 Chapter 1 Energy Fundamentals, Energy Use in an Industrial Society

8. The Principle of Energy Conservation is
   a. a possible means for extending our fossil fuel reserves
   b. inconsistent with the motion of a pendulum
   c. a law of physics often violated in thermodynamic systems
   d. a law of physics with no known exceptions
   e. obeyed in chemical reactions but not in nuclear reactions.

9. In the United States, fossil fuels account for about ___% of our energy use. (Do not consider direct use of solar energy in this context.)
   a. 30
   b. 50
   c. 85
   d. 91
   e. 99

10. Kinetic and potential energies are energies of ___ and ___, respectively.
    a. springs and gravity
    b. solids and liquids
    c. position and motion
    d. motion and position
    e. force and power

11. The total mass energy of one pound of anything is about:
    a. 9×10^{16} J
    b. 4×10^{16} J
    c. 4×10^{19} J
    d. 3×10^{8} J
    e. 1.36×10^{8} J
    f. 1.36×10^{11} J

12. One horsepower for one hour represents how many joules?
    a. 44,760
    b. 2,685,600
    c. 33,390,960
    d. 746
    e. 3,413
    f. none of the above

13. $\dfrac{(4.8 \times 10^{9}) \times (3.6 \times 10^{5})}{2.8 \times 10^{10}} =$
    a. 2.76×10^{5}
    b. 3.00×10^{35}
    c. 6.17×10^{4}
    d. 9.42×10^{24}
    e. 8.40×10^{45}
    f. 49.8×10^{5}

14. Classify the following terms according to whether they represent energy (E), power (P), or neither (N).
    a. calorie   ___
    b. horsepower   ___
    c. joules/sec   ___
    d. joule·sec   ___
    e. kilowatt/hour ___
    f. watt   ___
    g. Btu/hr   ___
    h. kilowatt·hour ___
    i. Btu   ___
    j. horsepower/day ___

CHAPTER 2

The Fossil Fuels

(*Source*: (c)David Scott Smith)

2.1 Introduction

As we saw in Chapter 1 (Table 1.1), a total of about 98 QBtu of energy is used every year in the United States. Of this amount, about 84 QBtu, or 86%, comes from fossil fuels. Our reliance on coal, natural gas, and petroleum is overwhelming. In order to achieve some understanding of the problems related to our use of these nonrenewable energy sources we will explore how each fuel was formed, the history of its use and resource development, projections for its use in the future, and the total U.S. resource. The story of each of the major fossil fuels is different, as is the impact of its use on the environment. Two other types of fossil fuels: shale oil and tar sands, which are presently underutilized, will also be discussed because they could be significant resources in the future.

2.2 Petroleum

We use a definition of petroleum that excludes solid forms of crude oil but does include liquid crude oil and natural gas liquids as well as the common fuels propane and butane. The latter two fuels are commonly transported and stored as liquids. Gasoline, kerosene, heating oil, and diesel fuel are all included in our definition of petroleum.

The energy input to the fossil fuels came to the earth hundreds of millions of years ago as radiant energy from the sun where nuclear fusion reactions provided the energy for the radiant solar output. This radiant energy was then captured in plant life on earth through the process of photosynthesis. The highly complex process of photosynthesis fixes carbon from the carbon dioxide in the air into carbohydrates when solar radiation interacts with carbon dioxide in the presence of an enzyme (biological catalyst) and water. Hydrogen from the water is also fixed into the molecular structure of the carbohydrates. In this way the electromagnetic energy of the solar radiation is stored as chemical energy in the carbohydrates. Following this, animal life dependent on plant life evolved.

Our petroleum comes from plant and animal life in the ancient seas of hundreds of millions of years ago. The organic material accumulated on the bottom of these seas and decomposition occurred through the action of bacteria. Eventually sediments of sand and mud on top of these organic deposits created conditions of high temperature and pressure that resulted in loss of oxygen. In this way a variety of hydrocarbon molecules were created, some of them solid, some of them liquid, and some of them gas.

The liquid and gaseous hydrocarbons could travel rather large distances in the porous sedimentary rock. Eventually, under the influence of the geologic structures and gravity, some of the hydrocarbon material accumulated in geologic traps such as shown in Figure 2.1. These reservoirs now lie beneath dry land as well as offshore under the oceans.

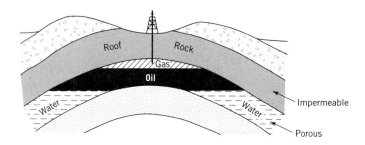

Figure 2.1 A typical geologic formation in which oil and natural gas are found. The gas is found above the oil, and the oil above water, all interspersed in a porous formation.

The profession of locating petroleum deposits has become increasingly more sophisticated as the undiscovered resources have become less abundant. In one method of geophysical exploration, sound waves from an explosion or from vibrating weights are sent down into the earth. When these sound waves encounter the interface between layers of rock with different acoustical properties they are partially reflected back from deep underground to the surface where they can be read by sensitive receivers called geophones. In this way the exploration geologists can determine if an underground geological formation is likely to contain a significant reservoir of petroleum. The only real proof of an oil or natural gas deposit must be obtained by drilling.

The first well drilled in a new area is called an exploratory or wildcat well. Even with the remarkable advances made in locating oil by the geophysicists, only about one commercial well is found in the United States for every nine wildcat drillings. After a petroleum deposit has been found by drilling, further wells are drilled to determine the extent of the reservoir. Then the total amount of oil that can be commercially extracted from the new area is estimated and added to what is called the *proved reserve*.

The amount of petroleum now present underground is a fixed number for practical purposes. Essentially all of what we now have dates back to from 10 to 250 millions of years ago and the amount to be formed within the history of man is completely negligible. What is available for use is a nonrenewable source of energy whose magnitude is reasonably well determined.

2.3 *History of the Production of Petroleum in the United States*

"Colonel" Drake started the whole thing off in 1859 at Titusville, Pennsylvania. Before Drake's time there were only a small number of oil springs or seeps coming to the surface that provided a few gallons of oil per day. Drake ignored advice against the venture, and in spite of economic problems he drilled and found a substantial amount of oil at a depth of 69 feet which he then pumped to the surface. The main problem at that time was to find suitable containers for the oil because the barrels cost more than the oil was worth. News of the strike spread rapidly and led to the drilling of many additional wells and an abundance of *rock oil*, as it was called. Refining of the oil produced kerosene which had a ready market for heating, cooking, and lighting. In 1862, 3 million barrels of oil were produced from more than 75 wells in Pennsylvania.

Although the Pennsylvania oil fields were rather short-lived, oil was also soon found in Ohio and Indiana, and more in California and Texas. By 1909 the United States was producing 500,000 barrels per day, more than was being produced by all other countries combined.

There could have been a decline of the new oil industry at about the time electric lights were replacing kerosene lamps, but just then the automobile was also beginning to replace horses. The need for gasoline soon gave crude oil a new importance. The graph in Figure 2.2 shows the amount of oil produced in

34 Chapter 2 **The Fossil Fuels**

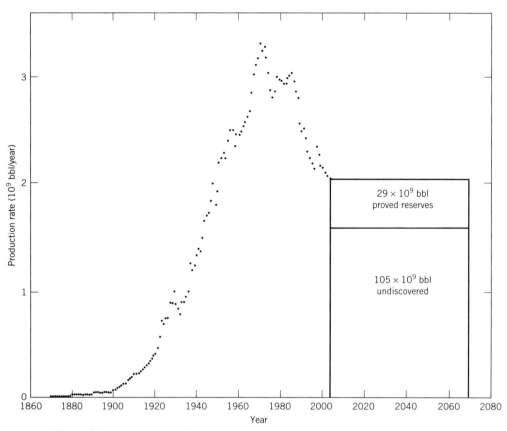

Figure 2.2 Annual rate of petroleum production in the United States, including Alaska. Data through 1996 are from the American Petroleum Institute. Data from 1997 through 2003 are from the U.S. Energy Information Administration. The two data sets do not quite match due to differing definitions of petroleum. The rectangle at the right has an area representing the 134 billion barrels estimated to be remaining for future production. The proved reserves are from Table 2.2. The undiscovered recoverable petroleum is from Table 2.1.

the United States each year since 1860 along with an estimate of the oil remaining for future production. Annual production climbed fairly steadily for 110 years until 1970, when a peak of about 3.3×10^9 barrels per year was reached. This was the year that the Prudhoe Bay field was brought on-line through the Alaskan pipeline. Since that time, production has fallen off in an irregular manner. Since 1948 the United States has imported more oil each year than it has exported. We now import considerably more than 50% of the oil we consume.

Drilling for oil in the United States has been reduced in the last 25 years because the chances of producing a paying well have become less with the passage of time. Apparently the rich oil fields have already been discovered. The decline of drilling is also related to the fact that the price of a barrel of oil is determined by abundant production in such places as Saudi Arabia. For example, 6600 oil wells were drilled in 2004 in the United States, compared to 44,000 in 1981, the peak year. In 2003, 155 million feet of total (oil, natural gas, and dry hole) footage were drilled in the United States, compared to 408 million in 1981. In 2003, 1000 drilling rigs were in operation, compared to 4000 in 1981. In North America, development and exploration costs in 2004 had risen to $11/bbl from $5/bbl in 1999.

2.4 Petroleum Resources of the United States

How much economically recoverable oil is left in the United States? Many people have been trying to answer this question over the years. With our extensive use of automobiles and a rising imbalance of payments in international trade, it is indeed an important question.

In discussing this subject, we use an important concept, Q_∞, the total resource that is available over all time from the beginning of the oil industry to the exhaustion of the resource. The total area under a curve fitted to the data of Figure 2.2 would represent Q_∞ (petroleum) graphically. Some qualification such as "economically recoverable" must be considered when trying to decide on the total resource because there will always be some petroleum left underground that takes more energy or costs more to produce than it is worth. One interesting way of estimating the amount of a natural resource is to examine both its rate of discovery and rate of production as a function of time. For petroleum one finds that there has been a typical delay of about 11 years between discovery and production.

Figure 2.3 shows a plot of *cumulative* discovery and production for a finite resource versus time. Over the lifetime of the resource a line representing the produced quantity must start at zero and ascend to Q_∞. At any point in time on this graph, the slope of the line will represent the production *rate* such as is shown in Figure 2.2. There is no general rule by which the production rate line must be

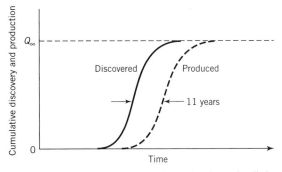

Figure 2.3 History of cumulative discovery and production of a finite resource consumed to depletion. For oil in the United States, production has generally followed discovery by about 11 years.

a smooth, symmetric, bell-shaped curve that is described by any particular mathematical function. Such a smooth curve is only a convenient description of a complex process but it does quite well. A number of external factors, such as cost of drilling, price for a barrel of oil, politics, or wars, can modify the shape of the production rate curve. Eventually, however, the curve must go up, and in some fashion come down, and the area under the production rate curve will represent Q_∞.

Using the concepts previously discussed, as early as 1956 a well-known petroleum geologist, M. K. Hubbert, made some reasonable, but startling, predictions based on his analysis of U.S. petroleum data. He predicted a Q_∞ of 165×10^9 barrels of oil for the United States exclusive of Alaska. This was at a time when others were predicting values many times larger. He also predicted that the U.S. oil production rate would peak between 1966 and 1971 and then fall back toward zero in spite of large efforts and technical advances in the industry. These predictions were not widely or cordially accepted by many others in the field. The common opinion among the experts was far more optimistic. We see in Figure 2.2 that the production rate actually peaked in 1970. Hubbert was right.

Estimating the total petroleum resource has been an ongoing endeavor of the oil companies as well as of government agencies such as the U.S. Geological Survey (USGS). The results of several very early studies are shown in Table 2.1, along with the results of a three year USGS study published in 1995. The latter study was based on detailed data from eight regions consisting of 71 separate areas including Alaska and offshore resources, and it was considerably more optimistic than some of the other studies, owing mainly to an increase in the growth of reserves. In brief, the mean values of the study give 20×10^9 barrels in proved reserves and a total of 113×10^9 barrels as the remaining U.S. resource. More recent data compilations are given in Table 2.1, Table 2.2, and Figure 2.2. As of the end of 2002, a total of 190×10^9 barrels of crude oil had been produced in the United States. When the future resource of 134×10^9 barrels is included, a value of 324×10^9 barrels is obtained for Q_∞. Although this is higher than Hubbert's early prediction, he did not include offshore or Alaskan resources.

What is the significance to us of the remaining U.S. resource of 134×10^9 barrels? The United States is now consuming about 6.7×10^9 bbl/yr of petroleum, so at that rate the oil resource would be completely exhausted in 20 years. It may last longer than that because we are now importing appreciably more than half of what we consume, but on the other hand, with a growing population and more miles traveled per year, our consumption rate will tend to increase. Other factors that enter into the picture are the use of natural gas as a motor fuel, the increased use of higher efficiency vehicles, the stability of oil supplies from the Persian Gulf area, the potential for shale oil and tar sands, and enhanced secondary and tertiary recovery of underground petroleum.

Petroleum geologists believe that after a well stops producing by normal extraction processes, about 70% of the petroleum is left behind in the ground. In the primary extraction process oil either comes to the surface under natural ge-

Table 2.1 Estimates of Undiscovered Recoverable Oil and Natural Gas in the United States

	Oil (Billion Barrels)	Natural Gas (Trillion Cubic Feet)
Oil Companies		
Company A (Weeks, 1960)	168	—
Company B (Hubbert, 1967)[a]	21–64	280–500
Company C (1973)[b]	55	—
Company D (1974)	89	450
U.S. Geological Survey		
Hendricks (1965)	346	1300
Theobald et al. (1972)	458	1980
McKelvey (1974)	200–400	990–2000
Hubbert (1974)	72	540
USGS (1975)	61–149	322–655
USGS (1989)	58	399
USGS (1995)	93	939
National Academy of Sciences		
National Research Council (1975)	113	530
U.S. Energy Information Admin.[c]	105	682

[a]Exclusive of Alaska.
[b]Estimated discoverable between 1973 and 1985.
[c]*Annual Report*, 2003.

ological pressures or it is pumped to the surface. About 15 to 20% of the oil underground can be recovered in this way. Secondary recovery involves pumping of either water or gas into the surrounding area to force more oil toward the well. This process can produce perhaps another 15% of the petroleum underground.

Tertiary, or enhanced recovery, techniques now being tested at a number of U.S. oil fields are based on overcoming the two effects that lock the oil in: interfacial tension (surface tension) and viscosity. Oil is generally not found in underground pools of free liquid oil, but rather in layers of rock or sand of varying porosity. If the high viscosity of the oil is the main problem, carbon dioxide gas can be injected under pressure. When the gas dissolves in the petroleum, the viscosity is reduced, and the oil flows more freely. Steam injection is also used to heat the oil and thereby reduce its viscosity. It is also possible to inject air or oxygen and burn a fraction of the oil in place. This heats the general area and reduces the viscosity of the remaining oil.

If surface tension is the problem and the oil is locked into the pores of the rock, detergents can be injected and then the rock can be flooded with water

Table 2.2 Major Oil-Producing Countries

Country	Production[a] (10^3 bbl/day)	Proved Reserves[c] (10^9 bbl)	No. of Producing Wells[d]
Non-OPEC			
Canada	2,243	16.9	54,061
China	3,406	23.7	72,255
Egypt	623	3.6	
Mexico	3,339	16.0	
Norway	2,861	10.1	833
Russia	7,979	69.1	41,192
United Kingdom	2,123	4.5	
United States	5,731	29.4	521,070
Total Non-OPEC[b]	40,907		
OPEC			
Algeria	1,587	11.3	
Indonesia	1,190	4.4	
Iran	3,744	130.7	1,120
Iraq	1,101	115.0	1,685
Kuwait	2,176	96.5	790
Libya	1,415	36.0	
Nigeria	2,148	34.4	
Qatar	767	15.2	
Saudi Arabia	9,031	262.7	1,560
United Arab Emirates	2,351	97.8	
Venezuela	2,144	78.0	15,395
Total OPEC	27,654		

[a]A seven-month daily average for 2003. U.S. Energy Information Administration, October 2004.
[b]Totals are for all countries, not just those listed. World total is $68,561 \times 10^3$ bbl/day.
[c]*BP Statistical Review of World Energy*, June 2004. Data include crude oil, gas condensate, and natural gas liquids. World total is $1,146 \times 10^9$ bbl.
[d]*Oil and Gas Journal*, December 23, 2002.

containing polymers. There is no one technique applicable for all situations and there is generally a long period of trial and error. Enhanced recovery is expensive and can have unfortunate environmental consequences. One estimate of the total yield that may be realized by enhanced recovery in the United States is in the range of 20 to 75 billion barrels of oil.

2.5 World Production of Petroleum

The major oil-producing countries of the world are listed in Table 2.2 along with their current production rates, proved reserves, and number of wells. The total world production in 2003 was $68,561 \times 10^3$ bbl/day. The countries belonging to the Organization of Petroleum Exporting Countries (OPEC) account for $27,654 \times 10^3$ bbl/day, or 40% of the total.

It can be seen that the United States has a very large number of wells producing a relatively modest amount of oil. The average production from a single well in the United States is only about 11 bbl/day. In Saudi Arabia the production per well is over 500 times greater, about 5800 bbl/day. The comparatively small per-well production rate for the United States is indicative of the fact that we are now getting our domestically produced oil from relatively small reservoirs as compared with those producing oil in the Middle East.

In 1969, M. K. Hubbert analyzed *world* oil production to estimate when it would peak. For an extreme lower value of 1350×10^9 bbl estimated for Q_∞ (world), he predicted that world oil production would peak at 24×10^9 bbl/yr in 1990. For an upper limit estimate of 2100×10^9 bbl for Q_∞(world) he predicted that production would peak in the year 2000 at 37×10^9 bbl/yr. In either case, 80% of the world's oil would be used up in a time span of about 60 years.

A more recent analysis in 1993 by Edwards used the general method of Hubbert, but it was based on a longer period of observed oil production and came up with a somewhat more optimistic picture. Edwards' analysis predicted that world oil production will peak between the years 2010 and 2020 in the range of 74 to 98×10^6 bbl/day (27 to 36×10^9 bbl/yr). His estimate of the world's oil that is left and that is economically recoverable is 1792×10^9 bbl with proved reserves of 1092×10^9 bbl. With a cumulative production to 1993 of 719×10^9 bbl, this would mean a Q_∞(world) of 2511×10^9 bbl, somewhat larger than Hubbert's upper estimate from 1969.

There is growing evidence, based on the diminishing rate of discoveries, that a peak in the world production of oil is near. From that point forward, production will continue to decrease in a time of increasing world demand. The year at which the peak will occur is surprisingly insensitive to the value of Q_∞. For example, according to one recent estimate, increasing the worldwide Q_∞ from 2×10^{12} bbl to 3.3×10^{12} bbl only moves the year of peak production from 2007 to 2010.

2.6 The Cost of Gasoline in the United States

Gasoline in the United States is relatively inexpensive, comparable in price to some bottled water. In 2005, we hear complaints about having to pay $3.00 a gallon from drivers who remember paying only 23 cents a gallon. It is true that the price has risen, but not so much if you compare the price in constant dollars. In constant 2005 dollars, the cost of gasoline per gallon was about $2.10 in 1947 and over $2.80 in 1980. There has been a real price increase since the mid 1990s, but gasoline is now not much more expensive than it was 50 years ago. Gasoline is less expensive in the United States than in any other major industrialized country. With the exception of Canada, the gasoline prices in other industrialized countries are two or more times greater than in the United States. With our petroleum resources being rapidly depleted, and with an increasing number of cars on the road, what is going on?

There are two major forces at play. While the resource size and production are down in the United States, there is still tremendous production in countries such as the former Soviet Union, Saudi Arabia, Iran, and others. Oil production is an essential source of income for these countries, and competition has kept the international price low. OPEC has tried to limit production to drive prices higher, but this has generally failed because the revenues were needed and agreed-to limits were not respected. The price of oil worldwide has been kept down by this more-than-ample production. Because the United States imports more than 50% of its petroleum, we are highly vulnerable to disruption in foreign production and delivery.

The other factor in play is our tax rate. For various reasons, taxes on a gallon of gasoline in the United States are lower by a factor of 6.5 than in most industrialized countries. In 2002, we were levying a tax of $0.40 on a gallon of gasoline while other countries averaged around $2.80 per gallon. Canada had a tax of $1.00 per gallon.

It is not very popular to complain that a tax is too low, but the low cost of gasoline has some negative effects in terms of pollution, crowded highways and city streets, hastened resource depletion, imbalance of foreign trade, and so forth. Efforts to enact even a modest tax increase are beaten down by perhaps the most powerful lobby around. Restaurant and motel owners, fast food chains, oil companies, land developers, highway contractors, car manufacturers, tire manufacturers, service stations, and many others have profits linked directly to the number of cars on the road and hence to the cost of gasoline.

A higher tax would not necessarily bring the per mile cost of driving above what we have experienced in the past decades because technical advances have made automobiles considerably more fuel efficient. In 1973 an average new American passenger car got only 13 miles per gallon; by 2000, this number had risen to 28 miles per gallon.

2.7 *Petroleum Refining*

The crude petroleum taken directly from the ground is not directly usable for most applications. It is a mixture of hundreds of chemical compounds called *hydrocarbons* and some impurities such as sulfur. The crude petroleum needs to be treated to generate products that are essential for an industrialized society.

The first step in the process is *fractional distillation*. The crude oil is vaporized by heating it to about 400°C; the vapors are then introduced at the bottom of a fractionating column or tower. The vapors condense at various collection points going up the tower as the temperature for condensation decreases. Figure 2.4 shows the general arrangement. Gasoline, with the lowest boiling point, condenses and is drawn off at the top, whereas heating oil, which condenses at a higher temperature, is drawn off lower down. The gases that do not condense leave from the top of the column. The residue that comes from the bottom of the column contains asphalt and heavy fuel oil. Table 2.3 shows some of the products of fractional distillation, their molecular size, boiling point range, and typical uses.

The mix of products from a distillation column is not normally the same as the mixture demanded by the marketplace. For this reason, it is desirable to alter the products of fractional distillation to make them more useful. The usual deficiency is in the lighter molecules that are usable in gasoline, such as octane (C_8H_{18}), and the excess is in the heavy molecules of paraffin and tar. The oldest method of breaking up the heavy molecules into lighter ones is called thermal cracking, which uses high temperatures and pressures. A more efficient

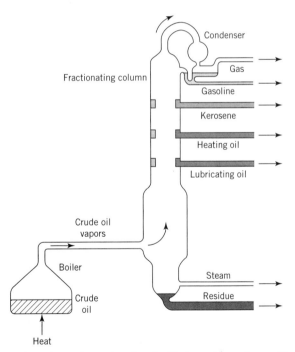

Figure 2.4 A fractionating column for the distillation of petroleum. The temperature of the column decreases going up from the bottom to the top so that the less volatile components are condensed near the bottom and the more volatile components are condensed toward the top of the column.

Table 2.3 Products of Petroleum Distillation

Fraction	Molecular Size Range	Boiling Point (°C)	Typical Uses
Gas	C_1–C_5	−164 to 30	Gaseous fuel
Petroleum ether	C_5–C_7	30 to 90	Solvent, dry cleaning
Gasoline	C_5–C_{12}	30 to 200	Motor fuel
Kerosene	C_{12}–C_{16}	175 to 275	Stove, diesel, jet fuel
Fuel oil	C_{15}–C_{18}	Up to 375	Furnace oil
Lubricating oil	C_{16}–C_{20}	Above 350	Lubrication
Greases	C_{18} and up	Semisolid	Lubrication
Paraffin (wax)	C_{20} and up	Melts at 52–57	Candles
Pitch and tar	High	Residue in boiler	Roofing, paving

Source: Spencer L. Seager and H. Stephen Stoker, *Chemistry: A Science for Today* (San Francisco: Scott, Foresman, 1973), p. 299.

process for producing higher octane gasoline is catalytic conversion. In this process petroleum vapor is passed over an alumina–silica mixture or certain types of clay that act as catalysts in bringing about the chemical changes. A photo of a typical refinery is shown in Figure 2.5.

Figure 2.5 A gas and oil refinery in Scotland, showing a complex array of modern chemical technology. Several fractional distillation towers are visible in this view. (*Source*: Mark A. Leman/Stone/Getty Images)

The reverse process of joining together light hydrocarbon molecules and forming heavier ones is also used and is called polymerization. Natural gas and other gases from the cracking process are made into high-octane fluids in this way.

The alkanes form an interesting series of hydrocarbon molecules. They follow the sequence of C_nH_{2n+2}, where n is an integer which can be chosen to be 1, or a higher number. The first eight members of the sequence are listed in Table 2.4.

Table 2.4 The Alkane Series of Hydrocarbons

n	Molecule	Name	Primary Use
1	CH_4	Methane	Natural gas
2	C_2H_6	Ethane	Natural gas
3	C_3H_8	Propane	Bottled gas
4	C_4H_{10}	Butane	Bottled gas
5	C_5H_{12}	Pentane	Gasoline
6	C_6H_{14}	Hexane	Gasoline
7	C_7H_{16}	Heptane	Gasoline
8	C_8H_{18}	Octane	Gasoline

The heat of combustion for these hydrocarbons ranges from about 53,000 Btu/kg for methane to 45,000 Btu/kg for octane.

2.8 Natural Gas

In recent years natural gas has become a favorite fossil fuel that many believe has great prospects of filling our energy needs in the coming decades. Natural gas delivered to the consumer is somewhat variable in composition. It is always principally methane, CH_4, typically 80 to 95%, and usually contains ethane, C_2H_6, along with small amounts of heavier hydrocarbons. There are several reasons for predicting that our use of natural gas will increase relative to other fuels.

(a) There have been improvements in the technologies for exploration and production and the resource base has increased.

(b) Relative to other fossil fuels, natural gas releases less harmful combustion products per unit of energy produced.

(c) There have been a number of new applications of natural gas such as motor fuel for transportation that increase the demand.

(d) On a Btu basis, compared to oil and electric energy, natural gas is a less costly source of energy.

Let's examine this last point as an example. In the United States, in September 2004, a typical cost for natural gas delivered to residential customers was $13.28/1000 ft^3. A typical cost of a kWh of electricity was about $0.089.

Example 2.1

Using the price information given in this chapter and the energy equivalents listed inside the book cover, find the ratio of the energy cost of electricity to that of natural gas.

Solution

Natural gas:

$$\frac{\$13.28}{10^3 \text{ ft}^3} \times \frac{10^3 \text{ ft}^3}{(1.035 \times 10^6 \text{ Btu})} = \mathbf{\$12.83/10^6 \text{ Btu}}$$

Electricity:

$$\frac{\$0.089}{\text{kWh}} \times \frac{1 \text{ kWh}}{3413 \text{ Btu}} = \mathbf{\$26.08/10^6 \text{ Btu}}$$

Relative cost of energy:

$$\frac{\text{Electricity}}{\text{Natural gas}} = \frac{\$26.08}{\$12.83} = \mathbf{2.03}$$

Thus in applications where heat is the main goal, there can be an obvious advantage to using natural gas. Household appliances such as clothes driers, water heaters, ovens, furnaces, and so forth are less expensive to operate using natural gas rather than electricity. A word of caution is needed, however, when furnaces and water heaters are considered. A natural gas furnace or water heater has an efficiency that can range from about 50 to 95%. Much of the heat energy lost escapes up the chimney. Ordinary electric resistance heating commonly used in water heaters or for space heating in homes is 100% efficient, offsetting some of the price advantage of gas heat. Electric heating does not use a chimney. On the other hand, the electric energy delivered to homes may come from a fossil fuel–burning power plant having an efficiency of only about 35%. In this case, it is questionable whether electric energy should be used for heating rather than burning the fuel directly in the home's furnace, with the electric energy reserved for higher uses such as electric motors, microwave ovens, lights, and televisions where fossil fuels cannot be used directly. If the electric energy used for heating comes from windpower, hydropower, or nuclear power, then this choice-of-fuels argument is not a consideration.

When one compares the price of natural gas to that of gasoline, using 2004 prices, natural gas is found to be slightly less expensive.

Example 2.2

Use your estimate of the cost of gasoline, and the cost of natural gas energy from Example 2.1, to find the ratio of energy costs for these two fuels.

Solution

Gasoline:

$$\frac{\$1.85}{1 \text{ gal}} \times \frac{1 \text{ gal}}{1.25 \times 10^5 \text{ Btu}} = \mathbf{\$14.80/10^6 \text{ Btu}}$$

Relative cost of energy (using natural gas price from Example 2.1):

$$\frac{\text{Gasoline}}{\text{Natural gas}} = \frac{\$14.80}{\$12.83} = \mathbf{1.15}$$

The lower cost of energy from natural gas is one reason why compressed natural gas is finding its way into powering taxis, city buses, and other vehicles, and it is being considered for extensive use for private automobiles. Other reasons are reduced emissions and engine wear. The cost of natural gas for large users such as industry and electric utilities is much lower than for residential consumers. The large-user cost in 2004 was only about $6/10^6$ Btu, less than half of what residential users pay.

2.9 The History of Use of Natural Gas

The long history of the use of natural gas dates back to the sixth century B.C. in China and Japan. In those countries gas was transported to lighting fixtures through bamboo pipes. However, those early uses were not common. More widespread use occurred in 1821 in Fredonia, New York, where a pipe provided gas for 30 burners from 27 feet underground. Although abundant sources of natural gas were found later, particularly in Kansas and Texas, their development was held back for many years by the fact that there was no system of pipelines in place from source to user. By 1923 the annual use of natural gas in the United States was only 1×10^{12} ft^3 (1 tcf). In those early years, much of the natural gas emerging from oil wells was wasted by venting it into the atmosphere or flaring it at the source because of the lack of a delivery system.

During World War II, and particularly in the years following the war, there was a rapid expansion of the natural gas industry in the United States. An extensive pipeline system was laid from the Texas panhandle to the northeastern United States. This was later extended to the west coast. Development of large diameter welded steel pipes and improved pipe-laying techniques made the rapid expansion possible. The economic advantage of natural gas over other fuels, as well as its convenience and cleanliness, drove widespread adoption of this fuel. Oil, coal, or coal gas (manufactured gas) were at a disadvantage relative to natural gas.

Figure 2.6 shows the annual production rate of natural gas in the United States. Production increased very rapidly from 1900 to 1973, and then fell until 1986 when it began to rise again. The production curve also approximately rep-

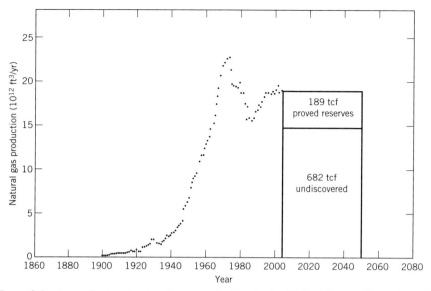

Figure 2.6 Annual rate of natural gas production in the United States. Data through 1996 are from the American Petroleum Institute. Data from 1997 through 2003 are from the U.S. Energy Information Administration. The rectangle at the right has an area representing the 871 tcf of natural gas estimated to be remaining for future production. The proved reserves are from Table 2.5. The undiscovered recoverable natural gas is from Table 2.1.

resents the consumption rate, since very little natural gas is exported and since only about 17% of our natural gas is imported, mostly by pipeline from Canada. In recent years, increased concerns about air pollution from coal have made natural gas the fuel of choice for many new electric power plants. A substantial fraction of the power plants now under construction in the United States will use natural gas as a boiler fuel to produce steam to drive turbines. This is a proven technology with many years of experience behind it. Also, there have been advances in newer types of gas turbines that use the combustion of natural gas to drive turbines directly, without the need for a steam-producing boiler. Steam-driven power plants typically have an efficiency of about 35% for converting the energy in the fuel to electricity. A gas turbine system, on the other hand, can have an efficiency of 45% or higher when used with a combined cycle that utilizes the excess heat from the gas turbine to generate steam to power a second turbine. On a capacity basis, such plants cost about half as much to build as a conventional coal-fired plant equipped with adequate pollution control devices. The use of natural gas as a fuel entirely avoids air pollution due to sulfur and particulates.

In addition to increased use in power plants, natural gas is also coming into use as a motor fuel in cars, trucks, and buses. For this use, in order to get sufficient vehicle range, the natural gas must be compressed, under pressure in the vehicles' fuel tanks. There are still limitations in the natural gas distribution and storage systems for motor fuel use, but these problems are yielding to the powerful forces of lower cost and reduced air pollution.

2.10 The Natural Gas Resource Base in the United States

The possible advantages of using natural gas more extensively, particularly in the transportation sector, will never be realized if we do not have the natural gas. In contrast to oil, the natural gas that we use comes mainly from domestic production. In 2002, a volume of about 23 tcf (23×10^{12} ft^3) of natural gas was consumed in the United States. About 20 tcf was produced domestically, and nearly 4 tcf came by pipeline from Canada. Smaller amounts also came by pipeline from Mexico and by liquefied natural gas transport ships from Trinidad, Algeria, and Nigeria, as well as minor imports from other countries. The available statistics do not always quite add up correctly for any given year, but they are close.

The cumulative production of natural gas in the United States, including Alaska, from the beginning of the industry through 2002 is about 1037 tcf. The proved reserves at the end of 2003 are about 189 tcf. Dividing the 2002 consumption rate (about the same as the 2003 rate) into the 2003 proved reserves gives us about 8 years left at the beginning of 2004, ignoring imports and undiscovered reserves. Also, it is apparent that at the beginning of 2004 the remaining proved reserves amount to only 189/(1037 + 189) = 15% of the known domestic natural gas consumed and in proved reserves. Ignoring the undiscovered resource, we have now consumed about 85% of our natural gas.

The actual situation is not that alarming, although it is clear from Figure 2.6 that the maximum production rate was reached more than 30 years ago, and that we can now expect a generally declining rate of production for the future. As with oil, the real question is how much economically recoverable natural gas is still in the ground and not yet discovered. From Figure 2.6, it is unlikely that a single bell-shaped curve could be constructed to fit the existing data sufficiently accurately to obtain Q_∞ in the spirit of a Hubbert-type analysis. Nevertheless, analysts have examined the data and made projections of future production by complex means, sometimes involving composites of more than one bell-shaped curve. A conclusion presented recently by the U.S. Energy Information Administration is that, as of 2003, there remain 682 tcf of natural gas still to be discovered. An optimistic view is that, at current rates of consumption, this would be good for 30 years. However, this number will be shortened considerably if the consumption rate increases. Also, as one nears the end of a resource, it is unlikely that an adequate production rate can be maintained.

Q_∞ for natural gas in the United States can be estimated by adding the amount consumed to the beginning of 2004 (1060 tcf), the proved reserves at that time (189 tcf), and the undiscovered resource (682 tcf) to obtain a Q_∞ of 1931 tcf. The amount we have already consumed is about 55% of this Q_∞.

48 Chapter 2 **The Fossil Fuels**

There have been some new and encouraging developments in the search for natural gas, related to coalbed methane. Methane (CH_4) is the major component of natural gas. It was produced at the same time that coal was formed hundreds of millions of years ago. Methane has been a hazard in the underground coal mining industry because it is toxic to the miners and can lead to disastrous underground explosions. Now the problems of extracting the methane from the coalbeds are being solved and the gas is beginning to reach the marketplace. One of the problems has been the large amount of water in the coal seams that must be pumped to the surface before substantial quantities of methane can be extracted. In 2000 about 1.2 tcf of coalbed methane was produced.

One must use some caution in quoting fossil fuel reserves to an accuracy of four significant figures because the numbers cannot be that accurate. For example, the USGS has estimated a 95% chance that the remaining gas resource is greater than 926 tcf and a 5% chance that it is greater than 1260 tcf. Their best estimate, 1102 tcf, is near the middle of this range.

Occasionally one will hear statements to the effect, "We have 50 years' worth of natural gas and we are finding more all the time." There are several pitfalls with these kinds of generalizations. First, people often forget that the calculation was done using the present rate of consumption. If natural gas is going to be used more widely for transportation and other uses, by an increasing population, the rate of consumption will go up thereby decreasing the remaining lifetime. Second, the fact that more natural gas is being discovered is already accounted for in the calculation. What we have now in proved reserves, with our present rate of consumption and no new discoveries, is good for only about eight years.

2.11 The Natural Gas Resource Base for the World

There is an impressive amount of natural gas in the reserves of the countries of the world. The proved reserves of the United States are about 3% of the total. Countries with some of the larger reserves are listed in Table 2.5. The total for all countries is about 6400 tcf.

Exploration for natural gas is less advanced than for oil. In fact, much of the known natural gas resource is a by-product of oil exploration. This situation is rapidly changing as the economic and environmental benefits of natural gas are providing an impetus for oil companies to search specifically for natural gas in likely geologic formations.

In addition to the conventional resources of natural gas that have been considered in this discussion, the potential of unconventional resources cannot be discounted. One example would be the natural gas hydrates, or methane clathrates, known to be present on deep seabeds worldwide. This resource has not yet been exploited to a large degree for various technical reasons, but the potential quantity of methane is immense. The U.S. Geological Survey reports, "The worldwide amount of carbon bound in gas hydrates is conservatively esti-

Table 2.5 Major Natural Gas Proved Reserves (2004)

Country	Reserves (tcf)
Russia	1680
Iran	940
Qatar	910
Saudi Arabia	231
United Arab Emirates	212
United States	189
Algeria	160
Nigeria	159
Venezuela	148
Iraq	110
Australia	90
Indonesia	90
Malaysia	75
Norway	75
Netherlands	62
Canada	59
Egypt	59
Kuwait	55
China	53
Libya	46
India	30
Pakistan	27
Mexico	15

Source: *Oil and Gas Journal*, January 1, 2004.

mated to total twice the amount of carbon to be found in all the fossil fuels on earth." The role of this huge store of methane as an eventual contribution to atmospheric carbon dioxide must be carefully assessed.

One of the major hindrances to wider use of natural gas on a worldwide basis is the lack of extensive pipeline systems to bring the fuel to consumers. In many countries these pipelines do not exist and they will be expensive to construct. The United States now imports more than half of the oil it uses, but this degree of importing natural gas fuel will be difficult to achieve because Canada and Mexico are the only likely sources of pipeline gas for the United States.

While it is technically possible to transport natural gas as a liquid (LNG) at low temperature ($-260°F$ at atmospheric pressure), the process and the specialized ships are expensive and there is a potential for explosions. Nevertheless, imports of LNG are increasing and now amount to about 10% of natural gas imports. The transport on ships of natural gas in gaseous form at normal temperatures is not a reasonable alternative; the amount of energy transported in this less dense form is far too small to make the process practical. The liquefied gas contains about 600 times more energy per cubic meter than does the gaseous form.

2.12 The Formation of Coal

The history of coal formation differs significantly from that of oil and natural gas. Coal has its origin in the abundant plants that started to grow 350 million years ago. There were huge swamps then with giant ferns, shrubs, vines, trees, and algae that grew and then fell into decay. The carbon in the organic material accumulated in layers at the bottom of the swamp. Initially the dead plant material was decomposed by aerobic bacteria, yielding carbon dioxide (CO_2), methane (CH_4), and some other gases. Later the layers of what was to become coal were completely covered so they were no longer exposed to air, and the decay proceeded further by anaerobic action.

The first stage of coal formation produces peat. In time, the peat or humus type material was heated and compressed by the overburden and carbon rich coal was formed. Anthracite is the oldest form of coal. It is the hardest and cleanest burning, with a carbon content of up to 95%. Bituminous coal is about 300 million years old and is about 50 to 80% carbon. Lignite, a third type of coal, is about 150 million years old and its carbon content ranges up to 50%. Some lignite will show traces of its fossil origins. The lowest grade of coal-like material is peat; it provides a low-quality fuel in many parts of the world as well as serving as organic material for gardeners.

Coal occurs in stratified deposits interlain with soil and rock in sedimentary basins. The average depth beneath the earth's surface is about 300 ft, and the average thickness of the coal strata (or seams) is about 2 to 8 ft. In parts of the West, however, seams 100 ft thick are being mined. The largest deposits are in Wyoming, North and South Dakota, Montana, and New Mexico. The western coals are mainly subbituminous and have a typical heat content of about 9000 Btu/lb. Table 2.6 shows the characteristics of some coals of the United States.

2.13 Coal Resources and Consumption

The size of the coal resource is more readily determined than that for oil or natural gas because the coal beds tend to occur quite near the surface over large areas that are well documented. Coal is found in significant amounts in three major regions in the United States: the Appalachian Basin (parts of West Vir-

Table 2.6 Analysis of Some Coals of the United States

Rank	Source	Percent Carbon	Percent Sulfur	Btu/lb
Anthracite	Pennsylvania	88	0.9	13,300
Semianthracite	Arkansas	79	1.7	13,700
Bituminous	Pennsylvania	57	1.4	13,870
	Illinois	49	0.9	11,930
	Michigan	50	1.2	11,780
	Kansas	53	4.3	12,930
	Utah	51	0.8	12,760
	Washington	50	0.5	12,250
Subbituminous	Wyoming	46	0.6	10,750
Lignite	North Dakota	27	0.6	6,750
	Texas	16	1.3	7,140

ginia, Pennsylvania, Ohio, and eastern Kentucky), the Illinois Basin (Illinois, western Kentucky, and Indiana), and the northern Great Plains and Rocky Mountain region (Montana, Wyoming, Colorado, North Dakota, and New Mexico.) The minable coal in the United States is 71% bituminous and subbituminous, 28% lignite, and only 1% anthracite. Almost all of the anthracite deposits are in Pennsylvania and are accessible only by underground mining. The bituminous coal is mainly east of the Mississippi in Illinois and West Virginia. The very large coal deposits are subbituminous and lignite and they are located in the Rocky Mountain region. The coal in this region is low in sulfur and much of it is obtainable from surface mines. Sulfur dioxide is perhaps the pollutant of most concern that results from burning coal in electric power plants. Ash-producing substances in the coal are also important as they add useless weight to the unburned coal and require expensive removal after burning. More importantly, they contribute to the particulate emissions during the burning process.

There is a vast amount of coal in the world (see Table 2.7) and a substantial fraction of the proved reserves (about 25%) is located in the United States. On a Btu basis there is about eight times as much energy in the coal reserves of the United States as there is in either the remaining oil or natural gas.

As we saw in Figure 1.1, in the middle of the 19th century wood was the major source of energy in the United States. As the industrial revolution proceeded, coal quickly replaced wood. There are as much as 13,000 Btu in a pound of coal, and about 8,000 Btu in a pound of wood. Abundant sources of wood became limited and it required enormous manual labor to harvest the quantities needed to power steam engines. As the United States and the other industrialized coun-

Table 2.7 Estimated Worldwide Coal Proved Reserves

Country	Amount (10^9 tonnes)	Percentage of World Total	R/P ratio (years)
United States	250	25	260
Former Soviet Union	230	23	≥500
Europe	138	14	300
China	115	12	70
Australia	82	8.3	240
Africa	55	5.6	230
South and Central America	22	2.2	350
North America (not incl. U.S.)	7.7	0.8	110
World total	984	100	192

Proved reserves are the amount recoverable from known deposits under existing economic and operating conditions.
The reserves-to-production (R/P) ratio is the number of years that the remaining proved reserves would last if production continues at the present level.
Source: BP Global Reports, December 2004. Reserves data from World Energy Council.

tries turned increasingly to steam engines to power trains, ships, mills, and factories, coal became the fuel of choice. By 1900 more than three times as much energy was derived in the United States from coal as from wood. The development of coal-powered steam electric plants to furnish the growing demand for electricity accelerated the consumption of coal. The annual production of coal is shown in Figure 2.7. Expansion of the coal-powered railroads provided the needed transportation of coal from the mines to the consumers. The major use of coal now is for production of steam in electric power plants. In 2000, 958 million tons of coal was consumed in the United States. Electric utilities used more than 90% of this.

In 2000, 1073 millions tons of coal was produced in the United States and 58 million tons of this was exported. There was no general increase in coal production over the 50 year period from 1920 to 1970. However, in the past 35 years, coal production has increased considerably to double what it was in the previous decades. Some of the failure for coal production to increase even more rapidly in this time period is due to concerns about the environmental impact of surface mining, the effect of CO_2 emissions on global climate, and health effects of SO_2 and particulate emissions. Federal regulations on SO_2 and particulate emissions now require a large investment in mitigation equipment, and there is no practical way known to reduce CO_2 emissions except by reducing the rate of burning carbon-based fuels. All of these factors have held back what would have been even larger increases in coal consumption.

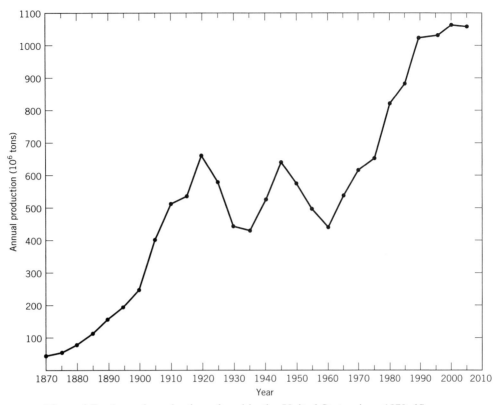

Figure 2.7 Annual production of coal in the United States since 1870. (*Sources: Historical Statistics of the United States: Colonial Times to 1970*, Washington D.C.: U.S. Department of Commerce, Bureau of the Census, 1975; and U.S. Energy Information Administration, *Annual Energy Review*, 1989, 1996, 2003.)

In discussing coal, it is important to distinguish between *proved reserves* and the total *resource*. The term *resource* is usually used to describe the total amount of coal known to be in the ground. The *proved reserves* is a smaller number describing the amount that is recoverable under existing economic and operating conditions. In the United States the coal *resource* is some six times larger than the *proved reserves*.

The proved reserves of coal in the United States are about 250×10^9 tonnes.[1] If we continue to consume 958×10^6 tonnes per year, the resource would last 261 years, an impressively long time. However, it is not credible to think that the consumption rate will remain constant for such a long time. Also, it could turn out that this enormous fossil-fuel resource will never be used as serious global warming becomes a reality.

---

[1] The *tonne* is the metric ton, equal to 1000 kilograms. It is also equal to 1.102 ton, or 2205 pounds. One ton is the same as 2000 pounds.

54 Chapter 2 **The Fossil Fuels**

In addition to the three conventional fossil-fuel resources just discussed, there are two nonconventional sources that should be mentioned: shale oil and tar sands.

2.14 *Shale Oil*

About 50 million years ago, the Green River Formation, the most extensive deposit of oil shale in the United States, was covered by two large tropical lakes. Organic matter from the surrounding hills was deposited along with sedimentary material on the lake bottoms to a depth of about 3000 feet over a period of many millions of years. The combined materials from plants and possibly some aquatic animals formed a carbon-bearing mudstone, or marlstone, on the lake bottoms. This is the source of shale oil. Through geological changes in the earth's crust, these deposits were brought up to their present elevations. Erosion by rivers cut into the formations, leaving much of the oil shale now exposed.

The oil shale, or marlstone, contains an organic substance called *kerogen*. Kerogen is a solid hydrocarbon, waxlike substance that vaporizes when heated sufficiently. The shale oil that results from condensing these vapors can be further refined to produce gasoline, fuel oil, and coke.

There is a long history of shale oil being just on the edge of becoming an economically recoverable and vast source of fuel. The resource is astonishingly vast; the Green River Formation contains between 600 to 2000 billion barrels of oil compared to a Q_∞ for petroleum in the United States of about 324 billion barrels. If shale oil were made available, the petroleum resource problem might not be entirely solved, but it would postpone the oil shortage crisis for a long time. The reality is that aside from small outputs from various pilot projects, shale oil has yet to play a serious role in the petroleum market. Significant production on the order of 100,000 barrels per day may not be possible for many years.

Figure 2.8 shows the location of the Green River Formation in Utah, Wyoming, and Colorado. Areas with oil shale that yields 25 gallons per ton or more and with deposits 10 feet or more in thickness are indicated with the bold cross hatching.

Table 2.8 shows the size of the oil shale resource in Colorado, Utah, and Wyoming under two different criteria: (a) layers at least 10 feet thick with 25 or more gallons of oil per ton of shale and (b) layers at least 10 feet thick with 10 to 25 gallons of oil per ton of shale. A rich oil shale sample is shown in Figure 2.9.

The removal of the overburden to get at the shale has been an expensive process and has hindered the development of the resource. Because of this difficulty, it has been estimated that only about 80 billion barrels of oil from the richest potential of 600 billion barrels may be economically recoverable.

A fundamental factor holding back the development of oil shale is related to the energy density of the material, exclusive of the overburden. If we consider only the better oil shale having 25 gallons of oil per ton of shale, this amounts

Shale Oil 55

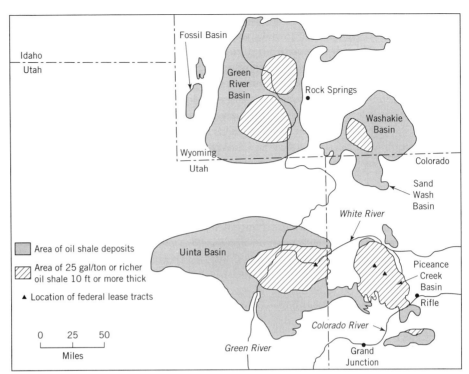

Figure 2.8 Oil shale deposits in the Green River Formation of the western United States.

to an energy density of about 3.5 million Btu per ton. For coal the energy density is typically about 8 times greater, 27 million Btu per ton. This large difference in energy density presents an enormous disadvantage for shale oil as long as abundant and cheap coal is available on a competitive market. There are well-established processes for making both liquid and gaseous fuels from coal, but at some cost in energy.

Among other difficulties encountered in attempts to produce liquid fuel from oil shale is the problem of disposing of the spent shale. In the retorting (heating)

Table 2.8 Oil Shale Deposits in the Green River Formation

	Billions of Barrels of Oil in Place			
	Colorado	Utah	Wyoming	Total
>10 ft thick, >25 gal/ton	480	90	30	600
>10 ft thick, 10 to 25 gal/ton	800	230	400	1430
Total: >10 ft thick, >10 gal/ton	1280	320	430	**2030**

Figure 2.9 A rich sample of oil shale found in western Colorado. (*Source*: Steve Northrup/Black Star)

of the shale to drive out the hydrocarbons the solid material expands by 35%, this means that the waste can't simply be disposed of by returning all of it to its original location. This presents a problem in finding a disposal site. Also, there is a large water requirement in shale oil processing; typically 3 gallons of water per gallon of oil produced is needed. This is a particular problem for the Green River Formation because it is in one of the most arid regions of the country. Some estimates indicate that the available water is only sufficient to support the production of a million barrels of oil per day, about 5% of our national oil consumption.

Attempts have been made to overcome some of these difficulties by developing an *in-situ* conversion process (ICP). In this process subsurface heaters convert the kerogen to high quality oil and hydrocarbon gases. The ICP process reduces significantly the problems of open-pit or subsurface mining, leftover tailings, and water use. The company that has worked on development of this process has indicated that the process could be competitive once steady-state production is achieved. To this date, the ICP process has not produced commercial quantities of shale oil.

The size of the shale oil resource is impressive, but the problems involved in bringing it to the marketplace are extensive.

2.15 Tar Sands

Tar sands are deposits of sand impregnated with a thick, tarlike hydrocarbon substance which is a very viscous crude oil called bitumen. Because of the high viscosity of the bitumen, it will not flow like normal liquid petroleum into wells from which it can be pumped to the surface. Instead, the tar sands must be mined as a solid material and transported to a processing plant where the bitumen is extracted by steam or hot water. The bitumen then can be refined by the same methods used for heavy crude oil.

Known deposits of tar sands are sufficient in size to make an impact on the petroleum resource picture. The Athabasca deposit near Fort McMurray in northeastern Alberta, Canada, is by far the largest deposit in North America. The Athabasca deposit contains up to a third of the world's known liquid hydrocarbon energy resources. Although large-scale efforts have been underway since 1966 to mine and extract oil from the deposit, it is only recently that costs have been controlled to the point that the resource can begin to be exploited in a major way and yield a profit. It is estimated that 300 billion barrels of oil are economically recoverable from the Athabascan deposit.

At Fort McMurray, huge shovels scoop up bites of tar sands from an open-pit mine and put the tar sand into trucks for transport to separators. The tar sand is put into rotating drums and mixed with hot water and caustic soda that separate the sand and water from the bitumen. Further purification takes place in a centrifuge and then the bitumen is broken down (cracked) into naptha, kerosene, and heavy fuel oil. In 2003 an average of 920,000 barrels/day were produced, and by 2010 it is expected that production will increase to 1.8 million barrels/day. The production cost in 2002 was somewhat less than $20/bbl. It takes two tons of tar sands to produce one barrel of oil. Thus the energy density in terms of Btu/ton is similar to that of oil shale but far less than for coal.

2.16 Summary

The fossil fuel resource presents a broad range of complex problems and potentialities. The resource in all its facets is made up largely of a vast store of carbon atoms that originated in the ancient atmosphere of many millions of years ago. These atoms are now there for our use. When we cause them to combine with oxygen in our engines and stoves they return to their original form as the carbon dioxide of today's atmosphere, yielding energy in the process. The hydrogen component of the hydrocarbon fossil fuels likewise has been held for our use since the hydrogen atoms were taken one by one from ancient water. We now make them combine again with oxygen to give us new water and useful energy. Of course, the energy of this closed process came originally from the sun, never to return but to find its way eventually into the background thermal energy of the universe.

The fossil fuel resource, in the forms of petroleum, natural gas, and coal, that we have consumed eagerly for the past many decades, especially in the United States, now has a visible end. The remaining fuels, along with other resource information, are summarized in Table 2.9. Whereas once the nation had very little worry about its ever-increasing appetite for gasoline and natural gas, we are now forced to plan for other energy sources. New discoveries in envi-

Table 2.9 Conventional Fossil Fuels in the United States

Type	Remaining[a]	Energy Content (QBtu)	Percent
Petroleum	134×10^9 bbl	780	9
Natural Gas	871 tcf	900	10
Coal	275×10^9 ton	7300	81

[a]The remaining petroleum and natural gas includes both the proved reserves and the undiscovered recoverable resource shown in Figures 2.2 and 2.6. For coal, the remaining amount is the proved reserves recoverable from known deposits under existing economic and operating conditions shown in Table 2.7. (Note: 1 tonne = 1.10 ton.)

ronmental science are making it apparent that even if the fossil-fuel resource were not limited, our prodigious burning of fossil fuels carries with it a threat against the stability of the global climate, and there might now be powerful cause to curtail use of fossil fuels.

Key Terms

Photosynthesis
Proved reserve
Resource
Primary recovery
Secondary recovery
Tertiary recovery
Enhanced recovery
Fractional distillation
Thermal cracking
Catalytic conversion
Polymerization

Methane
Peat
Lignite
Bituminous
Anthracite
Oil shale
Marlstone
Kerogen
Retorting
Bitumen
Tar sands

Suggested Reading and References

1. Hubbert, M. King. "Energy Resources." *Resources and Man.* National Academy of Sciences—National Research Council. San Francisco: W. H. Freeman, 1969.
2. Hubbert, M. King. "The Energy Sources of the Earth." *Scientific American,* **224** No. 3 (September 1971), pp. 60–70. (Also in *Energy and Power,* a Scientific American book. San Francisco: W. H. Freeman, 1971.)

3. Hall, C. A. S., and Cleveland, C. J. "Petroleum Drilling and Production in the United States: Yield Per Effort and Net Energy Analysis." *Science,* **211** (February 6, 1981), p. 576.

4. Yergin, Daniel. *The Prize.* New York: Simon and Schuster, 1991.

5. Priest, Joseph. *Energy: Principles, Problems, Alternatives.* Reading, MA: Addison-Wesley, 1991.

6. Menard, H. William. "Towards a Rational Strategy for Oil Exploration." *Scientific American,* **244** No. 1 (January 1981), pp. 55–65.

7. Aubrecht, Gordon. *Energy.* Columbus, OH: Merrill Publishing Company, 1989; 2nd Ed., Prentice-Hall, 1995.

8. 1995 National Assessment of United States Oil and Gas Resources. *U.S. Geological Survey Circular* **1118,** Washington D.C.: U.S. Government Printing Office.

9. Edwards, John D. "Crude Oil and Alternate Energy Production Forecasts for the Twenty-first Century: The End of the Hydrocarbon Era." *AAPG Bulletin*, **21,** No. 8 (August 1997), pp. 1292–1305.

10. Campbell, Colin J., and Laherrere, Jean H. "The End of Cheap Oil." *Scientific American* (March 1998).

11. George, Richard L. "Mining for Oil." *Scientific American* (March 1998).

12. Anderson, Roger N. "Oil Production in the 21st Century." *Scientific American* (March 1998).

13. Fouda, Safaa A. "Liquid Fuels from Natural Gas." *Scientific American* (March 1998).

14. Appenzeller, Tim; photographs by Leen, Sarah; The End of Cheap Oil; *National Geographic* **205 6** (June 2004), pp. 80–109.

15. Parfit, Michael; photographs by Leen, Sarah; After Oil, Powering the Future; *National Geographic* **208 2** (August 2005), pp. 2–31.

Questions and Problems

1. In Example 2.2 it is shown that gasoline costs more per Btu than does natural gas. Why then do we power our cars and trucks predominantly with gasoline rather than natural gas?

2. Why are fossil fuels not being created in nature fast enough to keep up with our use of them?

3. Why is two-thirds of the petroleum left in the ground after the primary extraction process comes to an end? Can this problem be overcome?

4. If there were originally 30 billion barrels of recoverable oil in Alaska, for how many years could the United States supply its present demand from that source alone?

5. On a worldwide basis, how much remaining oil is there per person? How does this compare to the remaining oil per person in the United States?

6. What is oil shale and how can it be converted into useful fuels for transportation and heating?

7. If your home requires one million Btu of heat energy from the furnace on a cold winter day, what is the cost in dollars to heat the house with natural gas, assuming a 60% furnace efficiency?

60 Chapter 2 **The Fossil Fuels**

8. How many pounds of western subbituminous coal would be needed to heat the house in problem 7 with a 60% efficient furnace?

9. What measures would you recommend to be enacted by the U.S. Congress to promote the conservation of fossil fuels?

10. How many tons of oil shale are required each day to supply a 10,000 bbl/day retort, if the shale yields 25 gallons per ton?

11. How much energy, in foot-pounds, is needed to raise one barrel of oil 25,000 feet? How much is this energy in terms of the heat energy content, in equivalent barrels of oil? Assume that the oil weighs 7.0 lb/gal.

12. It has been estimated that a person can perform continuous manual labor at a power of 50 watts for an 8-hour working day. How many pounds of coal contain the energy equivalent of the useful physical labor a person can perform in this time period? How many gallons of oil?

Multiple Choice Questions

1. A bacterial colony starts growing in a jar at about 11:00 A.M. The size of the colony doubles each minute, and the jar is just full at 12:00 noon. At what time was the jar 1/8 full? See Appendix A.
   a. 11:07
   b. 11:15
   c. 11:30
   d. 11:36
   e. 11:45
   f. 11:56
   g. 11:57
   h. 11:58

2. Coal, oil, and gas represent approximately ___%, ___%, and ___%, respectively, of the U.S. fossil fuel energy use.
   a. 27, 46, 27
   b. 25, 20, 55
   c. 90, 5, 5
   d. 35, 35, 30
   e. 52, 38, 10

3. In the United States the total energy consumed per year per person is the equivalent of about ___ barrels of oil.
   a. 5.8
   b. 58
   c. 580
   d. 5800

4. The declining order of importance for the various energy sources in the United States is
   a. natural gas, oil, coal, nuclear, hydro
   b. oil, coal, natural gas, nuclear, hydro
   c. oil, natural gas, coal, hydro, nuclear
   d. coil, oil, natural gas, hydro, nuclear

5. A reasonable estimate for Q_∞ (petroleum) for the United States, including Alaska and offshore oil, is about
   a. 324×10^9 bbl
   b. 250×10^6 bbl
   c. 113×10^9 bbl
   d. 458×10^9 bbl

e. 55×10^9 bbl
f. 113×10^6 bbl
g. 165×10^8 bbl
h. unknown

6. The total annual energy consumption in the United States is about ___ QBtu, and this is the equivalent of about ___ tons of coal per year for each person.
   a. 80, 13
   b. 800, 130
   c. 8000, 1.4
   d. 98, 13

7. One of the problems of producing shale oil is the amount of water needed for aboveground retorting. The engineers designing this process estimate that about ___ barrels of water are needed for each barrel of oil produced.
   a. 300
   b. 30
   c. 3
   d. 0.3

8. In tar sands, the organic material that contains the oil and that surrounds the grains of sand is
   a. kerogen
   b. kerosene
   c. marlstone
   d. bitumen
   e. paraffin

9. The energy content of the recoverable conventional fossil fuels remaining in the United States are in the ratio of approximately ___ to ___ to ___ for petroleum, natural gas, and coal, respectively.
   a. 1, 1.1, 9
   b. 1, 0.5, 15
   c. 1, 0.75, 58
   d. 1, 0.2, 150
   e. 1, 3, 150
   f. 1, 0.2, 28
   g. 1, 1, 98
   h. 1, 1, 1

10. The coal that we burn in our power plants today represents solar energy trapped in organic molecules about ___ years ago.
    a. 300,000
    b. 3,000,000
    c. 300,000,000
    d. 3,000,000,000

11. As compared to coal, oil shale of good quality has about ___ as much energy content per ton of excavated material.
    a. one-tenth
    b. ten times
    c. one-hundredth
    d. one-half
    e. twice

12. The estimated total minable coal in the United States is about
    a. 0.5×10^9 tons
    b. 1.5×10^{12} tons
    c. 0.5×10^{12} tons
    d. 275×10^9 tons

13. The United States is currently importing about ___ barrels of oil each year at a cost of roughly 60 dollars per barrel.
    a. 1×10^3
    b. 1×10^6
    c. 4×10^9
    d. 4×10^6
    e. 6×10^9
    f. 6×10^6
    g. 9×10^9
    h. 9×10^6

14. Natural gas consists mostly of ___, and to some extent ___.
   a. propane, ethane
   b. butane, ethane
   c. propane, methane
   d. methane, ethane

15. The United States is using liquid petroleum at the rate of about
   a. 6.7×10^9 gal/day
   b. 6.7×10^9 bbl/day
   c. 6.7×10^9 gal/year
   d. 6.7×10^9 bbl/year

16. The oil that we burn in our cars today represents solar energy that was captured in organic molecules about ___ years ago.
   a. 200,000
   b. 2,000,000
   c. 200,000,000
   d. 2,000,000,000

17. At our present rate of use, the estimated total remaining recoverable petroleum of the United States would supply our needs for about ___ years.
   a. 20
   b. 8
   c. 38
   d. 72

18. The average percentage of the total petroleum left in the ground after primary and secondary recovery has been accomplished is about
   a. 30%
   b. 20%
   c. 10%
   d. 5%
   e. 70%
   f. 90%
   g. 50%
   h. 0%

CHAPTER 3

Heat Engines

(*Source*: Hulton Getty/Stone/Getty Images)

3.1 The Mechanical Equivalent of Heat

If we take a look at the energy units, it becomes apparent that what we think of as a rather small unit of heat energy, for instance one Btu, is in fact a very large amount of energy compared to what we see in a mechanical unit of energy, the foot-pound. One Btu is the same as 778 foot-pounds. The thought of lifting a one-pound weight 778 feet into the air with the energy released by the burning of only one match can certainly be appealing to those who do heavy lifting for a living. To realize this goal, we must first find a way to capture the heat energy of the fuel and turn it into mechanical energy.

The possibility of easing human labor by utilizing heat sources has been the driving force behind a long history of development of what we now call *heat engines*. This is the main subject of this chapter.

3.2 The Energy Content of Fuels

We go to a great deal of effort to obtain fossil fuels and other types of fuel for two basic purposes: to provide direct heating and lighting, and to power heat engines. The general pathways by which we do this are shown in Figure 3.1.

Heat engines and the thermodynamic principles that govern their operation will be discussed in detail later. First, let's look at how heat is derived from fuels. In simplified form, the burning of hydrocarbon fuels is merely the combining of the carbon and the hydrogen from the fuel with oxygen from the air. Burning of a hydrocarbon compound is the reverse of photosynthesis, and proceeds according to the basic chemical reactions:

$$C + O_2 \to CO_2 + \text{heat energy,}$$

and

$$H_2 + O \to H_2O + \text{heat energy.}$$

These two simple reactions are both included in the overall reaction formula for an actual fuel. For example, we may consider the burning process for heptane, C_7H_{16}, a colorless liquid constituent of gasoline.

$$C_7H_{16} + 11\ O_2 \to 7\ CO_2 + 8\ H_2O + 1.15 \times 10^6 \text{ calories per 100 g } C_7H_{16}$$

This example illustrates that when heptane and oxygen are completely combined in the burning process, each carbon atom unites with two atoms of oxygen to form carbon dioxide, and every two hydrogen atoms combine with one atom of oxygen to form water. Carbon dioxide and water are the only material products of the reaction, and the energy liberated is in the form of heat. In a sense, what is taking place is the recycling of carbon and hydrogen. By burning a hydrocarbon we put back into air and water what was once taken from air and water. The cycle can be continued indefinitely. This burning returns the carbon and hydrogen atoms to the molecular forms they had in the ancient world where the photosynthesis that formed the fuel took place.

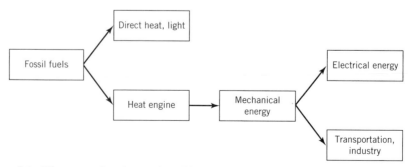

Figure 3.1 The general pathways by which we utilize energy from fossil fuels.

The number at the right in the formula, 1.15×10^6 calories per 100 grams, is the *heat of combustion* for heptane. Every fuel has a tabulated value for this quantity, often given in so much energy for so much mass of the fuel and sometimes in an energy amount for a volume amount. Some representative values are given in the Energy Equivalents table inside the front cover. The heat of combustion is the definite maximum amount of energy available from a fuel, which cannot be exceeded by even the most clever techniques of combustion or carburetion. For a fossil fuel, the heat of combustion can be considered a measure of the solar energy stored in the fuel since ancient times. For firewood or other biomass, the heat of combustion is the solar energy stored in the plant life during the growing process.

3.3 *The Thermodynamics of Heat Engines*

A *heat engine* is any device that can take energy from a warm source and convert a fraction of this heat energy to mechanical energy. The warm source is commonly a coal-fired steam boiler, the combustion chamber of an automobile engine, or even the combustion chamber of a Fourth of July rocket. Other warm sources for heat engines could be a solar-powered steam boiler, a natural hot spring, or a nuclear reactor.

Over the past few hundred years, people have been ingenious in devising heat engines to convert heat energy into work wherever a temperature difference can be found in nature (geothermal or solar) or can be created, usually by burning some fuel.

Whenever we observe a fuel-burning heat engine in operation, it's clear that not all of the heat energy from the burning of the fuel is being used to perform useful work. Our automobiles, locomotives, and motorcycles all are seen to dissipate waste heat energy to the surrounding air through their radiators, cooling fins, and exhausts. Coal-burning power plants for electricity generation are all connected to rivers, lakes, or cooling towers or have some other means to carry off the considerable waste heat. The same is true for nuclear power plants. Why is this? Have the designers carelessly permitted energy to become waste heat rather than useful work? Not entirely. There is a fundamental requirement, based on the laws of physics, that some fraction of the heat energy taken from the source must always be rejected, at a temperature cooler than that of the warm source, to the environment. This is an important but subtle point, not at all obvious without detailed mathematical analysis.

The result of such a study of a heat engine is often presented as a *Carnot efficiency*, named after a French engineer, Sadi Carnot (1796–1832), who put forth this idea in 1824.

The Carnot efficiency, a number always less than 100%, is the percentage of the energy taken from the heat source which is actually converted to useful mechanical work, under the important assumption of an ideal engine. The ideal engine has never been achieved; there are practical realities which keep us from ever achieving this ideal. Bearings do have friction, insulation is never perfect enough to stop all leakage of heat, and so forth. But the concept of an ideal engine does lead to an expression for the maximum efficiency of a heat engine if all practical engineering problems could be overcome. It is perhaps surprising that this ideal efficiency is always less than 100%; for it to be 100% one would have to have a source of energy at an infinitely high temperature or reject heat energy at degrees Kelvin.

Let's take a closer look at the operation of a heat engine. A diagram of energy flows, work done, and temperatures that applies to any heat engine is shown in Figure 3.2.

By the usual definition of efficiency,

$$\text{Efficiency} = \frac{\text{work done}}{\text{energy put into the system}},$$

for any heat engine, in terms of the heat energy flows, this becomes

$$\text{Efficiency} = \frac{Q_{\text{hot}} - Q_{\text{cold}}}{Q_{\text{hot}}},$$

where the numerator, which replaces work with the difference between energy amounts, is a result of the Principle of Energy Conservation, as shown in Figure 3.2. We have simply observed that the sum of the energy flow out ($Q_{\text{cold}} + W$) must be equal to the energy flow in (Q_{hot}).

We can also express the efficiency in terms of a percentage, using a minor algebraic rearrangement of terms in the above equation, to get

$$\text{Efficiency} = \left(1 - \frac{Q_{\text{cold}}}{Q_{\text{hot}}}\right) \times 100\%.$$

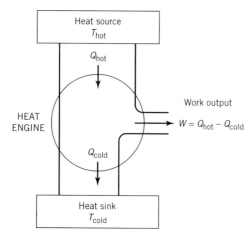

Figure 3.2 A thermodynamic diagram of a heat engine operating between a heat source and a heat sink at a lower temperature. The work output, because of conservation of energy, must equal the difference between the heat energy extracted from the source and that rejected to the sink.

The noteworthy contribution of Carnot was to show that for an ideal engine the ratio of two energy terms is identical to the ratio of two temperature terms:

$$\frac{Q_{\text{cold}}}{Q_{\text{hot}}} = \frac{T_{\text{cold}}}{T_{\text{hot}}},$$

where the temperatures of the heat source and heat sink are given on the absolute (Kelvin) scale. (The various temperature scales are discussed in the Appendix at the end of this book.) If we now follow along with Carnot and substitute the equivalent temperature ratio for the ratio of heat flows in the equation for efficiency, we get

$$\text{Efficiency (Carnot)} = \left(1 - \frac{T_{\text{cold}}}{T_{\text{hot}}}\right) \times 100\%$$

as the efficiency for an ideal heat engine. It is remarkable that this efficiency depends only on the temperatures of the two reservoirs between which the heat engine operates.

We can now calculate the Carnot efficiency for a realistic situation. For a coal-fired electric power plant such as shown in Figure 3.3, T_{hot} (the boiler temperature) would be about 825 K (552°C), and T_{cold} (the cooling tower) would be about 300 K (27°C). This leads to

$$\text{Efficiency (Carnot)} = \left(1 - \frac{300}{825}\right) \times 100\% = (1 - 0.36) \times 100\% = 64\%.$$

If this power plant were truly an ideal heat engine, its efficiency for producing mechanical energy from the energy of the fuel could be as high as 64%, meaning that 36% of the heat energy from the fuel must be wasted by rejecting it through the cooling tower to the surrounding atmosphere. It would be possible to use the waste heat for other purposes, such as space heating or greenhouses, but it cannot be used for producing mechanical work unless there is available another reservoir at a temperature significantly lower than 300 K. To make the Carnot efficiency as high as possible, it would be desirable to increase T_{hot} and decrease T_{cold}. In practice, however, the limit on T_{hot} is imposed by the materials from which the boilers can be constructed and the limit on T_{cold} is imposed by the availability in nature of large heat sinks at sufficiently low temperature.

3.4 Generation of Electricity

Now that we've seen how heat energy is derived from fuels and have gained an understanding of thermodynamic efficiency, let us examine how these ideas come together in the design of an electric power plant.

In 1831, in London, Michael Faraday (1791–1867) discovered that there exists in nature a phenomenon we now call *electromagnetic induction*. This was certainly one of the greatest discoveries of all time. It made the generation of electricity possible and also the transmission of electricity. Electric motors and all of our communications and computer systems were developed as a result of this discovery. Without an understanding of electromagnetic induction we would not now have a modern technological society, at least not in its present form. Faraday observed, with what now seems to be simple tabletop experimentation, that by moving a magnet near a loop of wire an electric current could be induced in the wire. This discovery quickly lead to the invention of electric generators and many other devices.

Our modern methods of electricity generation follow from Faraday's discovery. Generators, in their simplest form as shown in Figure 3.4, have coils of copper wire mounted on a rotating armature so that as the armature is forced to rotate, the coils are rotated through regions of magnetic field. The current induced in the coils according to the Faraday law interacts with the magnetic field to resist the motion of the coils through the field. Therefore it takes energy from some external source to force the rotation of the armature and move the coils through the magnetic field. Of course energy must be conserved in the overall process. The energy that appears in the form of electricity was earlier mechanical energy in whatever mechanism drove the armature to rotate at high speed.

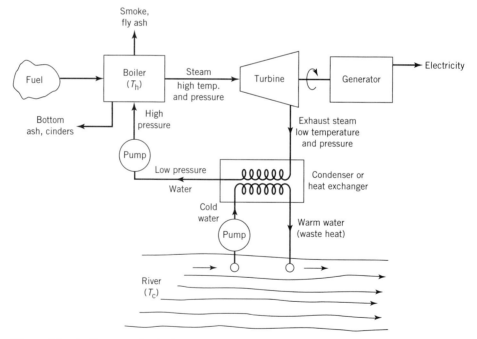

Figure 3.3 A diagram of a fuel-burning electric power plant. Here a river provides cooling water to the condenser, but lake water or a cooling tower could serve the same purpose.

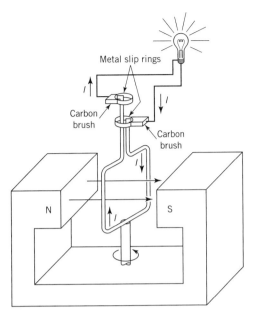

Figure 3.4 An elementary alternating current generator. A loop of wire is forced to rotate in a magnetic field. The induced alternating current enters the external circuit through contacts (carbon brushes) that rub against rotating metal rings, called slip rings, attached to the coil. The current generated, *I*, reverses in direction as the coil rotates. (Adapted from Cutnell and Johnson, *Physics*, John Wiley and Sons, 1989.)

Before that it may have been heat energy in a boiler, and before that it may have been chemical energy in a fossil fuel, before that solar energy, and so forth. The coils are connected in such a way that the induced electric current can be delivered to external loads through transmission lines.

Our electric power plants most commonly have the rotating coils of the generator mechanically connected to steam turbines or to water-driven hydroelectric turbines at large dams. In the first case, the steam comes from water heated either by burning fossil fuels or by a nuclear reactor. The flow of steam at high temperature and pressure is directed against the fanlike turbine blades to produce rotational motion. The main components of a typical electric power plant are shown in Figure 3.3. Plants of this type have been in use for over 100 years. As shown in Figure 3.5, engineering improvements have resulted in steadily increasing efficiency over the years. The actual efficiencies are only about one-third, meaning that two-thirds of the energy in the fuel winds up as waste heat. This is about double the waste that one would expect from the calculated Carnot efficiency. The difference is due to the realities of engineering, as mentioned earlier.

In summary, a standard large modern power plant that provides 1000 MW$_e$ of electric power typically releases 2000 MW$_t$ of thermal power as waste heat to the environment. The power from the fuel into the steam boiler is 3000 MW$_t$. It is generally true that for electric power plants of this type, three units of fuel must be burned to put one unit of electric energy into a home. A very large electricity generating plant is shown in Figure 3.6.

70 Chapter 3 **Heat Engines**

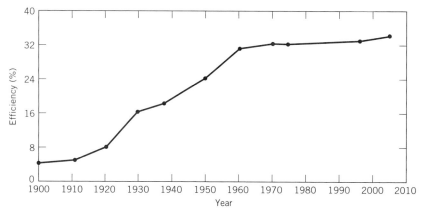

Figure 3.5 Typical efficiency of an electric power plant for converting chemical energy in the fuel into electric energy. The best new plants now achieve nearly 40%. (*Sources*: Delbert W. Devins, *Energy: Its Physical Impact on the Environment*, John Wiley and Sons, New York, 1982; and U. S. Energy Information Administration, *Annual Energy Review*, 2003.)

Figure 3.6 The Four Corners power plant near Farmington, New Mexico. This heat engine converts chemical energy from coal mined nearby into electrical energy that is supplied through transmission lines to the Los Angeles region at a distance of 500 miles. The waste heat is discharged to the environment through the cooling ponds. (*Source*: Michael Collier/Stock Boston)

3.5 Electric Power Transmission

In order for electric energy to be useful to society, it must be transported in some way from the power plants to factories, residences, and other locations of end use. Thomas Edison was responsible for the first electric power system, which was put into use in New York City in 1882. The local generating and distribution system used direct current (DC), in keeping with Edison's preferences. As the distribution system became larger and serviced customers further away from the generating plants, it was realized that alternating current (AC) offered greater flexibility in changing the voltage at different points in the system with transformers, which do not work with DC.

There is good reason for wanting to operate transmission lines at higher voltage than produced by the generators or provided to the end user. Energy is lost in all transmission lines (except for superconducting lines, which are not yet in widespread use) owing to their inherent resistance to the flow of electric current. In simple terms, the loss is proportional to the line resistance (in ohms) multiplied by the square of the current (in amperes). The power delivered by the line is proportional to the product of the voltage times the current. If the line voltage is higher, for the same power transmitted the current is lower, and the resistive loss in the line is smaller. Line losses now typically amount to several percent of the generated power in large systems. The power lost goes into heating up the conductors.

Line losses can be reduced by increasing the diameter of the conductors or increasing the number of conductors, but there is a practical limit to that approach. Lord Kelvin (1824–1907) showed that when the annual cost of the conductors, including interest on the investment, taxes, depreciation, etc., equals the value of the electric energy lost each year, an optimum has been reached. Thus it is not economically sound to increase the size or number of conductors beyond that point.

For this reason, raising the voltage continues to be the main way in which transmission losses are reduced. Step-up transformers take the voltage from the generators at 13 to 25 kV to higher line voltages often ranging from 115 to 765 kV before connecting to long-distance transmission lines. In some situations, as between the Four Corners Power Plant in New Mexico and Los Angeles, the distance between the generating plant and the users can be hundreds of miles. Substations with step-down transformers are located at points along the transmission lines to reduce the voltage to values appropriate for local distribution. Substations for large consumers produce voltages of 4 to 13 kV, and smaller consumers such as residencies, businesses, and some industries receive 240 and 480 volts.

One often sees transformers on power poles or sometimes in metal boxes on the ground in residential neighborhoods. These devices reduce the voltage to the 240 volts most commonly brought into our homes, where it is further divided by means of wiring connections to the 120 volt, 60 cycle AC outlets common in

North American households. Other countries typically use higher household voltages of 220 and 240 volts at 50 cycles AC.

With the passage of time, the number of generating stations and substations as well as the number and length of the transmission lines have increased dramatically as electric energy has become the only realistic choice for lights, machinery, some heating, appliances, electronic devices, some transportation, etc. The system of generating stations, substations, and transmission lines is called the electric power grid. Most of the electric power companies of the United States and Canada are integrated into a single power grid for reasons of economy, availability of backup power for emergencies, and ability to trade electric energy. The integrated system is immense, with about 850 electric utilities and more than 3000 power plants. The North American grid now has more than 200,000 miles of high-voltage transmission lines, almost all of which use three-phase AC. A few high-voltage transmission lines use DC for various technical and economic reasons, but these are relatively uncommon. However, high-voltage DC transmission is increasingly found in system interconnects, where it relieves the difficulty of maintaining the phase relationships of AC systems.

The grid is protected by circuit breakers that sense overloads caused by short circuits (an unintentional connection between a circuit's conductors due to a failure of insulation or some other accidental condition which can produce a potentially damaging flow of current) or other faults. If the equipment causing the fault is not isolated quickly by the circuit breakers, the problems can propagate through large sections of the grid, causing extensive blackouts such as the one in August 2003 that spread over much of the Great Lakes region, eastern Canada, and New York City.

3.6 Practical Heat Engines

Starting 300 years ago, with the work on steam engines by Savery in 1698, followed by that of Newcomen in 1705, and the major advances by Watt in the 1770s (about the time of the American Revolutionary War), practical heat engines began to appear wherever there was the possibility of using them to replace the labor of people and animals. Over previous millennia there had been other ingenious approaches to labor-saving devices, but the heat engines could do work when and where wind or flowing water could not. They were quickly put to use in the mines, in the fields, on the roads and the railroads, and in factories, ships, and aircraft. All that was needed was fuel and the knowledge of how to convert heat energy to mechanical energy. Heat engines have steadily improved since they were first invented. The steam turbine, an excellent means of converting steam energy to rotational motion, was developed by Parsons in the 1880s, and it now is the basis for most of our electricity generation. New types of heat engines have evolved since then, and more will certainly appear in the future. Some examples are discussed in the following sections.

3.6.1 Steam Engines

The principle of operation for a steam engine is based on the fact that when water is boiled to steam at atmospheric pressure its volume expands about a thousand times. If the steam is confined, pressure builds up and the steam tries to expand with great force. This force can be exerted against a piston, as in Figure 3.7, or it can work against the blades of a turbine, as in Figure 3.8. When steam at high pressure is admitted to one side of a turbine, for instance, it will force the blade assembly to rotate along with the shaft on which the blades are mounted. This rotation is achieved most effectively if the exhaust side of the turbine is maintained at low pressure so that there is a large pressure difference across the turbine assembly, from steam entrance to exit. The low pressure condition is assured by the presence of a condenser, typically a chamber into which the exhaust steam is admitted and kept cool by a flow of water from a river, lake, or other source. At the low pressure and low temperature side of the turbine, the exiting steam condenses to liquid water. As long as steam continues to be admitted, and as long as the condenser is kept cool, rotational force will be exerted on the turbine. The mechanical energy of the rotating turbine can be coupled to other machinery, often to an electric generator. The operation of a piston-type steam engine is explained by similar principles of a pressure difference from the top to the bottom of a piston connected to a rotating crankshaft.

Since their invention, steam engines have steadily increased in efficiency. The earliest engines had efficiencies of less than 1% for converting heat energy

Figure 3.7 A locomotive powered by the force of steam against pistons. The pistons move inside cylinders, one of which is seen clearly in this photograph. There is a similar unit on the other side of the locomotive. The motion of the pistons is coupled by connecting rods directly to the drive wheels. (*Source*: John Jackson/The Image Works)

74 Chapter 3 **Heat Engines**

Figure 3.8 Two modern steam turbines of the type used in large electric power plants and ocean-going vessels. (*Source*: Courtesy Niagara Mohawk Power Corp.)

to mechanical energy, and they burned prodigious amounts of fuel. The addition of an external condenser increased the efficiency to 2%, and numerous other engineering improvements resulted in efficiencies above 10% by the year 1900, and now above 30%.

Steam engines belong to the broad class of *external combustion* heat engines. In engines of this type, the fuel is burned outside of the pressurized part of the engine, at a relatively low temperature, at atmospheric pressure, and in the presence of an abundance of air. All of these factors lead to relatively low emissions of carbon monoxide and nitrogen oxides. The amount of emitted sulfur oxides and particulates depends on the fuel being burned.

3.6.2 Gasoline Engines

Most of the earliest motorized vehicles were driven by external combustion engines. As described above, this is the case for all the steam engines and it is also true for a few other types of engines, such as the Stirling hot-air engine that has been considered for automotive use. The heat engines we now have in almost all of our motor vehicles are *internal combustion* engines. In engines of this type, a fuel, such as gasoline, is vaporized and mixed with air to form a combustible mixture inside a closed chamber as shown in Figure 3.9. This mixture is compressed to about 6 to 10 times atmospheric pressure, then ignited with an electric spark timed to fire at just the right instant. On ignition, the fuel burns very

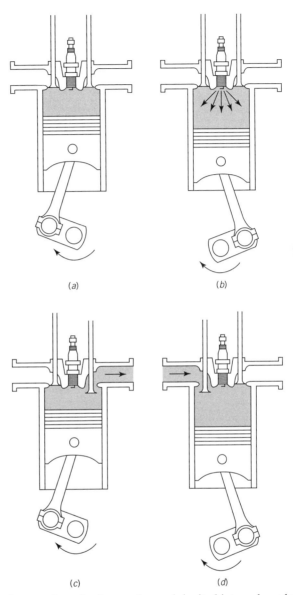

Figure 3.9 The four strokes of a four-cycle spark-ignited internal combustion engine: (a) compression, (b) combustion, (c) exhaust, and (d) fuel–air intake. This is the type of engine that powers almost all of our automobiles. (*Source*: Delbert W. Devins, *Energy: Its Physical Impact on the Environment*, John Wiley and Sons, New York, 1982.)

suddenly, in fact it explodes, forming hot gases predominantly of carbon dioxide and water vapor according to the chemical formulas given earlier in this chapter for the burning of hydrocarbons. The part of the air not reacting in the combustion process, mainly the nitrogen that makes up 80% of air, is also raised to a high temperature. The resulting hot gases (briefly above 1000°C) expand with great force against the piston, forcing it downward, causing the crankshaft to rotate. The maximum combustion pressure is in the range of 1000 pounds per square inch. About 25% of the chemical energy in the fuel can be converted to mechanical energy in a modern gasoline engine.

It's not easy to trace out an exact history of how the internal combustion engine evolved. Certainly it was recognized early in the 1800s that one could explode a fuel–air mixture above a piston, thus extracting work. The internal combustion engine took a great leap forward after the first extraction of petroleum from the earth in Pennsylvania by Drake in 1859. This meant that a volatile fuel—gasoline—became widely available, and by the turn of the 20th century, there were many manufacturers of automobiles powered by gasoline in the United States and Europe. Cars began to appear by the thousands. Internal combustion engines of various sizes also appeared in large numbers for pumping water, powering sawmills, driving electric generators, running farm tractors, powering boats and ships, and even operating washing machines.

3.6.3 *Diesel Engines*

Large trucks, locomotives, and farm tractors are predominantly powered by diesel engines. There are also diesel passenger cars on the road, but they are vastly outnumbered by gasoline-powered cars. The diesel engine is an internal combustion engine, in many ways similar to the gasoline engine. The main differences are that the diesel engine does not use electric spark ignition and it does not mix the fuel and air before admitting them to the combustion chamber. A diagram of a diesel engine is shown in Figure 3.10. The diesel engine is generally heavier and bulkier than a gasoline engine, it usually runs at a lower speed, and it may be slower to respond to the driver's need for more power. The primary advantage of the diesel engine is that it is significantly more efficient than a gasoline engine, converting greater than 30% of the fuel energy into mechanical energy. By the 1960s diesel engines had almost completely replaced steam engines on railroad locomotives. For more than a hundred years, locomotives had used reciprocating steam engines, but their efficiency of about 6% could not compete with the higher diesel efficiency.

During the compression stroke, the diesel combustion chamber contains only air. The compression ratio is quite high, producing a pressure more than 15 times atmospheric at the peak of the compression stroke. At this high pressure, the temperature of the compressed air has been increased to the ignition point for a fuel–air mix. Under the force of a high-pressure pump, a short burst of fuel is injected by spraying it into the combustion chamber, timed so that the fuel arrives near the peak of the compression stroke. The fuel mixes with the hot air and immediately ignites without the help of an electric spark. The combustion temperature is higher than in a spark-ignited gasoline engine, producing a higher thermodynamic efficiency, as we would expect from the Carnot analysis. Common diesel fuel is a middle-grade fuel oil, heavier than gasoline, but the diesel engine will operate well on a very wide range of fuels, including some of rather

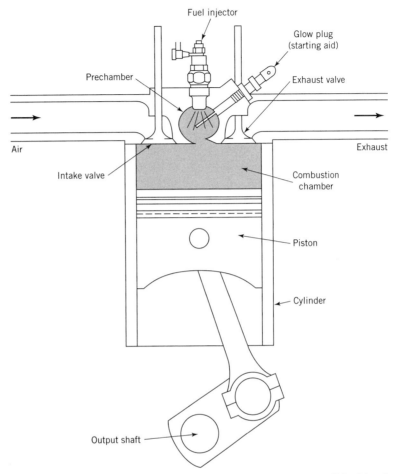

Figure 3.10 Cutaway drawing of a diesel engine. Ignition is accomplished by the high temperature produced by the compression of air. (*Source*: Delbert W. Devins, *Energy: Its Physical Impact on the Environment*, John Wiley and Sons, New York, 1982.)

low grade. There are legends of diesels operating on everything from coconut oil to distilled spirits.

Diesel fuel of the usual type, being heavier than gasoline, has about 10% more Btu per gallon than gasoline. This, along with the inherently higher efficiency, adds to the operating cost advantage of the diesel over the gasoline engine if the two fuels are available at the same per gallon cost. The carbon monoxide emissions of a diesel engine are very low, less than 10% that of the gasoline engine. This low CO emission is explained by the fact that the diesel combustion takes place with an excess of air (and oxygen) in the combustion chamber relative to the amount of fuel admitted. Any CO formed is quickly oxidized to CO_2. Disadvantages of the diesel include its greater noisiness, greater initial cost, greater weight, harder starting in cold weather, characteristic odor, and occasional emission of visible smoke and particulates. Because of the higher combustion temperature and pressure, there is greater emission of nitrogen oxides.

3.6.4 Gas Turbines

The gas turbine is one of the newer types of internal combustion heat engines. It is what powers jet aircraft and a few of our smaller electric power plants. Understanding how the gasoline and diesel engines operate is rather straightforward, but the operating principles of the gas turbine are more subtle. In the gas turbine, air is drawn in at the front and compressed by a fanlike compressor. That's the rotating part that we see when we look into a jet engine from the front. After this, the air is mixed with finely dispersed fuel and the mixture is ignited, causing the resulting heated gases to expand, much as in the description of the gasoline engine discussed earlier. The expanding gases move through a turbine as they exit through the exhaust at the back of the engine. This turbine is connected by a rotating shaft back to the compressor fan, giving it the power it needs. Because of the expansion which took place, the exhaust gases move out the back of the engine at higher velocity than the air had when it entered the compressor. This is what gives the jet engine its thrust. It is effectively pushing air and other exhaust gases backwards at high velocity. One could think of it simply as a fuel-powered fan. One difficulty in designing a gas turbine is associated with the need to ensure that the increased pressure in the combustion chamber does not force exhaust back out through the compressor instead of having it exit to the rear through the power turbine. This goal is accomplished by a careful design balance between the pressure increase produced by the compressor fan and the pressure drop across the exhaust turbine.

The gas turbine can be viewed as an internal combustion engine operating continuously and smoothly, in contrast to the piston engine where there are sequential pulses of power produced by the individual explosions of fuel above the pistons.

When the gas turbine is intended not to propel an airplane directly but to provide shaft power to operate an electricity generator, the rotating turbine shaft can be connected directly to the shaft of the generator as discussed in Section 3.8 of this chapter. A similar consideration applies for a jet-engine-powered helicopter. In such applications, the turbine is designed to give a larger fraction of its power to the rotating shaft than in the case of an aircraft jet engine where the shaft needs only enough power for the compressor fan, the rest of the energy going into kinetic energy of the exhaust gases. Present gas turbines have efficiencies of about 20 to 30% for converting thermal energy to mechanical energy. Turbines have the advantage of being light relative to their power (helicopter applications); they respond well to sudden power demand (helicopters and electric utility peaking power); and they're relatively inexpensive (public utility peaking power). In the latter application, the lower efficiency compared to conventional steam turbines is tolerable because of the lower capital cost and the intermittent nature of their use.

Development of the gas turbine has been held back by the need for materials which could withstand the high temperatures (above 1000°C) and high rotational speeds. Their recent success in industrial power plants is the result of extensive development work done for the aircraft industry.

Gas turbine power plants are now finding their way into cogeneration facilities where they are a good match to the need. Their relatively low efficiency for providing mechanical power is augmented by beneficial use of the waste heat energy. When both electricity generation and useful heat energy are considered, the combined efficiency can be quite high. For an example of such an installation see Section 3.8 on cogeneration.

3.7 Heat Pumps

The discussion of heat engines, illustrated by Figure 3.2, shows how it is possible to devise a system whereby heat energy is taken from a source, with part of this energy converted to work and the remainder being rejected as waste to a heat sink at a temperature lower than that of the source. Could this process be run backwards, thus removing energy from a cold place and delivering it to a warmer place? It is clear that the diagram of Figure 3.2 would still be consistent with all the laws of thermodynamics, including the Principle of Energy Conservation, if the directions of all three arrows indicating energy flow were reversed. This new diagram is shown in Figure 3.11; it corresponds to a heat engine run backwards and is a device that uses energy input in the form of work to cause the transfer of heat energy from the low temperature reservoir to another reservoir at higher temperature. Heat energy is thus pumped uphill in a temperature sense, and the quantity of heat energy delivered to the higher temperature reser-

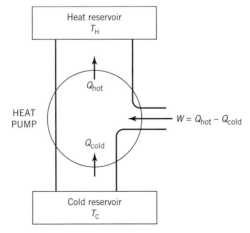

Figure 3.11 A thermodynamic diagram of a heat pump. A work input, W, is required to transfer an amount of energy, Q_{cold}, out of a cold reservoir and a larger amount, Q_{hot}, into a hot reservoir. Because energy is conserved, Q_{hot} must equal $W + Q_{cold}$.

voir can far exceed the work input. This remarkable phenomenon can and does make heat pumps highly attractive for space heating purposes. It is also the basis for refrigeration and air-conditioning.

It's evident from everyday experience that heat energy naturally flows only from high temperature to low temperature, never from low to high. Hot things never spontaneously get hotter than their surroundings and cold things never spontaneously get colder than their surroundings. To make either of these things happen, an energy input from some external agency is needed.

The idea of a Carnot efficiency applies to an ideal heat pump just as it does to an ideal heat engine used to extract work from a heat source. From Figure 3.11, using the Principle of Energy Conservation as a guide, we see that the energy flowing into the heat pump ($Q_{\text{cold}} + W$) must be equal to the energy flowing out, Q_{hot}. A commonsense measure of the effectiveness of the heat pump could be the ratio of the heat energy delivered (Q_h) to the place where it's wanted, divided by the work input (W). This ratio is the *coefficient of performance* (C.O.P.):

$$\text{C.O.P.} = \frac{Q_h}{W} = \frac{Q_h}{Q_h - Q_c} = \frac{1}{(1 - Q_c/Q_h)} = \frac{1}{(1 - T_c/T_h)} = \frac{T_h}{(T_h - T_c)}.$$

In arriving at this expression for the ideal C.O.P., we have made use of the Carnot relationship, $Q_c/Q_h = T_c/T_h$. We see that the C.O.P. for an *ideal* heat pump depends *only* on the temperatures (in degrees Kelvin) of the two reservoirs between which it operates. As with the Carnot efficiency of a heat engine, this coefficient of performance applies only to an ideal heat pump. It is a theoretical maximum value never attained in practice.

When a heat pump is used to transfer heat energy out of an enclosure, thus maintaining the enclosure at some desired T_c lower than the surrounding T_h, it is known either as a refrigerator or an air conditioner. The device is usually designated simply as a heat pump when its purpose is to transfer energy into an enclosure, such as a house, in order to maintain the house at some T_h warmer than the surrounding T_c. The same device can often be operated in either direction, providing warmth in winter and cooling in summer. For residential space heating purposes, it is becoming common to have a heat pump, powered by an electric motor, which extracts heat from the outside air (thus refrigerating the outdoors). The C.O.P. calculated in Example 3.1 suggests that 13.3 W of heating can be provided for every watt consumed by the heat pump. In practice, a C.O.P. in the range of 2 to 6 is typical for an air-to-air heat pump system, depending on the outside temperature. As the outside temperature drops (or as $T_h - T_c$ increases, meaning that the heat has to be pumped farther uphill), both the ideal and actual coefficients of performance diminish. Because of this loss of performance, electrically driven air-to-air heat pumps are most useful in moderate climates, losing their advantages over ordinary electric resistance heating when the outside temperature falls below about 15°F. Air conditioners and refrigerators sometimes carry the specification of energy efficiency ratio (EER). This is the rate at which heat energy is removed in Btu/hr divided by the rate at which energy is consumed by the appliance in watts.

In addition to the air-to-air heat pumps discussed in this section, it is common to use the ground at several feet of depth, or surface water such as a river, for the cold reservoir. In these cases the heat pump would be designated ground-to-air or water-to-air if used for space heating. In cold climates the ground is usually warmer than the outside air, thus giving a higher C.O.P. for the ground-to-air heat pump system.

Example 3.1

Calculate the ideal coefficient of performance for an air-to-air heat pump used to maintain the temperature of a house at 70°F when the outdoor temperature is 30°F. This example is not entirely realistic because ideal mechanical devices cannot be achieved.

Solution

$$T_h = 70°F = 21°C = 294 \text{ K}$$
$$T_c = 30°F = -1°C = 272 \text{ K}$$
$$\text{C.O.P.} = \frac{T_h}{(T_h - T_c)} = \frac{294 \text{ K}}{(294 \text{ K} - 272 \text{ K})} = \frac{294}{22} = 13.3$$

Thus, for every watt of power used to drive this heat pump, 13.3 watts are delivered to the hot reservoir (the interior of the house), and 12.3 watts are extracted from the cold reservoir (the outside air). In practice, the C.O.P. for such a situation would be much less favorable. It would probably be about 3.

The actual mechanism by which a heat pump operates can vary from device to device, but a typical system is illustrated in Figure 3.12. In this system a compressor driven by an electric motor compresses Freon gas to raise its temperature and pressure. The gas then flows through a radiator-type heat exchanger where it is cooled by a flow of room-temperature air and condensed to a liquid, still at high pressure, thus giving up its heat to the room. On passage through a small orifice (an expansion valve) into a region of lower pressure, the liquid expands into a gas, becoming very much colder in the process. This extremely cold gas then passes into a second heat exchanger located outside, where it is warmed to the temperature of the outside air, thus extracting heat from the outside air. The gas then passes into the compressor to repeat the cycle.

The overall effect of heat pumps is to make it possible to use electrical energy for space heating, and also water heating, with an overall (heat pump plus power plant) efficiency about the same as if the fuel burned in the power plant had been burned directly in a home's furnace. The primary fuel could be coal

82 Chapter 3 **Heat Engines**

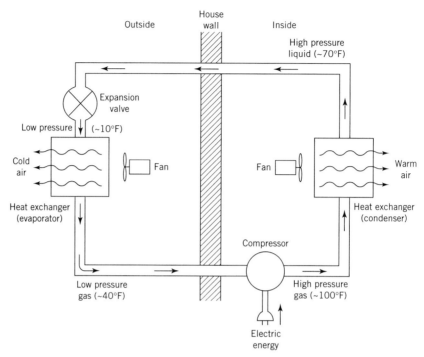

Figure 3.12 An electrically driven heat pump using Freon as a working fluid. In principle, the system becomes an air conditioner if the fluid flow direction is reversed. In practice, the reversal of function is more complex.

burned in a large plant, or the electricity could come from nuclear, geothermal, wind, or hydro, which are all difficult or impossible to use at home. Also, a home heated by an electric heat pump does not need connection to a gas line, or fuel storage, or even a chimney. These are all cost savings. The drawbacks to heat pumps center around their initial cost and concerns about long-term reliability as compared to electric resistance heating, as well as their reduced coefficient of performance in extremely cold weather.

3.8 Cogeneration

It's apparent from what we've seen earlier in this chapter that operation of a heat engine for any purpose is necessarily accompanied by the rejection of heat energy, often in large amounts. For large power plants this waste heat is generally dissipated into the atmosphere through use of a cooling tower or into an adjacent body of water such as a cooling pond, a river, a lake, or even the ocean. Not only is there expensive equipment involved in the process of throwing the heat away, but there are often concerns about the environmental effects of the heat added to the atmosphere or the body of water. From the energy standpoint alone, we must be concerned about the waste of heat energy. A new coal-fired electric power plant will typically have an efficiency of 38% for converting the energy from the burning of the coal into electricity, which means that 62% of the heat energy from the coal is rejected to the environment.

There is no fundamental reason why the rejected heat from a heat engine cannot be used in some beneficial way. The rejected heat is said to be of lower quality than that which powered the heat engine, meaning that it is more limited in application than energy of higher quality. We see a simple example of beneficial use of waste heat energy whenever we use the heater in an automobile. Here, waste heat energy from the engine, which would otherwise be thrown away, is used to provide a comfortable passenger compartment at no added cost in dollars or to the environment. Many other possibilities for the use of such low-quality heat come to mind. They include space and water heating in homes, institutions, and factories, as well as industrial process heat used in manufacturing. There can be several steps in the use of rejected heat—a cascading of applications—until the final temperature is equal to that of the atmosphere.

There is a long history of using waste heat from large electric power plants for space heating in cities, but this approach does not work well when the plants are so large that they must be situated far away from where the waste heat is needed. There is too much heat lost in transporting the steam or hot water over long distances. More recently, there have been advances in putting smaller, decentralized electric generating plants near the point of use, even right in the middle of a university campus. This is the case at the University of Colorado in Boulder. Here an electricity-generating plant powered by gas turbines provides electricity to the campus, with the rejected heat used for space heating and even space cooling (see Section 4.7). The plant operates so cleanly and quietly that most people who walk by it every day aren't even aware of its function. To them it's just another campus building. At times this plant generates electricity in excess of the university's needs, and the excess amount is put into the public utility's transmission lines for use elsewhere. The university is paid for the electricity delivered to the public utility. Some details for this installation are given in Table 3.1, where it can be seen that the overall efficiency for beneficial use of

Table 3.1 Cogeneration Plant, University of Colorado, Boulder

Fuel	Natural gas
Engine	2 Mitsubishi industrial gas turbines
Generating capacity	32 MW$_e$
Capital investment	$41,000,000
Construction started	1990
System lifetime	40 to 50 years
Efficiency for producing electricity	34%
Overall efficiency	70%

84 Chapter 3 Heat Engines

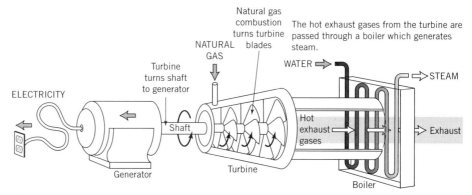

Figure 3.13 A small cogeneration plant that uses the combustion of natural gas to drive a gas turbine coupled to an electric generator. The hot exhaust gases boil water to steam for use in space heating and cooling. (*Source*: Exxon Corporation.)

fossil fuel energy is about 70%, significantly higher than in the usual electricity-generating plant.

Construction of this new facility was encouraged by implementation of the 1978 Public Utilities Regulatory Policies Act (PURPA). The three primary benefits afforded to qualifying facilities under this act are as follows:

1. Public utilities are required to purchase electricity from and provide backup service to cogenerators at nondiscriminatory rates.
2. Cogenerators are exempt from regulation as public utilities.
3. Electric utilities are required to interconnect and operate in parallel with cogeneration facilities.

Certainly as energy becomes more expensive, cogeneration approaches will find increasing application throughout our society. An example of a cogeneration plant powered by a gas turbine is shown in Figure 3.13.

Key Terms

Mechanical equivalent of heat
Heat of combustion
Gram molecular weight
Oxidation
Ideal engine
Efficiency
Fahrenheit, Celsius, Kelvin temperatures
Carnot efficiency

Electromagnetic induction
External combustion
Internal combustion
Steam engine
Steam turbine
Gasoline engine
Diesel engine
Compression stroke
Power stroke

Gas turbine
Heat pump
Heat pump, air-to-air
Heat pump, ground-to-air
Heat pump, water-to-air
Coefficient of performance
Cogeneration
PURPA

Suggested Reading and References

1. Krenz, Jerrold H. *Energy, Conversion and Utilization*, second edition. Rockleigh, NJ: Allyn and Bacon, 1984.
2. Devins, Delbert W. *Energy: Its Impact on the Environment.* New York: John Wiley, 1982.
3. Ross, M. H., and Williams, R. H. *Our Energy, Regaining Control.* New York: McGraw-Hill, 1981.
4. Gellings, Clark W. and Yeager, Kurt E., Transforming the Electric Infrastructure; *Physics Today* **57 12** (December 2004), pp. 45–51.

Questions and Problems

1. A typical room temperature is 68°F. What are the corresponding Celsius and Kelvin temperatures?
2. The basic equation of carbon burning is $C + O_2 \rightarrow CO_2$ + 95 kilocalories per mole (1 mole of carbon is 12 grams). From this equation calculate the number of Btu from burning one ton of coal, assuming it is pure carbon.
3. (a) Starting from the Principle of Energy Conservation, explain the reasoning that gives us the expression for thermodynamic efficiency, η, in terms of the temperatures T_c and T_h. You may assume the Carnot relation, $Q_c/Q_h = T_c/T_h$, for an ideal heat engine.

   (b) What is a heat engine?
4. An inventor claims to have developed a wonderful new heat engine that operates with a relatively cool (and therefore nonpolluting) flame at 150°C and discharges waste heat to the environment at 20°C. His promotional literature advertises that 45% of the fuel energy is converted into useful work. Calculate the maximum efficiency that can be expected for such an engine and compare it to the inventor's claim.
5. Because individual natural gas–fueled water heaters in homes do not have air pollution control devices, it has been proposed that water in homes be heated with electrical energy generated at a distant power plant that is equipped with effective air pollution control devices. The power plant burns natural gas and produces electrical energy with 40% efficiency; the electric transmission line between the plant and the home loses 10% of the electrical energy as heat. All that remains goes into heating the water. In contrast, the water heater gets only 60% of the fuel energy into the water. Which scheme uses more natural gas to heat a given quantity of water?

6. Why are fossil fuel–burning electric power plants situated near lakes or rivers or provided with cooling towers?
7. List the relative virtues of gasoline, diesel, and turbine types of internal combustion engines.
8. (a) This morning, a 1-kW electric heater was switched on for one hour. Trace back as far as you can the various forms the energy had before it was used for heat. Assume a coal-fired power plant.
   (b) Now trace the path the electric energy will take after it has been used for heat. What will be its ultimate destiny?
9. An electrically powered heat pump can deliver more energy than it draws from the power line without violating the Principle of Energy Conservation. Explain how this can happen.
10. How many tons of CO_2 are produced for each ton of methane burned?
11. A heat engine in each cycle extracts 50,000 Btu of thermal energy and rejects (or releases) 20,000 Btu of thermal energy.
    (a) How many Btu of work are done each cycle?
    (b) What is the efficiency of this engine?
12. Show that the combination of a 40% efficient power plant with a heat pump having a coefficient of performance of 4.0 would actually deliver 60% more heat energy than if the fuel were used directly to heat a house with 100% efficiency.
13. Considering problem 12, in the interest of conserving the fossil fuel energy resource and reducing pollution, should it be unlawful to burn fossil fuels directly in the home or factory for heat?
14. A heat engine operating between a geothermal steam source at 210°C and a river at 20°C achieves an efficiency of 20%. What percentage of its theoretical maximum efficiency is it achieving?
15. A refrigerator has an energy efficiency ratio (EER) of 10. For each unit of input energy (W), drawn from the electric power company, how many units of heat energy (Q_c) are removed from the cold box, and how many are delivered to the room (Q_h)?
16. A heat pump delivers 8 Btu of heat energy to a house for each 6 Btu of heat energy that it draws from its low temperature reservoir. What is its coefficient of performance?

Multiple Choice Questions

1. For an ideal heat engine, according to Carnot:
   a. $\dfrac{T_h}{T_c} = \dfrac{Q_c}{Q_h}$
   b. $\dfrac{T_c}{T_h} = \dfrac{Q_c}{Q_h}$
   c. $\dfrac{T_h}{Q_c} = \dfrac{T_c}{Q_h}$
   d. $\dfrac{T_c}{Q_c} = \dfrac{Q_h}{T_h}$

2. For a heat pump in normal operation:
   a. $Q_h > Q_c$
   b. $Q_c > Q_h$
   c. $W = Q_h$
   d. $W > Q_h$

3. One hundred degrees Fahrenheit is the same as ___ degrees Kelvin.
   a. 38
   b. 68
   c. 273
   d. 311

4. One gram molecular weight of octane, C_8H_{18}, has a mass of ___ grams.
   a. 8
   b. 18
   c. 26
   d. 114
   e. 146
   f. 224
   g. 1.15×10^6
   h. 6×10^{23}

5. How many tonnes of coal must be burned per day by an electric power plant that delivers 500 MW_e continuously if the efficiency is 30%?
   a. 1412
   b. 4650
   c. 8560
   d. 9416

6. If an ordinary household refrigerator is left operating in a closed, perfectly insulated room with the refrigerator door standing open, after a long time, the temperature in the room will _____
   a. go up
   b. go down
   c. remain constant
   d. fluctuate

7. A temperature of 98.6 degrees Fahrenheit corresponds to ___ degrees Celsius and ___ degrees Kelvin.
   a. 346, 73
   b. 73, 346
   c. 393, 120
   d. 120, 393
   e. 328, 55
   f. 55, 328
   g. 310, 37
   h. 37, 310

8. The maximum efficiency of a heat engine that has steam injected into it at 850°C and that rejects the steam at 225°C is _____
   a. 74%
   b. 56%
   c. 44%
   d. 20%

9. What is the Carnot efficiency for a heat engine that has steam injected into it at 1000°C and exhausted at 450°C?
   a. 86%
   b. 21%
   c. 67%
   d. 43%

10. For an ideal heat engine operating between thermal reservoirs at T_h = 1000°C and T_c = 110°C, the efficiency is _____
    a. 10%
    b. 29%
    c. 71%
    d. 90%
    e. 92%
    f. 95%
    g. 98%
    h. cannot be determined from information given

11. A typical efficiency for a coal-burning electric power plant is _____
    a. 1%
    b. 2%
    c. 5%
    d. 10%

e. 25%
f. 33%
g. 66%
h. 90%

12. A heat engine_____
    a. pumps energy "uphill"
    b. is identical in function to a heat pump
    c. can produce work from a temperature difference
    d. can occasionally violate the first law of thermodynamics
    e. is only an idea not yet achieved

13. Compare an ideal heat pump operating between T_h = 100°F and T_c = 0°F to one operating between T_h = 100°F and T_c = 45°F. In the second case, the Coefficient of Performance is ___ times larger than for the first.
    a. 0.9
    b. 1.8
    c. 3.6
    d. 7.2
    e. 14.4
    f. 18

14. One mole of any chemical compound is equal to _____
    a. 6×10^{23} gram molecular weights
    b. 6×10^{23} grams per mole
    c. a mass in kilograms the same as Avogadro's number
    d. one gram molecular weight
    e. 1.6×10^{19} atoms

15. The total rate of energy use in the United States is now such that each person has the equivalent of about ___ horsepower working continuously.
    a. 0.015
    b. 0.15
    c. 1.5
    d. 15
    e. 150
    f. 1500

16. The methane in natural gas burns according to $CH_4 + 2 O_2 \rightarrow CO_2 + 2 H_2O$. How many tons of CO_2 are produced for each ton of methane that is burned?
    a. 2.75
    b. 3.7
    c. 12
    d. 36
    e. 44
    f. 93

17. Which one of the following is not a heat engine?
    a. an internal combustion auto engine
    b. a 100 horsepower electric motor
    c. a 2000 horsepower jet aircraft engine
    d. an 800 pound diesel truck engine
    e. a Saturn V rocket

18. The Principle of Energy Conservation is
    a. frequently violated by practical devices
    b. one of the laws of thermodynamics
    c. an exception to Carnot efficiency

d. correct only on the Kelvin temperature scale
   e. correct only on the Celsius temperature scale
   f. a violation of the laws of thermodynamics
19. The Coefficient of Performance for a good heat pump is _____
   a. a negative number
   b. equal to the number 1.0
   c. larger for smaller temperature difference
   d. smaller for smaller temperature difference
   e. not a numerical quantity
   f. meaningless
20. A power plant having an efficiency of 30% burns 7200 tonnes of coal per day. The coal produces 2×10^4 Btu/kg. What is the electric power output of the plant in kilowatts?
   a. 2.75×10^3
   b. 7.25×10^2
   c. 6.25×10^7
   d. 5.72×10^6
   e. 6.75×10^4
   f. 5.27×10^5
21. Cogeneration _____
   a. is a vague dream not yet proven to be practical
   b. would increase a nation's demands on fuel resources
   c. might work for a small factory but not for a large institution
   d. offers considerable promise of reducing a nation's energy waste
   e. will probably increase thermal pollution of rivers

CHAPTER 4

Renewable Energy Sources I: Solar Energy

(*Source*: ©Barth Falkenberg/Deck House, Inc.)

4.1 Introduction

Fortunately, there are some sources of energy that are almost inexhaustible. Included in this category are geothermal and tidal, and solar in all its aspects: ocean thermal gradients, ocean currents, ocean waves, biomass, wind, hydroelectric, and direct use of solar radiation. While these sources may be inexhaustible in the sense that they can furnish energy for all time within our consideration, the *rate* at which energy may be drawn from them can be quite limited. For example, we may assume that the tides will go in and out forever, but the amount of energy that the tidal basins can furnish each year is severely restricted by the geography of coast lines.

With the exception of geothermal and tidal, all of the known renewable energies derive their energy from solar radiation. Tidal energy comes from the gravitational pull of the moon and sun on the oceans, and geothermal energy originates in the radioactive decay of long-lived isotopes within the earth.

According to some scenarios, the energy resources for nuclear energy technology, such as deuterium for fusion reactors and uranium for breeder reactors, could extend the nuclear energy program for many thousands of years. The problems and promises of nuclear energy will be treated separately in Chapter 6. Ocean thermal, ocean currents, oceans waves, and tidal energy are interesting concepts but they are not making a significant contribution to the energy needs of the United States today.

Energy consumption from renewable sources for 2003 is shown in Table 4.1. The renewable 6.15 QBtu per year, out of a total of 98.2 QBtu, is only 6.3%. This leaves much room for expansion. The fraction provided by renewables has actually decreased from 7.0% in 1990. The present numbers are also shown in Figure 1.5 and Table 1.1.

While breaking our dependence on the limited fossil fuel resources is the main motivation for switching more and more to renewable energy sources, there are also substantial environmental reasons for doing so. With the exception of biomass, it is possible for there to be no emissions from the use of the renewable resources. It is interesting to note that the use of biomass at a rate in equilibrium with its growth involves a closed cycle. The CO_2 emitted into the atmosphere in the burning process is exactly compensated by the removal of CO_2 from the atmosphere during the growing of the biomass. However, burning of biomass can emit other gases and particulate matter into the atmosphere.

In examining the role of renewable energy sources, we will first look at direct use of the solar radiant energy.

Table 4.1 U.S. Renewable Energy Consumption in 2003 in QBtu[a]

Energy Source	QBtu	Percent[b]
Conventional hydroelectric power	2.779	2.83
Geothermal energy	0.314	0.32
Biomass	2.884	2.94
Solar energy	0.063	0.06
Wind	0.108	0.11
Total	**6.15**	**6.3**

[a]Hydroelectricity generated by pumped storage is not included in renewable energy.
[b]Based on total energy consumption of 98.156 QBtu in 2003.
(*Source*: U.S. Energy Information Administration, *Annual Energy Review*, 2003.)

4.2 Energy from the Sun

The energy the earth receives from the sun is in the form of electromagnetic radiation. This form of energy includes visible light, infrared, ultraviolet, x-rays, and radio waves. As discussed earlier in Chapter 1, all forms of electromagnetic radiation travel through space at the speed of light, c (3.00×10^8 m/s), and each type of electromagnetic radiation is characterized by a frequency, f, and a wavelength, λ, where the product $f \times \lambda$ always is equal to c.

Figure 4.1 shows the spectrum of power versus wavelength for the electromagnetic radiation we receive from the sun. This figure is for the sun directly overhead with no cloud cover. The most intense solar radiation is in the wavelength region of visible light. Intensity is reduced for any wavelength in the infrared region, and is very weak in the ultraviolet region. The total power received, however, is about as great in the entire infrared region as in the visible region. The spectrum at the earth's surface, shown by the solid line, is reduced from its intensity at the top of the atmosphere, shown by the dashed line, because of absorption and scattering properties of the atmospheric gases.

The earth, at 93×10^6 miles (1.49×10^{11} meters) from the sun, makes a complete revolution about the sun once per year. The earth itself rotates on its north–south axis once every 24 hours, which, of course, accounts for our night and day. The axis of rotation is inclined at an angle of 23.5 degrees from the perpendicular to the plane of its orbit about the sun. In the summer the Northern Hemisphere is thus tipped toward the sun so that it receives the sunlight more directly on its surface; in the winter it is tipped away from the sun so that the sunlight comes in at a lower angle. This accounts for the longer and warmer days of summer and the shorter and colder days of winter. At the north pole the sun is in the sky 24 hours a day for a period of six months centered around June 21, but it is not seen at all for the other six months of the year.

The variations in solar intensity over day and night, summer and winter, as well as changing cloud cover complicate the use of solar energy. For many uses, it is necessary to provide some sort of energy storage system for the times of reduced sunlight. The power density at the top of the atmosphere on the side of the earth directly facing the sun is 2 cal/min · cm^2. This number is known as the Solar Constant, and is given in these rather unusual units because it is then easy to remember as the number 2. In fact, the Solar Constant is not quite constant because the earth's orbit is not perfectly circular; it is about 3% stronger in the Northern Hemisphere winter and 3% weaker in the Northern Hemisphere summer. Of course the *average* value on the earth's upper atmosphere is considerably smaller than the Solar Constant because in computing the average, one must consider that one side of the earth is always dark, and the solar intensity is small at the high latitudes. Taking these considerations into account, the result is that the average intensity on a horizontal area of the earth's upper atmosphere is only 1/4 of the Solar Constant, or about 1/2 cal/min · cm^2.

The solar radiation impinging on the upper atmosphere does not all reach the earth because it is partially absorbed and scattered by the atmospheric

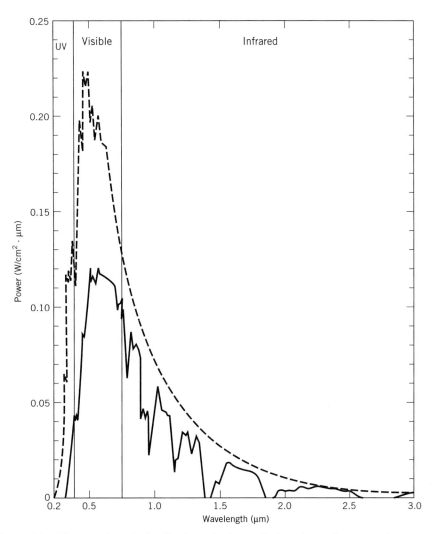

Figure 4.1 The wavelength distribution of solar radiation above the atmosphere (dashed line) and at the earth's surface (solid line). The Solar Constant is given by the area under the dashed curve. The sharp dips in the solid line are due to absorption of certain wavelengths by various atmospheric gases, including water vapor and carbon dioxide. (*Source*: Adapted from *On the Nature and Distribution of Solar Radiation*, Watt Engineering, Washington, D.C.: U.S. Government Printing Office, Department of Energy HCP/T2552-01, 1978.)

gases and the clouds (see Figure 4.2). The values shown in this figure are for average weather and are also averages over all seasons and latitudes. On average, then, we see that 47% (17% + 24% + 6%) of the solar power incident on the upper atmosphere reaches the ground and can be utilized in some fashion.

94 Chapter 4 Renewable Energy Sources I: Solar Energy

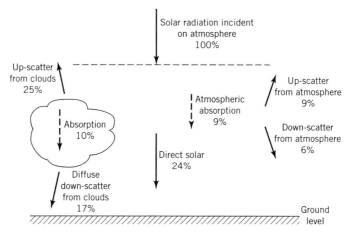

Figure 4.2 Absorption and scattering of solar radiation in the atmosphere. The values shown are for average weather, and are averaged over all seasons and latitudes.

To put available solar power into more useful units, the average on a horizontal area of 1 meter by 1 meter at the earth's surface is then

$$\frac{1/2 \text{ cal}}{(\min \cdot \text{cm}^2)} \times \frac{1 \text{ min}}{60 \text{ sec}} \times \frac{4.184 \text{ J}}{1 \text{ cal}} \times \frac{10^4 \text{ cm}^2}{1 \text{ m}^2} \times 0.47 \times \frac{1 \text{ W}}{1 \text{ (J/sec)}} = 164 \text{ W/m}^2$$

of horizontal surface area.

That is the average value for a 24 hour day. The average for an 8 hour day centered about noon is much larger. An optimistic, but still reasonable, estimate could be about 600 W/m² of horizontal surface area averaged over a typical year for an average latitude and for average weather conditions. It may be helpful in designing a solar space heater or water heater to have this number in terms of Btu and feet. The necessary conversion can be accomplished using the conversion factors given inside the front cover of this book. Thus,

$$600 \text{ W/m}^2 \times \frac{1 \text{ J/sec}}{1 \text{ W}} \times \frac{9.49 \times 10^{-4} \text{ Btu}}{1 \text{ J}} \times \frac{1 \text{ m}^2}{(3.28 \text{ ft})^2} \times \frac{3600 \text{ sec}}{1 \text{ hr}} = 190 \frac{\text{Btu}}{(\text{ft}^2 \cdot \text{hr})},$$

or

$$8 \text{ hr/day} \times 190 \text{ Btu/(ft}^2 \cdot \text{hr)} = 1520 \text{ Btu/ft}^2,$$

is the solar energy incident on a horizontal area for an 8-hour day for the average weather, latitude, and season.

Such values of the solar radiation are often called the *insolation*. As indicated earlier, the insolation varies greatly from time to time and from place to place. Tabulated values exist for numerous locations in various solar energy handbooks. Figure 4.3 shows how the annual mean insolation varies over the United States. The solar energy received by most of Colorado, for example, is in the range of about 5 to 5.5 kWh/m² per day, as an annual average. This would

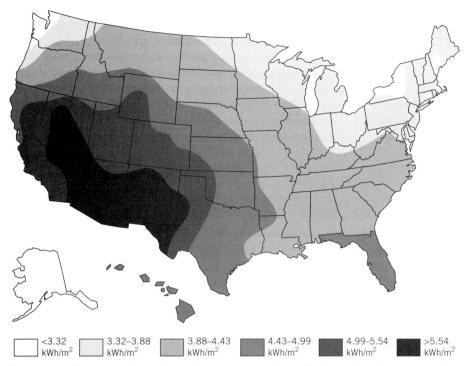

Figure 4.3 The mean daily solar radiation received on a horizontal surface, averaged over the year, for the United States. This figure includes both the direct and diffuse components of the solar radiation. (*Source*: National Renewable Energy Laboratory.)

correspond to a *power* of 610 to 680 W/m² for an 8-hour day. In the British system of units, the average energy received each day in Colorado would be in the range of 1600 to 1750 Btu/ft².

Let's now estimate the gross amount of solar energy incident on the United States in a year. To do this, we will need the known area of the United States, 3.615×10^6 square miles. Then, using the value of 1520 Btu/ft² per day estimated for 8-hour days,

$$3.615 \times 10^6 \text{ mi}^2 \times \left(\frac{5280 \text{ ft}}{1 \text{ mi}}\right)^2 \times 1520 \frac{\text{Btu}}{\text{ft}^2 \cdot \text{day}} \times \frac{365 \text{ day}}{1 \text{ yr}} = 5.6 \times 10^{19} \frac{\text{Btu}}{\text{yr}}.$$

As we saw in Chapter 1, 98 QBtu (98×10^{15} Btu) of energy was consumed in the United States in 2003. This is about 600 times less than the solar energy received. So, if we could find a way to do it, we would need only 0.16% of the solar energy received to furnish all of the energy needs of the nation. On a more personal scale, a 1000 ft² roof of a house receives the energy equivalent of 12 gallons of gasoline each day. The number we estimated, 600 W/m², for the annual eight hour per day average at a temperate latitude, is reduced to about 300 W/m² for the coldest months of the winter and increased to about 1000 W/m²

Table 4.2 Insolation in Btu/ft² per Day on Various Surfaces for Winter Months under Clear Sky Conditions at 40° N Latitude

	Perpendicular	Horizontal	Vertical South	60° South
October 21	2454	1348	1654	2074
November 21	2128	942	1686	1908
December 21	1978	782	1646	1796
January 21	2182	948	1726	1944
February 21	2640	1414	1730	2176
March 21	2916	1852	1484	2174

Note: The perpendicular surface is steered so that it is perpendicular to the sun's rays. The vertical surface is south-facing. The 60° south surface is south-facing and slanted back 30° from the vertical. (*Source*: Adapted from Kreider, J. F., and Kreith, F., *Solar Heating and Cooling*. New York: Mc-Graw–Hill, 1977.)

for the summer months. Table 4.2 shows the insolation in Btu/ft² per day for various times of the year and for several orientations of a collector surface.

Figure 4.4 shows that orientation of the collector is an important factor in the design of a solar energy system. A collector will receive the most energy each day if it is mechanically steered so that it always points directly at the sun, but this is an expensive complication for most applications. A collector panel permanently tilted back at the correct angle will do much better on a winter day than a flat horizontal surface at the 40 degree north latitude used in Figure 4.4. The correct angle will depend on the application for which the collector is intended. Heating water in the summer for a swimming pool will call for a differ-

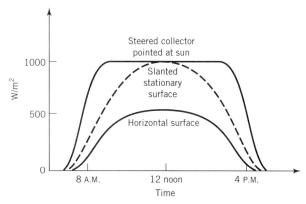

Figure 4.4 Solar power incident on three types of collectors for a typical winter day at 40° N latitude. The *energy* collected each day is proportional to the area under each curve.

ent angle than space heating in winter. At 40 degrees north latitude, on June 20, the noontime sun is at an angle of about 17.5 degrees south of vertical; on December 20, it is about 45 degrees lower in the sky which is at only 27.5 degrees above the horizontal.

There are two common general categories of solar devices: those that are intended to heat living or working spaces or produce hot water as an end product, and those that are intended to produce electric energy for various uses. A third, less common, category would be the powering of heat engines to drive irrigation pumps.

4.3 A Flat-Plate Collector System

It is quite common to utilize an *active system* for solar space heating or water heating. In an active system, a fluid is forced through a collector by an electrically driven pump or blower. The circulating fluid may be either liquid or air. The advantages of an active system are compactness of components, flexibility in the placement of the collector and storage units, and easy control. The disadvantage is the dependence on electricity. *Passive* solar heating is another approach and will be discussed in the following section of this chapter.

As an example of an active system, we will discuss a popular type based on circulating liquid with a flat-plate collector. There are numerous other options. The flat-plate collector is shown in Figure 4.5. It uses a black absorber sheet to maximize the absorption of the solar radiation. The tubes that carry the circulating fluid must be in good thermal contact with the absorber sheet, which is made of a metal that conducts heat readily, so that heat will be transferred efficiently by conduction to the fluid. Sometimes the tubes are soldered to the sheet, or they could be part of the sheet construction (tube-in-sheet). The fluid, which could be water in the warmer locations and possibly an antifreeze solution elsewhere, carries the heat in the system shown in Figure 4.6 to a heat exchanger. The two glass covers shown in Figure 4.5 let in the solar radiation and inhibit heat loss to the outside. The double glass, with an air space in between, is effective in limiting heat loss by conduction and the glass also serves to limit heat loss by thermal radiation.

At this point we introduce two well-known laws that govern thermal radiation from surfaces, often hot surfaces such as that of the sun. The same laws also apply to surfaces at ordinary temperatures such as are experienced with solar collectors. The first of these, *Stefan's law*, relates the radiated power to the temperature and type of surface:

$$\frac{P}{A} = \epsilon \sigma T^4,$$

where P/A is the power in watts radiated per square meter, ϵ is a number called the surface emissivity, σ is a number known as the Stefan–Boltzmann constant (it has a value of $\sigma = 5.67 \times 10^{-8}$ W/m² · K⁴), and T is the surface temperature

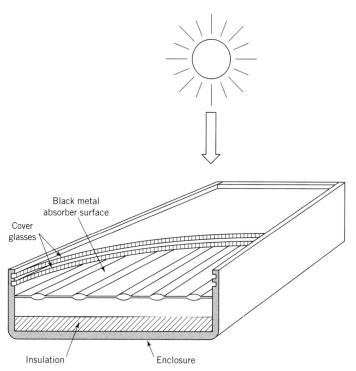

Figure 4.5 A cutaway view of a flat-plate solar collector with two cover glasses. A heat-transfer fluid is circulated through the tubular passages integrally formed into the metal absorber surface. (Not drawn to scale.)

in degrees Kelvin. Note that the radiated power increases rapidly with increasing temperature because of the fourth-power dependence. The emissivity has values ranging from near a maximum of 1.0 down to a minimum near zero. As a general rule, shiny metal surfaces have an emissivity at the lower end of this range—under 0.1. Dull surfaces will have values closer to 1.0.

The second important law is the *Wien displacement law*, which relates the wavelength of the most intense emitted radiation to the temperature of the surface. This law states that the wavelength (in microns) at the maximum intensity point of the spectrum of emitted radiation is given by

$$\lambda_{max} (\mu m) = \frac{2898}{T(K)}.$$

If we put into this equation the sun's surface temperature of 5800 K, we obtain a wavelength of

$$\lambda_{max} = \frac{2898}{5800} = 0.5 \text{ microns},$$

consistent with the peak of the spectrum shown in Figure 4.1.

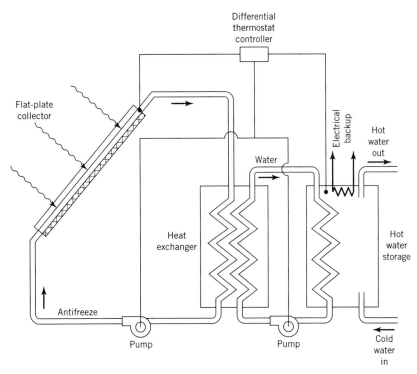

Figure 4.6 A circulating-liquid solar collector system that provides hot water for space heating and domestic use. In a typical installation the collector will be on the roof of a building with the other components in an inside utility area.

The sun's surface temperature of 5800 K determines both the total intensity and the wavelength distribution of the solar spectrum shown in Figure 4.1. The wavelengths in this spectrum are largely in or near the visible region, meaning that the radiation passes through the transparent glass covers into the collector with little loss.

The black collector surface inside the box operates at a temperature of perhaps 360 K, meaning that the thermal radiation which it emits is at a much longer wavelength than that from the sun. Whereas the wavelength of the solar spectrum ranges from a few tenths of a micron up to a few microns, with its peak at 0.5 micron, the wavelengths of the thermal spectrum emitted by the collector plate range from about 4 microns up to 24 microns, with a peak intensity at about 8 microns. From the Wien displacement law:

$$\lambda_{max} = \frac{2898 \; \mu m \times K}{360 \; K} = 8.05 \; \mu m.$$

It is known that glass is not transparent to infrared radiation in the wavelength region of 4 to 24 microns. Thus the thermal radiation emitted by the collector surface cannot escape from the solar panel. The infrared thermal radiation coming off the top face of the collector surface is absorbed by the glass, thus increasing its temperature. The glass then loses this added heat energy by conduction to the outside air and also by emitting thermal radiation, half of which goes back into the solar panel box and half of which is lost by radiation to the outside. The part of the thermal radiation from the glass which goes back to the collector surface has been effectively trapped within the solar panel. This effect has often been called the *greenhouse effect*, so named because of the mistaken belief that it accounts for solar heating of greenhouses used to grow flowers and vegetables in cold regions. However, experiments using greenhouse covers of plastic and other materials that are not opaque to infrared radiation have shown that they work about as well as those made of glass by reducing heat loss by conduction and convection. Nevertheless, the name *greenhouse effect* persists. The greenhouse effect as it relates to the carbon dioxide in the earth's atmosphere, however, is very real. Thermal radiation from the earth is inhibited from escaping by the atmospheric carbon dioxide, thus accounting in part for the earth's surface temperature and contributing to global warming as the atmospheric carbon dioxide content increases.

Figure 4.6 illustrates a system which uses flat-plate collectors to heat water for domestic use and space heating. This system uses an antifreeze fluid, rather than plain water, circulating through the panels to prevent damage when the outside temperature drops below freezing. The use of antifreeze requires a separate heat exchanger between the collector panels and the domestic hot water system to prevent contamination in the event of a leak in the pipes containing the circulating fluid.

With an active solar heating system, it is usually desirable to have a means of storing heat energy for use at night and on cloudy days. For the system shown in Figure 4.6, the hot water tank can be large enough to provide the storage. Another possibility would be to use a large insulated box of stones, when the circulating medium is air.

The efficiency of any solar heating system is less than 100% for several reasons. For the flat-plate collector shown, not all of the incident solar radiation is transmitted by the glass covers. The percent transmitted depends on the angle of incidence, the number of glass cover sheets, and the composition of the glass. For a single glass sheet, the transmitted fraction can be 90 to 95%. Some of the sunlight transmitted into the collector panel is absorbed by the collector plate and some is reflected. There are heat losses from the collector by conduction and radiation, which increase with increasing temperature. For a situation where there are two cover glasses, and a temperature difference of 100°F, the fraction of the incident solar energy that is usefully collected is typically about 50%.

Example 4.1

Estimate the collector surface area needed to heat 100 gallons of water a day from 50°F to 120°F when the daily insolation is 1000 Btu/ft^2. Assume an efficiency of 50%.

Solution

Recall that 1 Btu will heat one pound of water by 1°F, and a gallon of water weighs about 8 pounds. Then

$$\text{Heat energy needed} = 100 \frac{\text{gal}}{\text{day}} \times \frac{8 \text{ lb}}{1 \text{ gal}} \times 1 \frac{\text{Btu}}{°F \cdot \text{lb}} \times 70°F = 56{,}000 \frac{\text{Btu}}{\text{day}}$$

Because the efficiency is 50% there must be twice the needed amount, or 112,000 Btu, of solar energy incident on the collector each day. The area needed will be

$$\frac{112{,}000 \text{ Btu/day}}{1000 \text{ Btu/ft}^2 \cdot \text{day}} = \mathbf{112 \text{ square feet of collector}}$$

or about five panels, each 3 ft × 8 ft.

In the United States, the market for solar panels grew very quickly in the early 1980s because of increasing fossil fuel costs and a federal program of residential renewable energy tax credits. As a result of this program a homeowner could obtain solar heating equipment at no personal cost up to a limit of a $10,000 investment. When the tax credits were removed in 1985, and the fossil fuel prices came down at about the same time, the market for solar heating equipment shrunk very quickly. Many of the solar panels installed during the time of tax credits were for heating swimming pools.

Flat-panel solar collectors are now quite common in areas of the United States that have a reasonable number of sunny days each year. In 1994, 583,000 square feet of panels were installed at an average cost of $8.80/ft^2. As can be seen in Table 4.1, solar energy is still only a small portion of the total energy consumption of the United States, and flat panels are only a fraction of the solar energy devices in use. Published estimates have indicated that, even without tax credits, a flat-plate solar water heater can now be a reasonable investment in the United States. One such estimate shows that the simple payback time based on a $5,000 system is 27 years if one replaces or supplements a water heater using natural gas or oil, and 14 years if one replaces or supplements an electric water heater. The corresponding annual returns on investment are 3.7 and 7.1%, respectively. Of course, increasing energy costs would make the solar system even more attractive.

In some parts of the world, water for domestic use is heated predominantly by solar devices. On the island of Cyprus, over 90% of the homes use solar water heaters, and in Israel the fraction is more than 65%. In Israel the law requires solar water heaters for all buildings up to nine stories high.

4.4 Passive Solar

It doesn't take a textbook on energy to tell people to move into the sun when they're cold. From the earliest times, people in search of warmth have oriented their buildings toward the sun. The ancient cliff dwellings at Mesa Verde National Park are one example of primitive people building their homes to make full use of the sun. In more recent times, when central heating from burning fossil fuels became the norm, the lesson was often forgotten. Houses and apartment buildings are too often designed without considering the effects of the sun.

There can be an inherent appeal in designing a home that maintains itself at a comfortable living temperature but involves no burning of fuel, has no blowers or pumps, and is independent of the electric utility company. This is the passive approach. With the rising awareness of an impending fossil fuel shortage, and the likelihood of increasing costs, architects and homeowners are turning more toward passive solar techniques, where the building itself functions as the solar collector and storage unit.

A passive solar home can be seen in terms of three design elements: *insulation*, *collection*, and *storage*. The first element of a passive solar home is highly effective thermal insulation of all external walls, the roof, and the floor. Doors, windows, and vents must be designed to minimize heat loss. Without careful attention, far beyond the traditional practice, to controlling heat losses, the task of maintaining a building's warmth becomes overwhelming for realistic passive collection and storage schemes.

The next element of a passive system is that of solar energy collection. This can be achieved through the use of large windows on the south face of the house or perhaps a solar collector panel located on the south wall lower than the living space. In this arrangement, when the sun is shining, gravity alone causes the less-dense heated air in the collector panel to rise into the living space and returns the heavier cool air from the rooms through ductwork to the lower edge of the collector panel. The size of the collectors is matched to the calculated heat losses of the house and to the intensity of the incoming solar radiation, which varies from one location to another. In temperate or colder climates, windows and collectors are normally double-glazed to minimize the loss of heat by conduction.

The third element is a large amount of thermal mass inside the insulated walls of the house. The term *thermal mass* is used in discussions of passive solar heating for any material used so that it becomes warm under the influence of incoming solar energy and then cools down later, giving up its heat energy to the surroundings. Certain materials do this better than others, depending on their ability to absorb heat energy without large changes in temperature. Thermal mass is no more than heat storage material; it dampens rapid temperature changes in either direction. The amount of this mass must be great enough to store sufficient heat energy from a warm sunny day for use throughout the following cold night or even a cloudy day. The necessary amount is determined by balancing the heat energy it can store and release against the overnight heat losses from the house. In a passive system, the heat energy is stored at rather low temperature not many degrees above the normal temperature of the living space. This means that a very large thermal mass is needed. The thermal mass can be provided by the internal parts of building itself, such as a thick masonry floor, or even the soil under a thinner floor, or it can be in added components, specifically intended only for heat storage. The very best common material, in terms of stored Btu per cubic foot of material for each degree of temperature change, is ordinary water. Water stores 62 Btu per cubic foot per degree F. After water, a good common material, according to the same consideration, is iron. Scrap iron would do just fine, but iron has the disadvantage of being much heavier than other materials we might choose to use. Iron stores 54 Btu/ft$^3 \cdot$°F. Brick stores 25, loose stone 20, and concrete 22. Even wood is better than some of these materials, storing 25 Btu/ft$^3 \cdot$°F for oak.

Tanks of water, even recycled oil drums painted a dark color and situated in the direct sunlight just inside the windows, have often been used for thermal mass. The oil drum approach can be effective, but it's rather unaesthetic and perhaps better suited to a greenhouse than a living space. Various other schemes, including the Trombe wall shown in Figure 4.7, have been used. The intention here is that the massive concrete wall will be heated on the sunny side, which is painted black, so that the air between the wall and the windows is warmer than the air in the living space. Natural convection causes the warmer air to circulate through openings near the top of the wall into the living space while the cooler air moves through openings near the bottom of the wall into the region between the wall and the windows. Shutters can be used to close off the openings at night so the airflow direction does not reverse and cool the room. The wall also radiates heat into the room. Because the wall is so massive, once it has been heated by solar radiation during the day, it continues to warm the house well after the sun has set.

A less intrusive, but still effective, arrangement known as the *direct-gain method* is shown in Figure 4.8. Here the living space has a normal appearance. At the northern latitudes, such as in the United States, the vertical south-facing windows admit substantially more direct sunlight at midday during the winter than in the summer. As illustrated, an exterior overhang helps to keep out of the intense summer sun and still lets the winter sun shine directly onto the massive interior floor that provides passive heat storage.

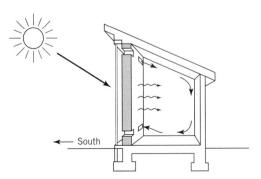

Figure 4.7 A typical Trombe wall installation. The massive concrete wall inside a glass window acts both as a collector and a heat storage medium. The room air circulates by natural convection as shown. Heat is also radiated by the wall into the living space. (*Source*: Adapted from J. Douglas Balcomb, *Passive Solar Space Heating*, Los Alamos National Laboratory, LA-UR-80-2555.)

104 Chapter 4 Renewable Energy Sources I: Solar Energy

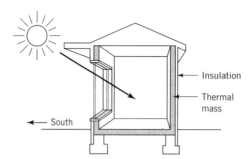

Figure 4.8 A home heated by the direct gain passive solar method. South-facing windows act as solar collectors. Sunlight enters the living space, the energy is converted to heat at absorbing surfaces, and the heat is dispersed throughout the space and to the various enclosing surfaces and room contents. The windows can be covered at night with movable insulation to reduce heat loss. A massive masonry floor and back wall serve for heat storage and prevent overheating. The exterior overhang helps to prevent overheating in the summer. (*Source*: Adapted from J. Douglas Balcomb, *Passive Solar Space Heating*, Los Alamos National Laboratory, LA-UR-80-2555.)

Example 4.2

(a) Calculate the amount of solar energy, in Btu, that would come in through a double-glazed vertical south-facing window of 100 ft² on December 21 at 40° N latitude on a clear day. Estimate that 75% of the incident light is transmitted.

(b) Unfortunately, heat also escapes through windows. If the conductive heat loss from inside to outside that day is 480 Btu/ft² (we will deal with calculating such heat losses in Chapter 7), how many Btu are lost by conduction that day?

(c) Find the net solar gain.

Solution

(a) From Table 4.2, 1646 Btu/ft² will be incident on the window that day. This number does not change by very much over the heating season. Estimating that 75% is transmitted, the result is

$$1646 \frac{\text{Btu}}{\text{ft}^2 \cdot \text{day}} \times 100 \text{ ft}^2 \times 0.75 = \mathbf{123{,}450 \frac{\text{Btu}}{\text{day}}}$$

(b) The heat lost by conduction that day is

$$480 \text{ Btu/ft}^2 \times 100 \text{ ft}^2 = \mathbf{48{,}000 \text{ Btu}}$$

(c) The net heat gain for one day for the 100 ft² window is

$$123{,}450 \text{ Btu} - 48{,}000 \text{ Btu} = \mathbf{75{,}450 \text{ Btu}}$$

A house in the United States insulated to the usual standards can typically require about 1 million Btu of heat energy on a cold winter day. For such a house, based on the above example, one would require an impossibly large number of very large south-facing windows, or some other source of heat such as a furnace. This finding gives emphasis to our listing of thermal insulation as the first element of a passively heated home. With exceptional attention paid to controlling heat losses, and provision of adequate thermal mass, the necessary window area required can be reduced to a practical number.

Passive heating of buildings is generally not included in tabulations of solar energy usage because every building gains some solar energy through its windows, even those that are not directly south-facing. Tabulating these gains has been difficult to do with sufficient accuracy. For this reason, among others, the actual importance of solar energy to the natural energy budget is understated in most accounts of energy sources.

4.5 Solar Thermal Electric Power Generation

An obvious application of solar energy is the production of electric power by using the sun to boil water to steam, which then could be used to drive an electric generator. The equations developed in Chapter 3 concerning the efficiencies of heat engines are directly applicable; one simply uses the sun to supply the heat energy to produce a suitable temperature, T_h, of the steam. A major problem in making use of a diffuse source of energy such as the sun is attaining the necessary high temperature of the working substance in order to achieve a reasonable Carnot efficiency. Except in special situations, this high temperature requirement means that solar collectors will have to be used that focus the energy falling on a relatively large area to a spot, or focus, much smaller in area. A number of different schemes have been proposed, and some are currently being tested. We shall examine several of these approaches to gain an understanding of the main ideas involved and to assess the practicality of generating electric power in this way.

We could also possibly make use of solar energy to generate electric power by utilizing the natural thermal gradients in the tropical oceans to drive immense heat engines, and we now commonly generate electricity from solar energy in the form of wind and in falling water. These somewhat specialized topics in solar energy are discussed in the following chapter, along with other alternative sources of energy.

There are two general categories of proven solar-thermal electric power generating devices. Both involve focusing the sunlight to attain high temperature. The first, often called the power tower, uses a large array of many reflectors to concentrate the light onto a single central receiver mounted on a tower within the array of reflectors. This results in a very high temperature at the receiver, producing a similarly high temperature of the working fluid that is circulated through the receiver to a heat engine, thus resulting in a high Carnot efficiency for converting solar energy to electricity. In a second general type of power system, many concentrating reflectors are also used, but each reflector focuses sunlight onto an individual receiver for that collector. The heated circulating fluid is then brought to a central location to drive a steam engine. The choice between the two systems is whether to use the reflected rays of the sun or a flow of heated fluid to bring the solar energy to the central site.

Before describing some examples of these systems, it will be useful to learn how concentrating reflectors work. Figure 4.9 shows a cross section of a parabolic surface and a few light rays from the sun. The sun is so far away that the light rays coming from it can be considered to be parallel. On striking the parabolic surface, which is usually either polished metal or metallized glass or plastic, the rays are reflected so that they all go through the focal point, F. A receiver located at this focus will thus receive the energy from many light rays. The distance of the focal point from the surface depends on the curvature of the surface. For a practical system, a mirror with a spherical surface is often cheaper and easier to construct and will work about as well as one with a strictly paraboloidal surface.

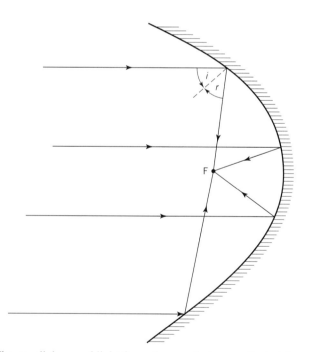

Figure 4.9 The parallel rays of light from the sun that are reflected from a shiny parabolic surface are brought to a focus at a point. For each ray, the angle of incidence (i) is equal to the angle of reflection (r). In this drawing, the angles are measured relative to a line drawn perpendicular to the surface at the point where the ray is incident.

Instead of using a reflecting surface to concentrate the sunlight, one could also use a transparent lens. Lenses are generally more expensive to fabricate, and because of losses due to reflections at the surface and absorption in the material, they are seldom used. A lens design called a Fresnel lens, however, does find some use because it has the focusing properties of a normal thick lens, but it can be made very thin and inexpensively from plastic.

In the large arrays of mirrors, flat—rather than curved—mirrors can also be used effectively. Of course they do not individually focus the sun's rays to a concentrated image; each mirror produces an image on the receiver about as large as the mirror itself. It is the superposition on the receiver of the images from many such mirrors that produces the high temperature.

The essential characteristic of a concentrating collector system is the concentration ratio (CR), the ratio of the net collecting aperture seen by the sun to the illuminated area of the receiver. A flat-plate collector obviously has a CR equal to 1, and the corresponding maximum temperature that can be achieved is about 100°C (373 K). At this temperature, with a $T_c = 20°C$, the Carnot efficiency is about 20%.

A cylindrical trough reflector may have a concentration ratio ranging up to about 50, and a corresponding maximum temperature of about 400°C. A parabolic or spherical dish reflector may have a concentration ratio in the range of 30 to 1000, and a maximum temperature of about 1200°C. There is no fundamental reason why higher concentration ratios and higher temperatures cannot be achieved. The economics of units with extreme concentration ratios are not favorable for driving heat engines, but other uses might justify these devices. The record for high temperature was set (over 3000°C) at Odeillo, France, where a 1 megawatt solar furnace was constructed by having the south-facing side of a building made into a huge paraboloidal reflector 40 m high and 54 m across. The standard power tower, with several hundred reflectors, or heliostats, steered so that they follow the sun and reflect the direct sunlight onto a central receiver throughout the day, can have a CR of up to several thousand, and temperatures of 1000 to 2000°C can be achieved.

Flat-plate collectors take advantage of both the direct and the indirect diffuse components of solar radiation, and they are normally mounted in a fixed orientation. Parabolic trough reflectors focus only the direct sunlight, and they can have their orientation steered to follow the sun. If the axis of the trough is horizontal, in an east–west direction, orientation adjustments may be necessary only a few times during the year. A reflector having a point focus for direct sunlight, such as a parabolic or spherical dish, or a system of heliostats, must be continually changed in orientation to accommodate changes in time of day and season.

There is a long history of heat engines driven by solar collectors. However, it is only within the last 30 years or so that the U.S. government, through the Department of Energy, has contracted with industry to construct systems of appreciable size that will produce 10 MW or more of electric power. Since the majority of the capital cost of the system is in the collectors, it is hoped that innovative designs and mass production techniques for the collectors will result from these pilot projects. As with many applications of solar energy, the problem is not to show that the device can work. It is necessary to demonstrate that an economically competitive system can be constructed, and that numerous technical and practical problems such as heat storage and equipment maintenance have been solved. Several typical systems will be discussed in this chapter to provide a perspective on what is involved.

4.5.1 Power Towers

The general idea of a solar power tower is to collect the light from many reflectors at a central point to achieve high power density and high temperature. The Sandia National Laboratory outside of Albuquerque, New Mexico, has been testing various heliostats and has constructed a 5 MW thermal facility that consists of 222 heliostats, each of which has 25 flat mirrors. The heliostats must continuously rotate to track the sun across the sky, so that the reflected light from all the mirrors falls on the receiver. The tower at Sandia is 61 m tall, and the receiver at the top has a peak incident power of 2.5×10^6 W/m², for a concentration ratio of 2630. This can be thought of as 2630 suns shining simultaneously on the receiver.

Based largely on the experience gained at Sandia, a 10 MW solar electrical power plant was constructed near Barstow, California. A photograph of this facility is shown in Figure 4.10, and a schematic view is shown in Figure 4.11. There are 1900 heliostats, each about 20 ft by 20 ft, surrounding a tower 295 feet tall. As the figure shows, an energy storage system is provided to permit the system to continue generating 7 MW of electric power without sunlight. The capital cost of such a power plant is much higher than that of a coal-fired power plant of the

Figure 4.10 A photograph of the solar-thermal power plant near Barstow, California, described in Figure 4.11. (*Source*: Peter Menzel/Stock Boston)

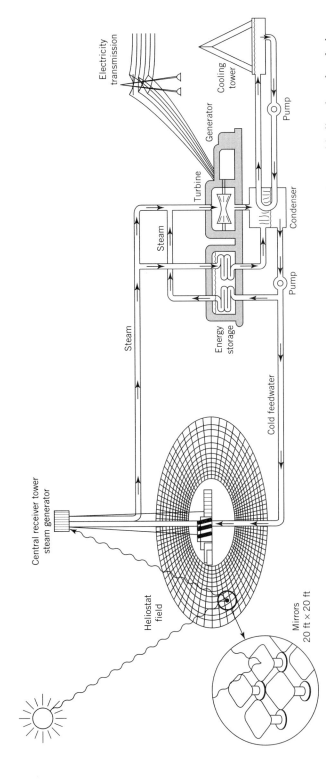

Figure 4.11 A schematic view of a 10 MW$_e$ solar-thermal power plant near Barstow, California. The receiver and boiler that absorb the sunlight reflected from 1900 heliostats are at the top of a 90 meter tower. The heliostats are each steered by computer control to reflect the sunlight onto the receiver. The steam from the boiler can be either delivered directly to the turbine and generator or to storage. The storage system can provide steam for 4 hours of generation at a level of 7 MW$_e$ without sunlight. (*Source*: Solar Energy Research Institute.)

same capacity, but of course there is no cost for fuel, ash disposal, or control of stack emissions. Even with these considerations, the solar-thermal generated electricity is more costly than that from conventional generating plants. As more power towers are constructed, it is expected that economies will be realized and that technological advances will be made that will bring the cost down, but it will be many years before the costs are competitive.

One way to reduce the cost of the solar-thermal electricity would be to use the waste steam from the turbine for industrial processes or space heating. A unit has been under study for use at the Fort Hood army base that combines 1 MW of electric power production with space heating and air-conditioning for the whole base.

4.5.2 Parabolic Dishes and Troughs

In some systems made up of parabolic dishes or troughs that focus the sunlight on a small receiver for each focusing device, the heated working fluid is circulated to a central location point and used there to drive a steam turbine. A field of collector devices of this type may consist of more than a hundred separate dishes or troughs.

In 1983 the first of a series of Solar Electric Generating Stations (SEGS) was installed by Luz International in the Mohave Desert in southern California. The SEGS I plant used solar troughs to produce steam for a conventional steam turbine electric generator. The reflecting troughs focus the sun's rays onto specially coated steel pipes containing a heat-transferring oil and housed in vacuum-insulated glass tubes. The troughs rotate about their long axis and use sensors and microprocessors to track the sun. The oil is heated to 700°F and is pumped to heat exchangers where it transfers its heat energy to water to produce superheated steam. Natural gas was used as a supplementary fuel for up to 25% of the needed heat energy so plant output could be adjusted to meet utility peaking requirements. The 13.8 MW electric energy output was sold to Southern California Edison Company. The cost of SEGS I was $6,000 per peak kilowatt rating and the overall efficiency was about 25%.

A federal tax credit of 25% was matched by another 25% from the state of California, and Luz International was able to build similar solar plants up through SEGS IX in 1991. They were able to reduce the cost of the plant to $3,000 per peak kW, and the cost of the electricity produced to 8 cents per kWh. With all nine plants running, 354 MW_e could be produced. The Luz company went bankrupt in 1991 after the tax credits were withdrawn, but the nine plants continue to operate. Much experience has been gained in the operation, maintenance, and cost-saving measures with the nine plants. Similar plants are now being considered for other sites in the southwestern United States.

A part of a system of parabolic reflector troughs is shown in Figure 4.12. These are components of a Luz solar steam-turbine electric generating station in California.

Figure 4.13 illustrates a technology that produces electricity from the high temperatures produced by parabolic dish reflectors. The device shown here is a

Figure 4.12 Parabolic trough reflectors at a Luz solar-electric generating station in the Mojave Desert of southern California. (*Source*: Hank Morgan/Photo Researchers, Inc.)

component of a system of thermoelectric generators, a technology which does not involve a circulating fluid, and which is not discussed in this text.

4.6 *The Direct Conversion of Solar Energy to Electrical Energy*

As we have discussed, the process of using the sun's energy to boil water to steam to power a heat engine that in turn rotates an electric generator is a usable but expensive and cumbersome scheme. Newer devices can now generate electricity directly from the sun's rays, thus avoiding many of the problems associated with large, complicated, mechanical equipment.

Figure 4.13 A parabolic dish reflector at a solar-electric power station in White Cliffs, Australia. A computer steers the dishes to ensure that they face the sun throughout the day. Each dish focuses sunlight onto a thermoelectric generator at its focal point, thus generating electricity directly from the high temperatures produced. (*Source*: John Mead/ Science Photo Library/Photo Researchers, Inc.)

In 1954, physicists at the Bell Telephone Laboratories discovered that a properly prepared wafer of silicon was sensitive to sunlight. Silicon is one of the most abundant elements in the earth's crust, but it is found in nature only in combination with other elements—most commonly as an oxide. The oxide known as silica, SiO_2, is what makes up sand and quartz. Silicon is inexpensive in its raw form but must go through expensive processing to make it into the pure crystalline form so valuable to the electronics industry. A crystal of silicon consists of a rigidly ordered three-dimensional array of single silicon atoms. The silicon used for the 1954 demonstration had to be extremely pure, and the silicon commercially available at that time had to be extensively refined to be usable. After the first laboratory discoveries, generation of useful amounts of electric power by silicon solar cells was soon demonstrated. Solar cells have since become a standard feature on space satellites to provide electric power and in many areas of ground-based technology.

Silicon solar cells belong to the general class of photovoltaic cells, devices capable of converting sunlight directly to electricity and commonly made of materials called semiconductors. A semiconductor is not quite an insulator, but it is also not a good conductor of electricity. Under normal circumstances, a voltage applied across a sample of crystalline silicon will produce almost no current; that is, the electrical resistance is very high. The silicon atom has four outer, or valence, electrons that participate in bonding each silicon atom to others in the crystal. Because all four of these electrons participate in this bonding, called covalent bonding, there are normally no free electrons available in a pure silicon crystal to conduct electricity. If one or more of the valence electrons could be caused to break away from its role in bonding, the crystal then could conduct electricity. Sunlight shining on the crystal can cause this effect near the surface within the penetration depth of the light.

Light is made up of little bundles of energy called photons. The amount of energy carried by a single photon depends on the frequency of the light. An electron struck by a photon can gain nearly the entire energy of the photon. This phenomenon is called the *photoelectric effect*, and its understanding by Albert Einstein led to his being awarded a Nobel prize in 1921. Einstein's work demonstrated that the energy of a photon is proportional to its frequency, and thus inversely proportional to its wavelength. If the light shining on a silicon crystal is of adequately high frequency or, equivalently, of sufficiently short wavelength, the photon is capable of freeing an electron from its bonding duties and bumping it up in energy to the point where it is free to participate in conducting electricity.

The maximum wavelength (corresponding to minimum photon energy) of light that will cause electrical conduction in silicon is 1.12 microns. About 77% of solar energy is at wavelengths less than 1.12 microns. If the energy of an incident photon happens to be greater than the amount needed to produce a conducting electron, the excess energy just goes into heating the crystal. The greater the intensity of light, the greater the number of photons arriving per second, the greater the number of conducting electrons produced per second, and hence the greater the electric current the crystal can conduct.

A photovoltaic cell is more than just a pure crystal of silicon. To produce electrical energy from sunlight one needs to have more than a material in a conducting state. The device must also produce a voltage to drive the current through an external load. This necessary driving voltage can be achieved by using a combination of silicon plus another element to obtain what is called a *doped semiconductor.*

It is common to use small amounts of arsenic, or antimony, or phosphorus, to dope a silicon crystal for the purpose of making a solar cell. An atom of arsenic has five valence electrons, and if an atom of arsenic replaces an atom of silicon in its normal crystalline lattice site, only four of these electrons are needed to provide the same covalent bonding that the replaced silicon atom had. The fifth electron, not needed in bonding, is free to participate in the conduction of electric current. Because the electron is negatively charged, such doped semiconductors with an excess of electrons are called *n*-type. Normally, only a very small amount of the doping material is added, perhaps about one part to a million parts of silicon.

If the doping atom substituting for a silicon atom has only three valence electrons, which is true for boron, aluminum, or indium, a vacancy is created in the normal electron site for a covalent bond. If this electron vacancy becomes occupied by an electron moving into it from a neighboring atom, the electron vacancy in effect has moved to the neighboring donor atom. This can happen over and over again, with the result being an effective motion of the vacancy through the crystal. This is equivalent to an electric current with a positive charge carrier. The vacancy (also known as a *hole*), being an absence of a negatively charged electron, can be thought of as having a positive charge. Thus doping with such atoms produces positive, or *p*-type carriers, and such a doped material is called a *p*-type semiconductor.

The means by which solar energy creates useful electric current in a solar cell is complicated, but the explanation can be reduced to a sequence of a few steps.

If a piece of *p*-type silicon is put into contact with a piece of *n*-type silicon, the junction between the two is known as a *p–n* junction, and this junction is crucial to the operation of a photovoltaic cell. In the region of the junction some of the electrons from the *n*-type material will move by normal diffusion into the *p*-type material and occupy electron vacancies. Each time this happens, an electron hole is left behind in the *n*-type material. So, accompanying the actual motion of an electron from the *n*- to the *p*-type material, a hole has effectively been moved from the *p*-type material into the *n*-type material. Because both materials were electrically neutral to begin with, the loss of electrons from the *n*-type material, and the loss of holes from the *p*-type material, leaves behind net positive and negative charges, respectively, in the two materials. The presence of these net positive and negative charges on either side of the junction creates an electric field across the junction. The electric field thus developed in the junction will force any free electron that happens to be in the junction region to move in the direction toward the *n*-type material. If there is also a hole in the junction region, it will effectively move in the direction toward the *p*-type material. Now, whenever solar radiation interacts with an electron in the junction region to give it some energy and liberate it from its atomic bonds and let it be free to move about, a hole is left behind at the original site of the electron. So solar radiation in this way creates electrons and holes in pairs in the *p–n* junction. The electric field there forces the electrons to move in one direction, and the holes in the other, through the silicon crystal and through an external electric circuit. Thus, electric current is made to flow through an external load, and solar energy is being converted into electric energy. The strength of the internal electric field is reflected in the voltage produced across the terminals of the solar cell. The maximum current that the cell can deliver will be determined by the rate at which incident photons raise electrons into the conducting state.

To manufacture a silicon solar cell one first has to obtain silicon of very high purity. Even though silicon is abundant in the earth's surface, it is relatively expensive to purify it sufficiently to make a semiconductor. After silicon of high purity is obtained, a crystal must be grown from it by any of several methods. One method involves slowly withdrawing a seed crystal from molten silicon. Some of the molten silicon cools and hardens against the seed crystal and takes on its crystalline structure. As the withdrawn silicon hardens, a crystalline ingot is formed up to several inches in diameter and up to several feet long. A *p*-type crystalline ingot is formed by doping the molten silicon with just the right amount of an element such as boron. The ingot is then sliced by sawing it into thin wafers. The *n*-type part of the solar cell is formed by thermally diffusing an element such as phosphorus into the surface of the wafer. It is important that the diffusion only goes into the wafer a fraction of a micron because light must penetrate through the *n*-type front surface of the crystal into the *p–n* junction region. To finish the solar cell, electrical contacts are made to the two surfaces, the crystal is packaged into an assembly, and an antireflection coating is applied to the front face. It is then ready for use. A schematic view of a solar cell is shown in Figure 4.14.

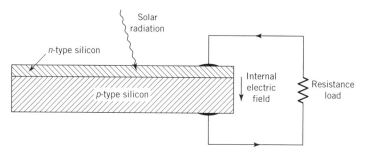

Figure 4.14 A solar cell made of *n*-type and *p*-type silicon. If the incident radiation has sufficient energy an electron can be taken from the valence to the conduction band. Because of the internal electric field, the electron will drift toward the contact on the *n*-type silicon to produce current in the external circuit and finally combine with a hole in the *p*-type silicon. In keeping with the usual convention, the current arrows shown are opposite to the direction of electron flow.

A typical individual solar cell can be about 2 inches in diameter and 1/16 of an inch thick. It will produce about 0.5 volt across the terminals when sunlight shines on it. Solar cells are rated at peak power (W_p) which is the electrical power output with 1000 W/m² of solar power density incident at a temperature of 25°C. A solar cell installation is shown in Figure 4.15. Forty or fifty solar cells are commonly grouped together and connected in series electrically so they produce an output voltage of about 20 to 25 volts. To achieve a larger output current, and thus a larger electric power, several such arrays of 40 or 50 cells can be electrically connected in parallel with each other. The assembly must be oriented toward the sun to optimize the solar collection, and it must be packaged to resist damage from the elements. One laboratory test of resistance to hail is to fire ice cubes at 100 miles per hour against the front surface. Modern solar cells pass this test without damage. If the solar cells are intended to put alternating current into the grid, a unit called an inverter converts the direct current into alternating current. The cost of these units must be included if one is estimating the total capital cost of solar energy.

Since solar cells were first developed in 1954, there has been enormous effort put into increasing their efficiency, developing new types of solar cells, and reducing the costs of manufacturing the cells. The central aim of all these efforts has been to bring down the cost of electricity generated from solar cells to the

Figure 4.15 A 115 kilowatt photovoltaic system at the Dangling Rope Marina in Utah's Glen Canyon National Recreation Area. This system eliminates the use of more than 65,000 gallons (250,000 liters) of diesel fuel for electric generators every year. (*Source*: ©National Park Service/Dept. of Energy/National Renewable Energy Laboratory)

point where they can compete with coal-fired power plants. By 2002 the average capital cost of single-crystal solar cells had been brought down to the range of $2.90 to $4.00 per peak watt of generating capacity. A peak watt is the maximum power that the cell will generate under ideal conditions, such as under a full noonday sun. The average power generated by a solar cell over a 24-hour day is of course much lower than its peak capacity. Some days, under heavy cloud cover, it will be close to zero. Because of this, one must be careful in making comparisons of solar power to other types of power generation that operate at constant efficiency over the 24-hour day in all weather and all seasons. It is clear that naive comparison of the peak power cost of solar cells to the cost of the steady-state power output of a coal-fired plant will lead to undue optimism about solar cells. A conventional coal-fired power plant has a capital cost of about $1.00 per watt of generating capacity, still very much less than solar power. Some of the considerably higher capital cost of photovoltaic solar electricity is offset because there are no costs for fuel, no emissions, lower maintenance costs, and operation in areas not served by power lines.

One of the avenues of attack on the cost of electricity from solar cells has been the development of less expensive ways of growing the silicon crystals. A process called edge-defined film-fed growth (EFG) is one of the more promising developments of this kind. In EFG a ribbon of crystalline silicon is drawn up through a die by capillary action from the molten silicon. This produces a ribbon of silicon of about the correct thickness and of reasonable quality so that solar cells having 10 to 12% efficiency can be made with far less expense than in the usual method of growing and slicing crystal ingots.

Amorphous silicon technology has made perhaps the greatest progress in the past decade or two. In this process the solar cell is made in a continuous-film deposition process with no attempt to have the film in the form of a single crystal. The advantage of this process is that it is relatively inexpensive and uses negligible amounts of silicon because the film is only about a micron thick. This type of solar cell is now widely used for watches, calculators, and walk lights.

Several materials other than silicon can be made into solar cells. Polycrystalline thin films of cadmium telluride, copper indium diselenide, and gallium arsenide are all being investigated, and significant progress has been made in lowering costs and increasing efficiency. Still another direction that has been taken is to stack three cells on top of one another in order to capture a wider portion of the solar spectrum. The top cell is designed to absorb the high energy (blue part) of the spectrum and the lower layers absorb the remaining part of the spectrum (mostly in the red).

The efficiency of solar cells for converting sunlight into electricity is far less than the 77% of the solar spectrum that has usable wavelengths. In fact, about 43% of the average absorbed photon energy merely goes into heating the crystal. It can be shown that the theoretically maximum possible efficiency of a silicon solar cell operating at a temperature of 0°C is about 24%. This maximum efficiency drops off rapidly as the temperature increases. At 100°C it is only 14%. In addition to the loss mechanisms considered in the foregoing discussion, there are other losses such as light being reflected off the front face and the internal electrical resistance of the cell. In practice, the efficiency of a solar cell is typically in the range of 14 to 17% (Table 4.3).

Another completely different method of making solar cells more competitive economically is to use them in conjunction with concentrating collectors. In principle, if a concentrating device such as one of the parabolic reflectors discussed earlier is used to focus the sunlight onto the solar cell, the effective solar intensity could be increased by 100 times, for example, and the size of the solar cell could be reduced by a factor of 100 for the same amount of electrical energy produced. This gain must be balanced against the added cost of the concentrator and tracking system. Also of concern is the fact that the solar cell will experience a much higher temperature than when used without a concentrating collector and the efficiency will drop accordingly. A system of circulating cooling water could be used to keep the temperature of the cells from rising too much. A publication by M. H. Ross and R. H. Williams has considered in some detail the design of a large solar system for a community where the electricity from the photovoltaic cells would provide a large fraction of the electrical needs, and the heated water from the cooling of the cells would provide a large fraction of the space heating and hot water needs of the community. Generators powered by diesel engines would supplement the solar electric power and the heat from the engines would also be used for space heating and water heating.

Production of solar cell modules in the United States increased from 35 MW_p in 1995 to 121 MW_p in 2002. Some of this production was exported, mainly to developing countries. Worldwide, 562 MW_p of solar cell modules were produced in 2002, with Japan leading the way at 251 MW_p. Solar cells are finding widespread and growing uses, including outdoor lighting, microwave repeaters, re-

Table 4.3 Solar Cell Efficiency and Manufacturing Cost

Cell Technology	Efficiency (%) (2003)	Manufacturing Cost ($/W_p$) (2002)
Single crystal silicon	14–17	2.90–4.00
Polycrystalline silicon	13–15	2.90–4.00
Silicon ribbon	13–15	3.00–4.00
Concentrators with silicon	22	4.00–6.00
Amorphous silicon	5–8	2.00–3.00
Copper indium diselenide	8–10	
Cadmium telluride	7–9	
Silicon film	8–10	

Source: Adapted from Paul Maycock, "The State of the PV Market," *Solar Today*, January/February 2004, p. 34.

mote pumps, satellites, navigational aids, as well as in residential grid-connected and off-grid systems. Every day we see them in new applications.

While the use of solar cells in remote applications and in special situations has been growing, the large-scale electric power applications have been few in number and are mainly test facilities and prototype systems. The Public Utilities Regulatory Policies Act (PURPA) enacted in 1978 required that public utilities purchase energy from various alternative energy sources at the utilities' normal cost. This act, along with tax credits, offered encouragement for the development of solar cell power stations. The largest of these was a 6.5 MW_p station at Carrissa, California, which was completed in 1985. There was another 1 MW_p station at Hesperia, California. These solar power plants were constructed by ARCO, the Atlantic Richfield Company, a major manufacturer of solar cells. Their construction and operation proved the feasibility of the solar cell power plant and led to many improvements. These plants are no longer functioning because of unfavorable costs. The solar cell division of ARCO was sold in 1990 to Siemens, a German electric equipment manufacturer.

Several other solar cell power plants have been sponsored by the Department of Energy. The largest of these is 2 MW_p for the Sacramento Municipal Utility District. In 1994 it produced 2195 MWh of electric energy for the district. It was never expected that the facilities would be cost-competitive with fossil fuel power plants, but much valuable experience has been gained.

What is the future of photovoltaic electricity in the United States? While it is encouraging to see the price reductions and the increase in market for solar cells, it should be noted that the 121 MW_p of total U.S. solar cell manufacture in 2002 is only a small fraction of one 1000 MW_e coal-fired plant. In order to make major inroads into the 65,000 MW additional power needs of the United States within the next decade, further improvement will be needed in the costs, efficiency, and reliability of solar cells. If the price could be reduced to under $1 per peak watt, the market certainly would open up very quickly. Many think that polycrystalline thin films offer the greatest hope for such reductions.

Another way of stating the problem is to note that the cost of wholesale electricity from a new coal-fired plant in the United States is in the range of $0.08 to $0.20/kWh, while electricity from photovoltaic power generation is between $0.50 and $1/kWh depending on the size of the system. Arguments are sometimes made that the total cost of coal-fired electricity should be increased to reflect the costs to society of the emissions of pollutants such as SO_2, and greenhouse gases such as CO_2. Even without corrections of this kind, solar cells seem destined to become a significant source of electric power in the 21st century.

When one looks back at the progress that has been made with solar cells in the past 25 years or so, it seems likely that continued work will lead to the manufacture of more efficient solar cells at lower cost. If the cost can be reduced to less than $0.50/$W_p$, photovoltaic generation of electricity should be competitive with gas- or coal-fired plants.

4.7 Solar Cooling

Where and when there is abundant sunshine there is often need for cooling. Because of this obvious fact air-conditioning has become a standard expectation throughout our country, especially in the Sunbelt where it makes the warm months comfortably livable. To meet today's demand for air-conditioning, a large and growing amount of electric power is being devoted to cooling homes, schools, shopping malls, manufacturing plants, theaters, and so forth.

The basic refrigeration process was introduced in Chapter 3. Today, almost all of our refrigerators and air conditioners have their operation based on a mechanical compressor driven by an electric motor. There is also a second method, well known for a long time and based on a technology with no moving mechanical parts. There is no need for an electric motor or for connection to a source of electricity. Fifty and sixty years ago, kitchen refrigerators using this system were widely sold and prized for their quietness. They were marketed under the brand name of *Servel*.

One of the well-established refrigeration processes that operates without moving parts is the Electrolux absorption refrigeration system. The Electrolux process came into common use after its invention by two Swedish engineering students in Stockholm. It needs only a source of heat energy to make the refrigerant go around the refrigeration cycle. Here is a simple summary of the complex process: heat is added to the refrigerant, ammonia gas dissolved in water, in a generator to force liquid ammonia into an evaporator section where it expands into a gas, thus producing cooling. The ammonia gas is then condensed back into a liquid in an air-cooled condenser. A flame from the burning of any of a number of fuels can be used to drive the process. The refrigeration coefficient of performance, also known as the refrigeration C.O.P., is a measure of the heat energy removed from the refrigerator relative to the energy put into the process. This number is much lower for absorption systems than for mechanical refrigeration systems, and is typically less than 1.0 for an absorption system, meaning that less energy is removed from the refrigerator box than is taken from the heat source. This disadvantage of the absorption system is offset by the substantially lower cost of fuel, on an energy basis, than the cost of the electricity for a mechanical system. If an absorption system were to use electricity as its heat source, its operation would be much more costly than that of a conventional mechanical system.

Solar energy, rather than a burning fuel, can provide the heat energy for the absorption refrigeration process. This generally requires a concentrating collector because the necessary temperature for running an absorption refrigerator cannot be achieved with a flat-panel collector. However, flat-panel collectors can be sufficient for the lesser degree of cooling needed for air-conditioning. Solar-powered refrigeration and air-conditioning are not yet common, or even available on the market, because of the intermittent nature of direct sunlight and because of the costs associated with the concentrating collectors. It would appear that development of standard manufactured units and pilot projects could advance the realization of the potential for solar cooling.

Simple solar ammonia-based coolers have been developed for use in primitive areas where food spoilage is a serious problem. In some areas it is not unusual for 25% of the food to be wasted due to lack of refrigeration. There is a compelling logic to the argument that we have the greatest need for solar refrigeration in the areas with the most intense insolation.

Key Terms

Geothermal	Fresnel lens
Electromagnetic radiation	Power tower
Wavelength	Heliostat
Infrared	Photovoltaic cell
Ultraviolet	Semiconductor
Solar constant	Photon
Active system	Photoelectric effect
Passive system	n-type silicon
Greenhouse effect	p-type silicon
Thermal mass	Amorphous silicon
Trombe wall	Electrolux process
Concentration ratio	

Suggested Reading and References

1. Brinkworth, B. J. *Solar Energy for Man.* New York: John Wiley, 1972.
2. Kreider, J. F., and Kreith, F. *Solar Heating and Cooling.* New York: McGraw–Hill, 1977.
3. Krenz, Jerrold H. *Energy, Conversion and Utilization,* second edition. Newton, MA: Allyn and Bacon, 1984.
4. McDaniels, David K. *The Sun: Our Future Energy Source.* New York: John Wiley, 1979.
5. Ross, Marc H., and Williams, Robert H. *Our Energy: Regaining Control.* New York: McGraw–Hill, 1981.
6. Weinberg, Carl J., and Williams, Robert H. "Energy from the Sun." *Scientific American,* **263,** No. 3 (September 1990), pp. 147–163.
7. Stone, Jack L. "Photovoltaics: Unlimited Electric Energy from the Sun." *Physics Today,* September 1993.
8. Parfit, Michael; photographs by Leen, Sarah; After Oil, Powering the Future; *National Geographic* **208 2** (August 2005), pp. 2–31.

Questions and Problems

1. Why are we still designing so many new homes with little regard for passive solar heating?
2. Why does your hand not feel the radiant heat energy from a fireplace if you place it behind a sheet of glass, even though you can see the fire very easily?
3. What steps could the federal government take to promote the further development and use of photovoltaic cells?
4. With the abundance of solar energy incident on our nation, what are the technical, economic, or political factors that must be overcome to break our reliance on fossil fuels?
5. Describe four situations where photovoltaic cells would have an advantage over other sources of electric energy.
6. What are some steps that could be taken by an individual or a family to lessen their dependence on nonrenewable energy?
7. Design an array of individual solar cells that would produce 120 volts DC. About how large an area of solar cells would be needed to produce 2 kW$_e$ of output power at noon on a clear day?
8. What collector area is needed for the system shown in Figure 4.6 to provide the necessary space heating for a home having 2000 ft^2 of floor area? Assume an average daily insolation of 1000 Btu/ft^2 and an efficiency of 50%. Also assume that the house requires 50 million Btu per thousand square feet of floor area for the 180 day heating season.
9. A car wash needs 1200 gallons of warm water a day heated from 50°F to 100°F. How large a solar collector would be needed to do this? The incident solar energy is 1100 Btu/ft^2 each day and the collector efficiency is 50%.
10. A passive solar home has energy stored in a concrete floor of 1,000 ft^2 area. How thick should this floor be to store 200,000 Btu with a temperature swing of 20°F?

Multiple Choice Questions

1. The Solar Constant, 2 calories per square centimeter per minute, is ___
    a. the annual amount of solar energy available at the top of the earth's atmosphere
    b. the solar power density at 93 million miles from the sun
    c. the average solar power density at ground level on earth
    d. extremely variable
2. The solar spectrum has its maximum intensity in terms of watts per square centimeter at a wavelength corresponding to ___
    a. the maximum of the ozone absorption spectrum
    b. the ultraviolet region

c. the visible region
d. a wavelength of about 10^{-7} meters

3. The spectrum of solar radiation above the earth's atmosphere has its maximum intensity in the _____
   a. ultraviolet region
   b. region of 100×100^6 meters
   c. visible region
   d. infrared region

4. Representative wavelengths for ultraviolet and infrared radiation, respectively, are _____
   a. 0.1 and 1 microns
   b. 1 and 10 microns
   c. 10 and 100 microns
   d. 100 and 1000 microns

5. You can easily feel the heat from the sun through a glass window, but behind a sheet of glass you do not feel the heat radiated from a fire in a fireplace. This is because _____
   a. glass is transparent to visible light and opaque to infrared
   b. the fireplace emits essentially pure ultraviolet radiation
   c. glass is opaque to visible light and transparent to infrared
   d. the sunlight is extremely infrared

6. Of the solar radiation incident on the earth's upper atmosphere, about ___, on average, is available at ground level.
   a. 1/8
   b. 1/4
   c. 1/2
   d. 100%

7. Your house foundation and basement floor are made up of 1000 ft³ of concrete. About how many Btu are required to raise the temperature of this mass from 0°C to 40°C? Given: 22 Btu/ft³·°F for concrete.
   a. 1,600,000
   b. 900,000
   c. 22,000
   d. 40,000
   e. 1,800
   f. 40

8. A flat-plate solar collector has a useful upper temperature limit of about ___°C.
   a. 2
   b. 10
   c. 20
   d. 100
   e. 200
   f. 1000

9. A useful rule of thumb for a flat-plate collector system is about ___ Btu per square foot per day of collected heat energy.
   a. 5
   b. 10
   c. 50
   d. 100
   e. 500
   f. 1000

10. In Boulder (latitude 40° N), a flat-plate collector should be tipped at an angle from the horizontal of about ___ in order to maximize the annual direct plus diffuse sunlight that it collects.
    a. zero degrees (flat on the ground)
    b. 90 degrees (upright)
    c. 55 degrees
    d. 15 degrees

11. At 40° N latitude a south-facing vertical surface, such as a window, has incident on it on a clear winter day roughly _____
    a. 157 Btu/m^2
    b. 200 Btu/cm^2
    c. 20 cal/min·cm^2
    d. 20 Btu/m^2
    e. 1600 Btu/in^2
    f. 1600 Btu/ft^2
    g. 1600 Btu/m^2
    h. 1600 Btu/cm^2

12. A reasonable size for overnight heat storage for a solar home is about ___ gallons of water.
    a. 10
    b. 50
    c. 1000
    d. 5000

13. A parabolic reflector _____
    a. can be used for solar collection systems only if the overall efficiency is less than 5%
    b. focuses both the direct and diffuse components of sunlight
    c. focuses only the diffuse component of sunlight
    d. focuses only the direct component of sunlight

14. Good photovoltaic cells have an efficiency of approximately _____
    a. 28 to 38%
    b. 14 to 17%
    c. 2 to 6%
    d. 38 to 58%

15. It has not been practical to enhance the performance of silicon photovoltaic cells by using lenses with high concentration ratios because _____
    a. the extreme magnification makes the sun's image too large to match the surface of the solar cell
    b. the Carnot efficiency cannot exceed the maximum
    c. the solar cell efficiency is reduced at the resulting high temperatures
    d. the lenses are more expensive per square centimeter than are the silicon solar cells

16. Only about ___ of the power in the solar spectrum is in a wavelength region to which silicon photovoltaic cells are sensitive.
    a. 47%
    b. 57%
    c. 67%
    d. 77%
    e. none of the above

17. The reason that photovoltaic cells are not generally competitive with fossil fuel–fired power plants is _____
    a. the cost of raw silicon is too high
    b. the expense of growing crystals and fabricating the cells is too high
    c. the efficiency is only 2%
    d. they cover too much land area

CHAPTER 5

Renewable Energy Sources II: Alternatives

(*Source*: PhotoDisc, Inc./Getty Images)

5.1 Introduction

Solar energy can fulfill our energy needs in many wondrous ways. Aside from the direct uses of solar radiation covered in the last chapter, there are also the less direct technologies of hydroelectricity, wind energy, biomass, ocean thermal gradients, ocean currents, and ocean waves for making use of energy from the sun. In addition to using the energy sources of solar origin, we have also learned how to exploit ocean tides and geothermal resources. These are all renewable energy sources that can be explored to see what potential they have for providing useful amounts of energy in an economically and environmentally sound way. The overall flow of energy to and from the earth is shown in Figure 5.1.

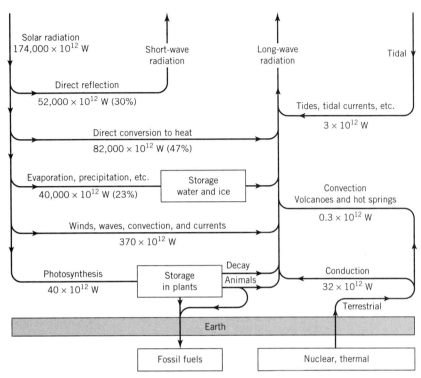

Figure 5.1 Natural energy flow (in units of power) to and from the earth. (*Source*: M. K. Hubbert, "Man's Conquest of Energy: Its Ecological and Human Consequences," in *The Environmental and Ecological Forum 1971–1972*. Washington D.C.: U.S. Atomic Energy Commission Publication TID-25857, 1972.)

5.2 Hydropower

Hydroelectricity is a well-established technology; it dates back to the beginnings of the electric power industry more than 100 years ago. About 7% of the electric power now generated in the United States comes from our damming the rivers as they flow to the ocean and then releasing the water to turn turbines connected to electric generators. The use of water power is much greater in some other countries; Norway obtains 99% of its electricity from water power, Nepal 95%, Brazil 93%, New Zealand 78%, Canada 58%, and Sweden 50% (half hydro, half nuclear).

Hydropower was important long before electricity generation became possible. About 2000 years ago, during the first century B.C., it was discovered that the force of moving water acting on a waterwheel could be used to ease human labor. Various types of ingeniously contrived waterwheel mechanisms were connected by rotating shafts and cogwheels to mills for grinding grain and sawing wood. By the 13th century, water power was used to operate hammers in the ironworks of western Europe, and by the 16th century the waterwheel was the primary source of industrial power in that part of the world. The steam engine eventually replaced water power in many applications, but in selected locations and for certain purposes, water power continued to be the preferred energy source. As our nation developed, mills were established at sites with reliable water flow of sufficient volume and velocity. Communities then grew up around these mills. Dams were built at many such sites to impound the flowing water and to even out the flow over the seasons. After electricity came into practical use, waterwheels were used to drive generators; it then became reasonable to locate hydroelectric plants at a distance from population centers and to transport the electricity over power lines hundreds of miles to the point of demand. Now we use water power almost exclusively for the generation of electricity, even if there is only a modest distance between the point of generation and the point of use.

Water power is, of course, a consequence of the natural cyclical transport of water between the earth's surface and the atmosphere. This hydrologic cycle, shown in Figure 5.2, involves the evaporation of the earth's surface water when it is heated by sunlight, followed by precipitation and the downward course of the water in rivers and streams under the force of gravity. In a sense the water is the working fluid in an enormous heat engine powered by sunlight.

The available energy of water stored at a height above a generator is potential energy in the earth's gravitational field. The solar energy that went into

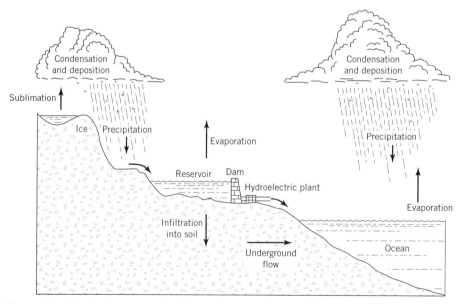

Figure 5.2 The hydrologic cycle. Electricity is produced in the hydroelectric plant by the action of water against a turbine connected to a generator. In this way the stored potential energy of the water in the reservoir becomes electrical energy.

lifting the water upward in this gravitational field can be made available again when, for example, the water falls downward onto a waterwheel. As each element of water of mass m falls freely a distance h, it attains a velocity v and a kinetic energy, $1/2 mv^2$, equal to its original potential energy, which is given by mgh. The acceleration of gravity, g, is 9.8 m/sec^2. Thus the potential energy available for conversion to kinetic energy is 9.8 joules per kilogram of water (or anything else) per meter of height above where the kinetic energy is to be utilized. The distance, h, is often called the *head*. Low-head hydroelectricity can be generated with h as small as 10 feet, whereas high-head hydroelectricity is generated with heads of hundreds up to greater than 1000 feet. The detailed design of the hydraulic turbines is different in the two cases.

It will be apparent on visiting any modern hydroelectric facility that the discussion given above about water falling freely through space before impacting on a turbine is not entirely realistic. Nowhere in a large hydroelectric plant will you see water falling freely through a large vertical distance. Nevertheless, the analysis in terms of potential energy of the water is still correct when the water emerges at high velocity through an opening near the bottom of the dam rather than over the top. The water coming out of the opening and impinging on the blades of a turbine has in effect fallen through a distance equal to that from the surface of the impounded water down to the exit. Modern hydroelectric installations convert the potential energy of water to electric energy at an efficiency of 80 to 90%.

Example 5.1

Calculate the flow rate of water (in liters/sec = kilograms/sec) required to provide 1 kW of electric power if the water falls a vertical distance of 90 m. Assume 80% conversion efficiency.

Solution

The potential energy in joules of a mass m at a height h is mgh, where $g = 9.8$ m/sec^2, m is in kilograms, and h is in meters. If the flow rate is 1 kg/sec, the power in the stream of water after a fall of 90 meters will be

$$\text{Power} = \frac{\text{energy}}{\text{time}} = \frac{mgh}{\text{sec}} = \frac{1 \text{ kg}}{\text{sec}} \times 9.8 \frac{\text{m}}{\text{sec}^2} \times 90 \text{ m}$$

$$= 882 \frac{\text{kg m}^2}{\text{sec}^3} = 882 \text{ J/sec}$$

$$= 882 \text{ W}$$

Here we have trusted that as long as metric units were used as inputs to this calculation, the result for power will be in J/sec = W, or we could note that 1 joule = 1 kg · m^2/sec^2.

If this 882 W is converted to electricity at 80% efficiency, the electrical power produced will be 0.8 × 882 W = 706 W at a flow of 1 liter/sec. To generate 1 kW (1000 W), a flow of

$$\frac{1000 \text{ W}}{706 \text{ W/(liter/sec)}} = \mathbf{1.42} \; \frac{\textbf{liters}}{\textbf{sec}}$$

will be necessary.

Hydroelectric power has many obvious advantages over conventional power plants driven by fossil fuels. There are no polluting emissions into the air or water, and no waste heat is rejected as thermal pollution. The operation of hydro plants relies only on renewable energy resources, and the plants have lifetimes of many decades and small maintenance requirements. They respond well to sudden changes in demand, making hydroelectricity well suited to matching peak loads. The dams can serve multiple purposes; water stored for irrigation, flood control, or a municipal drinking water supply can also power a hydroelectric plant as it flows to its other tasks. Figure 5.3 shows an example of a multipurpose installation.

In the late 1980s, the city of Boulder, Colorado, installed a system of six small hydroelectric plants on the water pipeline that brings municipal water down into the city from a watershed in the mountains at an elevation of several thousand feet above the city. The total power produced by these plants is about 6 MW, sufficient to supply about 8% of the electrical power used in the city. For many decades prior to the installation of these hydroplants, the excess water

Figure 5.3 Hoover Dam and Lake Mead on the Colorado River at the Arizona–Nevada border. This installation produces 2080 MW of electrical power as well as storing water for other purposes, including flood protection, irrigation, and recreation. (*Source*: Courtesy Bureau of Reclamation/U.S. Department of Energy)

pressure in the pipeline had been reduced through the use of special pressure-reducing valves. This had wasted the energy of the falling water. The installation of the hydroelectric plants eliminated the waste, and since the payback time for this system is 10 years, the city is making a profit on its investment. Through sales of electricity to the electric utility Xcel Energy, this hydroelectric system earned revenue of $1.7 million for the city in 2004.

Table 5.1 lists some of the larger hydroelectric plants in the United States. These, along with a number of smaller projects, amount to a total capacity of about 80,000 MW$_e$. It is apparent that although there has been rather little construction of hydroelectric plants in recent years, there was a flurry of hydroelectric facility development in the decades following the end of World War II in 1945. These trends can be seen in Figure 5.4, which shows the amount of hydroelectric energy produced in the United States from 1950 to 2003. Since 1974 there has been little overall increase. The large year-to-year fluctuations in hydroelectric generation seen in the graph over the last 20 years are largely due to variations in annual precipitation. Figure 5.4 also shows that the percentage of U.S. electric energy from hydroelectric plants has gradually declined from about 30% in 1950 to under 7% in 2003.

It may be noted here that the approximate average over recent years of 3×10^8 MWhr/yr of hydroelectric energy shown in Figure 5.4 is about the equal of 1 QBtu/yr (about 1% of the national energy consumption). Yet in Figure 1.6, Table 4.1, and elsewhere, the annual hydroelectric contribution is shown as close to 3 QBtu/yr (about 3% of the national energy consumption) An explanation

Table 5.1 Some Large Hydroelectric Projects in the United States

Project	River	First Year of Operation	Rated Capacity (MW$_e$)
Grand Coulee	Columbia	1942	7100
John Day	Columbia	1969	2500
Chief Joseph	Columbia	1955	2300
Moses–Niagara	St. Lawrence	1961	2160
Bath County, VA	Buck Creek	1985	2100
Hoover	Colorado	1936	2080
The Dalles	Columbia	1957	1870
Racoon Mt.	Tennessee	1979	1530
Glen Canyon	Colorado	1964	1300
McNary, OR	Columbia	1954	1130
Northfield, MA	Briggs	1971	1080

Sources: U.S. Energy Information Administration (2000); United States Bureau of Reclamation (2004).

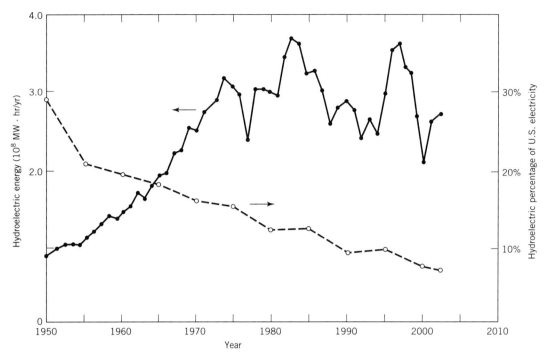

Figure 5.4 Electric energy from hydroelectric installations in the United States (solid line). The percentage of U.S. electricity provided by hydropower is shown as the dashed line. (*Source*: Data for 1950 to 1990 from *Annual Energy Review*, 1990; data for 1991 to 2003 from U.S. Energy Information Administration)

for this difference is given in Section 1.6 along with the discussion of Table 1.1. The same pattern appears for wind and some other renewables. What is often listed is the equivalent thermal energy it would take to produce the same amount of electrical energy in a fuel-burning or nuclear plant having an efficiency in the range of 30 to 35%. Hydroelectric, wind, and some other renewable generating technologies are not subject to the inefficiencies of the typical thermal-electric generating plant. Thus, to compare all energy inputs in terms of equivalent thermal energy, the electrical energy produced by these renewable technologies is multiplied by a factor of approximately 3 and converted to thermal energy units before it is entered into the overall energy compilations. This sometimes confusing practice is a convention of long standing in energy statistic compilations.

The hydroelectric potential in various sections of the country is shown in Table 5.2. The Pacific, which includes Alaska and the Northwest, has the greatest undeveloped capacity. Overall, about 50% of the potential capacity of the United States has been developed. For a variety of reasons—environmental, eco-

Table 5.2 Hydroelectric Potential in the United States, GW

Region	Potential	Developed	Undeveloped	% Developed
New England	6.3	1.9	4.4	30.1
Middle Atlantic	9.8	4.9	4.9	50.0
East North Central	2.9	1.2	1.7	41.3
West North Central	6.2	3.1	3.1	50.0
South Atlantic	13.9	6.7	7.2	48.2
East South Central	8.3	5.9	2.4	71.1
West South Central	7.3	2.7	4.6	36.9
Mountain	28.6	9.5	19.1	33.2
Pacific	64.4	38.2	26.2	59.3
Total	147.7	74.1	73.6	50.2

Source: Statistical Abstracts of the United States, 1995.

nomic, and political—it is unlikely that a significant portion of the undeveloped potential will be developed in the foreseeable future. The Wild and Scenic Rivers Act and the Endangered Species Act have inhibited development of some sites, and there are other problems:

1. Some hydroelectric installations have lifetimes limited to approximately 50 to 200 years because their storage volumes become steadily filled with silt washed downstream by the rivers that feed the reservoirs. This is a severe problem, with no solution in sight, for many of the world's largest hydroelectric facilities. It is less of a problem for other facilities fed by streams flowing over beds of rock rather than of soil. Once a reservoir has been filled with silt to the point that it is no longer useful for water storage, maintenance of the dam will be a continuing responsibility. Downstream areas must be protected from the sudden release of enormous volumes of silt that could flow downstream in the event of dam failure.

2. Other objections to hydroelectric power include the loss of free-flowing streams and the loss of the land flooded by the reservoirs. Native aquatic life of many forms is disturbed—particularly salmon in the Northwest. After spending most of their life far out at sea, native salmon return to the river where they were hatched, fight their way upstream, often against swift rapids, to spawn another generation of salmon. This natural cycle of life has been severely interrupted by the building of hydroelectric dams on many of the larger rivers. These dams produce intermittent stream flow in response to power demand, and they present physical barriers to the passage of all

aquatic life. The salmon population in the Pacific Northwest states, British Columbia, and parts of Alaska is now in serious decline. There have been various attempts to remedy this problem, such as by providing structures known as fish ladders to enable the salmon to work their way upstream around the dams, but this approach has not been entirely successful, and the problem continues. Plans to build more hydroelectric dams in this area of great undeveloped hydroelectric capacity have been held up by concerns over the future of native salmon populations.

3. Storage reservoirs are often situated upstream from major population centers. This presents a considerable risk in the event of dam failure, such as might be expected from an earthquake. Catastrophic dam failures occurred frequently throughout the 20th century. Between 1918 and 1958 there were 33 major dam failures in the United States with 1680 resulting deaths, an average of 42 per year.[1] From 1959 to 1965 nine large dams failed throughout the world. Dam failures involving loss of human life occurred in the United States in 1976 and 1977. We now have a number of population groups of more than 100,000 persons at risk from dam failure.

5.3 Wind Power

Since that distant time when our ancient ancestors first learned to rig sails on their rafts, wind power has been put to use. After this early discovery, the use of wind on fabric sails advanced technologically to the point where all the world's navies and transoceanic commerce were powered by the wind. The large sailing vessels of the 19th century could extract as much as 10,000 horsepower from the wind; no larger wind machines have ever been developed. Stationary wind machines now represent the main use of the wind's energy, and their early development also dates back many hundreds, perhaps thousands, of years. Windmills by the tens of thousands were operating in Europe during the 19th century, and several million were pumping water in the United States by the early 1900s. Starting about 1890, the use of windmills has been increasingly directed toward the generation of electricity and less toward their direct mechanical coupling to machinery such as mills or pumps. Today the development of wind machines is almost exclusively concentrated on electricity generation. Figure 5.5 shows a large wind machine constructed over 100 years ago.

In some ways wind power has an advantage over solar photovoltaic generation. The winds can blow day and night, sunny or cloudy, and are often at their strongest and most reliable during the coldest and darkest nights of winter when energy is needed the most. In the commonly overcast regions and high latitudes where solar power cannot be relied on, wind energy is often available. However,

---

[1]On March 13, 1928, more than 400 people died when the San Francisquito Valley in California was inundated with water after the St. Francis Dam failed.

Figure 5.5 A fully rigged American sailing ship of the early 19th century. Under good conditions this ship could extract thousands of horsepower from the wind. (*Source*: Corbis-Bettmann)

both technologies are intermittent in their power delivery and require energy storage for many applications.

The earth's winds are a direct consequence of solar energy. On both local and global scales, these winds are generated because the sun heats certain areas of the earth's surface and atmosphere more than others. The differential heating induces both vertical and horizontal air currents, with the patterns of the currents modified by the earth's rotation and contours of the land. The familiar land–sea breeze cycle is an example of how winds are produced. During the daytime, the sun shines on both land and sea, but the surface of the land becomes significantly warmer than that of the sea because it takes less energy to heat the materials of the earth's surface than it does to heat seawater. As the air over the warm land is heated, it rises and is replaced by the onshore breeze that moves air in from over the cooler sea. After the sun sets, the land and the air over it cools faster than the sea. Then the now warmer air over the sea rises and is replaced by an offshore breeze that moves the heavier cooler air out from over the land.

A small fraction of the solar energy incident on the earth's upper atmosphere is transformed into kinetic energy of the winds. This kinetic energy is dissipated into warming the atmosphere and against features of the earth's surface as fast as it is gained from solar heating, thus being maintained at a quite steady average amount. The rate at which solar energy goes into wind energy over the United States is about 30 times the rate of our present energy consumption.

The power in the wind increases rapidly with velocity; it is proportional to the third power of the velocity. This is explained as follows: each unit of air mass has kinetic energy proportional to the square of its velocity, and the amount of air mass moving past a given point (for instance, the location of a windmill) per unit time is proportional to the wind velocity. Thus the amount of wind power (energy/time) at any point in space varies as v^3. The exact relationship of power to velocity, of course, also involves the density and moisture content of the air. For average conditions, one may use the expression

$$P/m^2 = 6.1 \times 10^{-4} \, v^3$$

to find the power in kilowatts per square meter of cross section oriented perpendicular to the wind's direction, with the velocity v given in meters per second. This is the total power in the wind, which cannot all be extracted by any practical device.

A windmill built so that it too severely interrupts the airflow through its cross section will reduce the effective wind velocity at its location and divert much of the airflow around itself, thus not extracting the maximum power from the wind. At the other extreme, a windmill that intercepts an exceedingly small fraction of the wind passing through its cross section will reduce the wind's velocity by only a small amount, thus extracting only a small fraction of the power from the wind traversing the windmill disk. All practical windmills fall somewhere between the two extremes. An exact theoretical analysis of windmill performance shows that no more than 59% of the kinetic energy of the wind is recoverable as mechanical energy; modern windmills can attain an efficiency of perhaps 50 to 70% of the theoretical maximum.

Figure 5.6 shows the efficiency of several types of wind machines. The Savonius rotor has the shape of two halves of a barrel, cut lengthwise, with the two halves offset by something less than a diameter, and rejoined so that they face in opposite directions. They are mounted on a vertical axis through the region where the two halves are joined. A second vertical axis machine, the Darrieus rotor, is also represented on this plot. The vertical axis machines have the advantage that the generator is at ground level, and thus easy to service. Also, no mechanism is needed to keep them pointed into the wind.

Figure 5.6 shows that the ratio of the blade's tip speed to the wind speed is an important variable that must be considered in the design of wind machines. The most efficient type, the high-speed propeller, can typically attain 70% of the theoretical maximum efficiency, thereby utilizing about 42% of the power in the wind. The American multiblade windmill, used for pumping water in rural areas, utilizes a maximum of 30% of the wind's power, and the picturesque Dutch four-arm type extracts about 16%. The American multiblade type is well adapted to driving water pumps, because it has a relatively high starting torque, whereas the high-speed propeller type has a low starting torque but is most efficient at

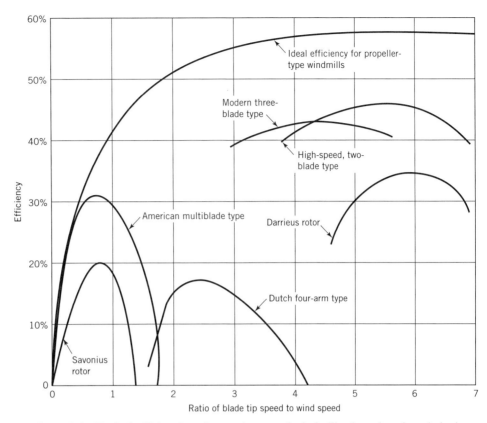

Figure 5.6 Typical efficiencies of several types of windmills plotted against their tip-speed ratio. The maximum efficiencies are seen to vary from about 16 to 46%. The ideal efficiency shown is a mathematical ideal, never to be achieved in practice. (*Source*: Basic data from R. Wilson and P. Lissaman, *Applied Aerodynamics of Wind Power Machines*, Oregon State University.)

the high rotational speeds suitable for small-scale electricity generation. When a windmill is used for generating electricity, one must consider two efficiencies: the efficiency for converting wind power to mechanical power and the efficiency for converting mechanical power to electric power. The product of these two numbers is the overall efficiency for converting wind energy to electric energy. The mechanical to electric efficiency is typically about 90%.

Example 5.2

Calculate the electric power produced per square meter of windmill disk area for a windmill operating at 70% of the theoretical maximum mechanical efficiency, with an electric generating efficiency of 90% when the wind velocity is 10 m/sec.

Solution

To get the electric power, we must multiply together the mechanical efficiency, the generating efficiency, and the power in the wind per square meter.

$$\text{Power (kW/m}^2\text{)} = (0.7) \times (0.59) \times (0.9) \times 6.1 \times 10^{-4} \, v^3$$
$$= (0.7) \times (0.59) \times (0.9) \times 6.1 \times 10^{-4} \, (10)^3$$
$$= 0.227 \text{ kW/m}^2 = \mathbf{227 \text{ W/m}^2}$$

Example 5.2 shows that with a wind velocity of 10 m/sec, a power output of about 200 W/m² can be achieved. This is far above the average performance to be expected year-round in the United States, because a wind velocity of this magnitude is experienced near ground only a small percentage of the time. Typical annual average *energy* output for a windmill can be expected to range from about 100 kWh/m² for relatively calm areas to approximately 500 kWh/m² for relatively windy areas. These numbers correspond to average levels of electric *power* production of only 11.4 W/m² and 57 W/m², respectively, over the entire 8760-hour year.

Example 5.3

Calculate the diameter of a windmill needed to supply the 5000 kWh of electrical energy needed annually by an American household. Assume average wind conditions in which 250 kWh/m² is produced annually by the windmill.

Solution

The area of the windmill disk can be obtained by dividing the energy needed by the energy produced per square meter.

$$\frac{5000 \text{ kWh}}{250 \text{ kWh/m}^2} = 20 \text{ m}^2$$

The diameter of the disk can be obtained from

$$\text{Area } (A) = \pi r^2 = \pi \left(\frac{d}{2}\right)^2 = \pi \frac{d^2}{4}$$

or

$$d = \sqrt{4A/\pi}$$
$$d = \sqrt{4 \times 20 \text{ m}^2/\pi} = \sqrt{25.5 \text{ m}^2} = 5 \text{ m} = \mathbf{16.5 \text{ ft}}$$

In choosing a site for a windmill, the constancy of a good wind throughout the day and throughout the year is more important than occasionally high peak speeds. Detailed wind measurements over a number of years are needed to help make a proper selection of a site. The windspeed profile with height above the ground shows that the speed generally increases substantially with height and depends on surface features such as buildings and trees. Ideally, it would be best to avoid such surface features and to situate the windmill at a very high level, like 300 meters, where the full potential of the wind is realized, but such a high structure is obviously impractical. Commercial wind turbine towers are about 50 meters or more high, high enough so that about 80% of the full potential is reached.

The wind turbines used today fall into two general classes: small wind machines for use by individual farms and households not connected to the power grid and larger machines generating bulk power for distribution by an electric power utility. In the years between 1850 and 1970 there were at times more than 6 million wind machines of less than 1 kW output in use in the United States. At first they were used mainly for pumping water, but later they also generated electricity. The smaller electricity-generating machines were common in the United States in the early decades of the 20th century, through the 1940s, in areas where electric power from a utility was not available. Extension of electric utility service into rural areas and increasing load demands by users have significantly reduced the use of home-type wind machines, but some wind-driven water pumps of the American multiblade type are still used in rural areas.

Following an early attempt in the 1940s, the 1970s saw serious efforts to develop and install large wind turbines connected to the power grid. Some very large individual wind generators of 1 to 3 megawatt capacity were installed at various places around the United States and in Europe, but later efforts concentrated on *wind farms* made up of large numbers of machines of moderate size, in the range of 50 to 600 kW. About half of the wind machines now installed in the California wind farms are of Danish design. Many of the newest wind farms in the United States and Europe are now using wind machines of 1 megawatt and greater generating capacity.

The U.S. wind energy generating capacity is rapidly increasing. In 1989 it was about 1400 MW_e and by 2003 it had grown to 6400 MW_e. In 2003 alone, about 1800 MW_e was added to the total. Since 2000, wind energy generating capacity has grown by nearly 25% each year. This record growth will not continue long without extension of the federal wind energy tax credit. As it now stands, there is a federal energy production tax credit of $0.018 per kilowatt hour of wind-generated electricity, adjusted annually for inflation, continuing for the first 10 years of operation for systems put into use by December of 2005.

An important factor in the remarkable growth of wind energy has been the ability of the utilities to fund renewable projects, mainly wind energy, by charging the increased costs to voluntary program subscribers. At least 350 utilities in 33 states now have announced or implemented "green" pricing programs. More than 6% customer participation has been achieved by some utilities, but the average is about 1%.

In spite of encouraging signs of growth, the contribution of wind energy to the total U.S. electrical energy consumption is still substantially less than 1%. Some analysts expect the wind energy contribution to triple over the next 10 years.

The reported power ratings for wind machines and wind farms should not be taken at face value because they present a far too optimistic picture of wind power in the United States. Most of our conventional generating capacity operates 24 hours per day, every day of the year. This is obviously not true of wind power. A kilowatt of wind power rated capacity is not the equal of a kilowatt of conventional capacity. The Pacific Gas and Electric Company, the buyer of the electric power produced by the California wind farms, estimates that the average output of a wind farm is only 20% of the rated capacity, meaning that a wind turbine in California typically operates at only one-fifth of its rated capacity.

What is the potential for having a much larger share of our electricity-generating capacity provided by the wind? Figure 5.7 shows the areas in the United States with various wind power densities (watts/m^2 of rotor-swept area) at 50 m above the ground. Wind power classes 3 (300–400 W/m^2 at 50 m altitude) to 7 (800–2000 W/m^2 at 50 m altitude) are generally considered suitable for wind power development. The classification scheme also takes into account the variability of the wind speed and the average density of the air. There are 625,000 km^2 in the contiguous United States that have class 3 or higher wind resources and which are within 10 km of electric transmission lines. This resource corresponds to a potential average 734,000 MW$_e$ from wind power (meaning that the rated capacity is much greater), and is based on coverage of the area by wind turbines spaced 10 rotor diameters apart in the direction of the prevailing wind and 5 rotor diameters apart perpendicular to this direction. If the wind turbines have a rotor diameter of 15 meters, for example, there would be one for every 11,250 square meters of land area within the region of high potential, or a total of more than 50 million wind turbines over the 625,000 km^2 area. The total installed capacity of all electric power plants in the United States (948,000 MW$_e$) is somewhat larger than what might be expected from this potential wind power resource.

In addition to the difficulties in financing, there are basic energy storage problems and the variability of the wind speeds that must be overcome by system design. Difficulties due to wind variability arise when the electric power from wind is more than 15 or 20% of the total installed system capacity, including the conventional modes of power generation. While the environmental problems of wind energy are not as severe as those of some other energy conversion technologies, they do exist. The differential pressure gradients around the wind turbines can be a problem for birds, and the noise affects both people and animals. The unsightly appearance of mile after mile of wind machines with the necessary transmission lines is of concern to some members of the public.

5.4 Ocean Thermal Energy Conversion

The world's oceans constitute a vast natural reservoir for receiving and storing the energy of the sun incident on earth. The oceans take in solar energy in proportion to their surface area, which is nearly three times that of land. Water near the surface of tropical and subtropical seas is maintained by this solar radiation at higher temperatures than the water at greater depth or at higher latitudes.

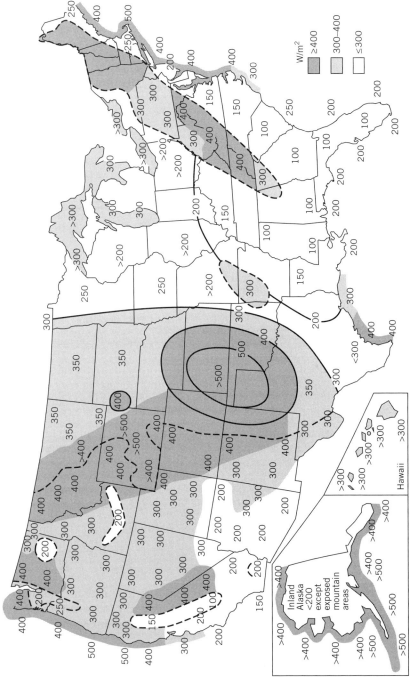

Figure 5.7 Annual average wind power density (watts per square meter) at 50 meters altitude. (*Source:* Figure supplied by the National Renewable Energy Laboratory.)

Some of the warm surface water is carried far from equatorial regions, either to the north or to the south, by ocean currents such as the great Gulf Stream or the Japanese Current. These movements result from the action of prevailing wind patterns on the surface of the ocean; the courses of these currents are shaped further by the effects of the earth's rotation (Coriolis effect) and by the shorelines of continental land masses.

It would be entirely possible to tap the energy in the ocean currents for electrical power by using gigantic underwater turbines anchored to the ocean floor and connected to generators—much as windmills extract energy from moving air. The ocean currents are attractive for this purpose because they flow at a relatively steady velocity, in contrast to the intermittent nature of the wind. This eliminates the need for energy storage. The energy available is enormous; the Gulf Stream has a flow a thousand times greater than the Mississippi River and a maximum velocity greater than 4 miles per hour. Serious proposals have been put forth to construct ocean current turbines with the electricity produced then cabled to shore. The proposed projects do not appear to be able to compete economically with conventional power generation, and they have not been advanced beyond the discussion stage.

There has been a larger effort to extract energy from the oceans through the use of heat engines that exploit the temperature differences between the warm tropical surface water at temperatures above 20°C and the colder water at depths of about 1000 m, which is only a few degrees above the freezing point. It has been demonstrated that a heat engine capable of driving an electric generator can be designed and operated between the warm surface waters that act as a heat source and the deep cold water that acts as a heat sink. The temperature differences are very steady in time, persisting over day and night and from season to season. This steadiness overcomes the energy storage problem encountered by some other solar energy sources. Figure 5.8 shows the regions of the ocean surface where thermal gradients are appreciable. Regions that are of particular interest to the United States are Florida, Puerto Rico, Hawaii, and other islands of the Pacific where electric energy is now obtained primarily from imported fossil fuels.

Two types of heat engines have been considered, and demonstrated, for ocean thermal energy conversion (OTEC). The first is a closed-cycle system of the type shown in Figure 5.9. In this engine heat from the warm seawater causes a working fluid such as ammonia to evaporate into a gas. The expanding vapor forces rotation of a turbine connected to an electric generator. The cold water brought up from the depths then cools a condenser which causes the ammonia vapor to condense back to a liquid which is then returned to the evaporator, and the process repeats itself. The overall process is similar to that employed in an ordinary steam-powered electric generating plant that uses water as its working fluid, as shown in Figure 3.3. The second type of OTEC engine is an open-cycle system where the working fluid is seawater. In this system the warm seawater from near the surface is brought into a chamber maintained at a partial vacuum by a vacuum pump. The warm seawater then boils into a vapor, and the vapor

Ocean Thermal Energy Conversion 141

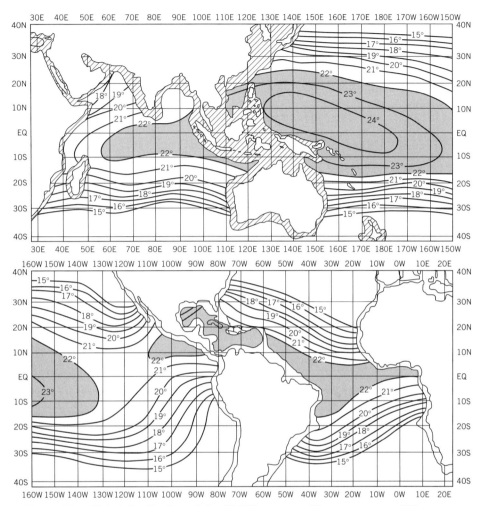

Figure 5.8 Global distribution of the OTEC resource. The temperature difference (degrees Celsius) is shown between the surface and 1000 meter depth. (Figure supplied by the National Renewable Energy Laboratory.)

drives a turbine connected to an electric generator. After passing through the turbine, the vapor is condensed back into a liquid in a heat exchanger cooled by the cold water brought up from deep below the surface. The condensed seawater is then discharged to the ocean. Unlike the closed-cycle system, there is no recycling of the working fluid, in this case, seawater.

It is estimated that a minimum temperature difference of about 17°C is needed for a practical OTEC heat engine to operate. With a temperature difference of 20°C, the Carnot efficiency is only about 7% (see Example 5.4), and the net efficiency to be expected in practice, about 2.5%. This small efficiency means that very large volumes of both warm and cold water must be circulated

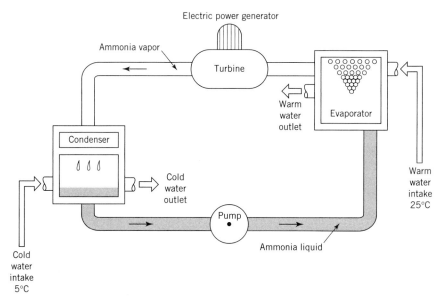

Figure 5.9 An OTEC heat engine using ammonia as a working fluid. The turbine is driven by the ammonia vapor and is connected to a generator to produce electricity. The warm water is drawn from the ocean surface; the cold water from a depth of 1000 meters. (*Source*: Figure supplied by the National Renewable Energy Laboratory.)

past the heat exchangers to produce useful amounts of power (see Example 5.5). It has been estimated that 25×10^6 liters/sec of both warm and cold water would be needed for 100 MW of electric output. For a 40 MW$_e$ plant a cold water intake pipe of 10 meters diameter (a good-sized traffic tunnel) would be necessary. This is about the same diameter as that of some of the water passages in the Hoover Dam hydroelectric facility.

Example 5.4

Calculate the thermodynamic efficiency, η, for an ideal heat engine operating between surface waters and water at 1000 m depth if the surface water temperature is 25°C and the deeper water is at 5°C.

Solution

$$T_c = 5°C = 278 \text{ K}$$
$$T_h = 25°C = 298 \text{ K}$$
$$\eta = (1 - T_c/T_h) = (1 - 278/298) = (1 - 0.933) = \mathbf{0.067}$$
$$\eta = \mathbf{6.7\%}$$

Of course, the efficiency of an *actual* heat engine will be less than that.

Example 5.5

(a) Calculate how much power (P_h) in MW is made available as input energy to a heat engine by the cooling of 1000 gallons of water per second by 2°C. Express your answer in watts. Assume 1 gallon = 3.8 kg. Recall that one calorie of heat energy is liberated by the cooling of one gram of water by one degree Celsius.

(b) If this heat engine operates between heat source and heat sink temperatures differing by 20°C, as in Example 5.4, what is the theoretical maximum power output?

Solution

(a)

$$P_h \text{ (cal/sec)} = \text{grams H}_2\text{O/sec} \times \Delta T \text{ (°C)}$$
$$= 3.8 \times 10^3 \text{ grams/gal} \times 10^3 \text{ gal/sec} \times 2°\text{C}$$
$$= 7.6 \times 10^6 \text{ grams} \cdot °\text{C/sec}$$
$$= 7.6 \times 10^6 \text{ cal/sec}$$

$$P_h \text{ (W)} = P_h \text{ (cal/sec)} \times 4.184 \text{ J/cal} = 31.8 \times 10^6 \text{ W}$$
$$= \mathbf{31.8 \text{ MW}}$$

(b) Using $\eta = 0.067$ from Example 5.4,

$$P_{out} = \eta \times P_h = (6.7 \times 10^{-2}) \times (31.8 \times 10^6) = \mathbf{2.13 \text{ MW}}$$

or about 1/500 of the electric power output of a large coal-burning or nuclear power plant.

Jacques A. d'Arsonval, a French physicist and engineer, first proposed the OTEC concept in 1881. In 1930, Georges Claude, a former student of d'Arsonval and inventor of the neon lightbulb, built and tested the first OTEC system. It was located in Matanzas Bay in Cuba and used the open-cycle system to generate up to 22 kW of electric power. Unfortunately, for this first trial system, more power was required to operate it than it produced. However, on receiving an award from the American Society of Mechanical Engineers in 1930, Claude said, "I affirm my faith in the realization of wonderful plants running ceaselessly throughout the year, unaffected by the seasonal scarcity of water in streams or variations in the cost of coal; and I hold that this is not the task of a remote future, but one of tomorrow."

Very little was done after the Claude experiment until the 1970s when the rising costs of fossil fuels prompted several countries, including the United States, to initiate design and economic studies of OTEC.

In the 1970s the U.S. Department of Energy financed the design of large floating OTEC power plants that were fixed in position by cables to the ocean floor. These plants were intended to provide electricity to islands. The electric transmission lines would run along the ocean floor from the OTEC plant to the island. A design concept for a 100 MW$_e$ OTEC power plant is shown in Figure 5.10. No plants of this size were actually built. A small test plant of 10 to 15 kW$_e$ was built and operated successfully for four months off the Kona coast of the island of Hawaii. An engineering test facility for a 1 MW$_e$ power plant was also built in that area, but no useful electric output was intended. Since the early 1980s there has been little support by the government for any program of power from the oceans, either by OTEC or by ocean currents.

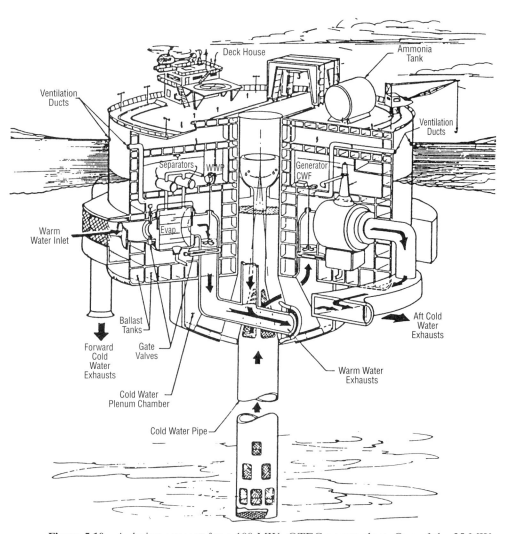

Figure 5.10 A design concept for a 100 MW$_e$ OTEC power plant. One of the 25 MW$_e$ power modules is shown in the cutaway portion. The platform has a diameter of 100 meters. Design by TRW Systems Group, Inc.

Several other possibilities besides direct electricity generation have been considered for OTEC plants. It would be possible to use the energy developed by an OTEC plant to generate hydrogen from water or to manufacture other energy-intensive products such as fertilizer. Once hydrogen is produced, it could be piped to shore for widespread distribution and use as a clean fuel. Also, various biological nutrient materials are upwelled from the ocean depths by the OTEC cold water intake pipes. These nutrients have been considered for supporting the growth of plants or animals. Another possible important use of OTEC energy would be the conversion of ocean water to fresh water to be piped to communities on shore.

5.5 *Biomass as an Energy Feedstock*

Solar energy has its most obvious role in the growing of plant life. Biomass formed from air and water through the influence of sunlight has been an important feedstock to our energy budget for a very long time. The chemical energy stored in the biomass is released for our use by combustion, either of the original biomass, or after its conversion to another form of fuel.

In the process of photosynthesis, the radiant energy from the sun drives a chemical reaction in which carbon atoms from the carbon dioxide (CO_2) gas always present in air, and hydrogen atoms from water (H_2O) are combined to form various carbohydrate compounds plus oxygen (O_2). We depend on this process for our food, fiber, wood, and innumerable other products from vegetable matter, and it is this same photosynthetic process that captured the sun's energy in living organisms hundreds of millions of years ago to provide the fossil fuels we use today.

As we saw in Chapter 4, the energy of the sun that reaches the earth is electromagnetic radiation with a spectrum that ranges from about 0.3 to 3 microns in wavelength. This corresponds to radiation from the near-ultraviolet through the visible to the infrared.

What are the details of the process by which plants use the energy of sunlight to form vegetable matter? To understand photosynthesis, we must use the quantum description of electromagnetic radiation. In this description, first understood about a hundred years ago, electromagnetic radiation is made up of quanta, or separate little bunches of energy, with longer wavelengths having less energy per quantum. The energy of a quantum or, equivalently, the energy of a photon (a quantum of electromagnetic radiation is also known as a photon) in the ultraviolet wavelength region is sufficient to break a chemical bond; this is not generally so, however, in the middle portion of the solar spectrum where solar intensity is at its peak and where most of the photosynthesis takes place. A photon in the visible portion of the spectrum, although not capable of breaking a chemical bond, has sufficient energy to raise an atom to an excited state. Many details of photosynthesis are not yet understood, but it is generally accepted that the excited atomic state may make it possible for bonding to take place between neighboring atoms, thus forming new compounds. The following generalized reaction expresses these ideas for the simplest carbohydrate:

$$CO_2 + 2\, H_2O + \text{light} \rightarrow CH_2O + H_2O + O_2$$

where 112 kcal of light energy is needed as input per mole of CH_2O formed. A mole is an amount equal to 6.02×10^{23} atoms or molecules, and one mole of a substance is the weight in grams equal to the molecular weight. For example, the mass number of oxygen is 16, so the mass number of the oxygen molecule, O_2, is 2 times 16, or 32. Hence one mole of O_2 weighs 32 grams.

The common carbohydrates are represented by the formula $C_x(H_2O)_y$, where x and y are numbers that distinguish one carbohydrate from another. Some of the simplest carbohydrates formed by photosynthesis are the sugars glucose, $C_6H_{12}O_6$, and sucrose, $C_{12}H_{22}O_{11}$. The photosynthetic reaction leading to glucose is

$$6\, CO_2 + 6\, H_2O + \text{light} \rightarrow C_6H_{12}O_6 + 6\, O_2$$

The light energy required to produce a mole of glucose is 674 kcal.

The photosynthetic reaction does not take place in a single step, but requires several steps in which the various components of the sugar molecule enter into the reaction. At least two types of pigments in the plant are involved. The most important is chlorophyll, which absorbs light in the red part of the visible spectrum (a wavelength of about 0.7 micron) and in the blue (a wavelength of about 0.4 or 0.5 micron). With these two parts of the spectrum absorbed, the reflected light appears green; hence this is the common color of plant leaves such as the eucalyptus trees shown in Figure 5.11. The light absorbed by the chlorophyll leads to the formation of oxidants and reductants. These help in the production of energy-rich adenosine triphosphate (ATP) which is important for the conversion of carbon dioxide to carbohydrates. Carotenoids, the second type of pigments, gather light energy and transfer it to chlorophyll. Phycobilins, another type of accessory pigments, are found in marine photosynthetic organisms.

The exact photosynthetic process is complex; it differs for different plants and involves a variety of enzymes and chemical steps. It is the pigments in a plant that allow it to adapt to the light it happens to receive. Below the surface of the sea, for example, the light is mostly green, and red algae thrive because they can absorb the green light, allowing photosynthesis to take place; green algae, on the other hand, which would reflect rather than absorb green light, cannot grow.

Respiration, essentially the opposite of photosynthesis, provides energy for the plant. This process of "burning" or oxidizing the carbohydrate molecule, releasing carbon dioxide, water, and energy, occurs continuously. It is basic not only for the survival and growth of the plant, but also for the nutrition of the animals that eat the plant and derive energy from the carbohydrates.

At what rate can vegetable matter be produced by photosynthesis on the surface of the earth? The answer to this question is important in assessing how many people can be fed and whether growing wood for fuel or grain for alcohol offers a significant source of energy for space heating and transportation. The

Figure 5.11 The energy on which life depends enters the biosphere in the form of light. The light energy is converted to stored chemical energy by photosynthesis. These are four-year-old eucalyptus trees on Bioenergy Development Corporation land on the island of Hawaii. (*Source*: ©Oak Ridge National Lab/Dept. of Energy/National Renewable Energy Laboratory)

amount of vegetable matter that can be grown certainly depends on the availability of sunlight to drive the conversion or CO_2 and H_2O into carbohydrates. In addition to the limitations of sunlight, there are also limits placed by the availability of appropriate land, temperature, climate, and nutrients in the soil, as well as by plant diseases and insects. Overlooking these complications for the moment, let us examine the overall efficiency with which biomass may be produced from sunlight.

In Chapter 4 the amount of solar radiation that finds its way to the surface of the earth was examined. The average solar energy per unit horizontal area and per unit time at the top of the atmosphere was found to be 0.5 cal/min · cm². This value is averaged over day and night and over all latitudes. For the purpose of calculating plant production, it is convenient to know the number of calories in a day per square centimeter. Given that 47% of the solar energy incident on the atmosphere reaches the ground, the energy available for food production averages

$$(0.5 \text{ cal/min} \cdot \text{cm}^2) \times (0.47) \times \frac{60 \text{ min}}{1 \text{ hr}} \times \frac{24 \text{ hr}}{1 \text{ day}} = 338 \frac{\text{cal}}{\text{cm}^2 \cdot \text{day}}.$$

On a typical summer day, a forest or field will have a somewhat higher value, about 500 to 700 cal/cm² · day. It's difficult to calculate from theory alone what fraction of this energy will end up as biomass. Only about 25% of sunlight incident on earth has the right wavelength to produce photosynthesis, and 60 to 70% of that will be absorbed by the plant leaves if there is dense foliage. The amount of energy stored per carbohydrate unit synthesized is about 5 eV. The photons driving the process are mostly from the red end of the visible spectrum where the energy per photon is 1.7 eV. Most measurements suggest that about eight photons of incident red light are required per carbohydrate molecule formed. On this basis, the energy input would be about 14 eV (8 × 1.7 eV) to have a stored energy of 5 eV. This corresponds to an efficiency of roughly 35% for the absorption process. Taking all these factors together (0.25 × 0.70 × 0.35 = 0.06) gives an overall efficiency of about 6% for converting solar energy to chemical energy stored in biomass.

A less theoretical, but perhaps more practical, approach to determining the efficiency of biomass production is to make direct measurements. Such measurements indicate that if there are 500 cal/cm^2 · day incident, the net potential plant production is about 71 g/m^2 · day. The gross production is 106 grams, but the respiration loss of 35 grams reduces this to a net yield of 71 grams.

Example 5.6

From the measured data given in the previous section, calculate the maximum efficiency with which solar energy is converted into biomass. Assume glucose ($C_6H_{12}O_6$) requiring 674 kcal per mole is the material produced.

Solution

First calculate the amount of energy stored per gram of glucose. Since carbon, hydrogen, and oxygen have atomic masses of 12, 1, and 16, respectively, a mole of glucose will have a mass of

$$(12 \times 6) + (1 \times 12) + (16 \times 6) = 180 \text{ grams/mole}$$

Next, find the energy stored per gram. Since glucose stores 674 kcal per mole, we have

$$(674 \text{ kcal/mole}) \times (1 \text{ mole}/180 \text{ g}) \times (10^3 \text{ cal}/1 \text{ kcal}) = 3744 \text{ cal/g}$$

Now we can find the energy stored per cm^2. With the known net production of 71 g/m^2 · day, this is equivalent to

$$(71 \text{ g/m}^2 \cdot \text{day}) \times (3744 \text{ cal/g}) \times (1 \text{ m}^2/10^4 \text{ cm}^2) = 26.6 \text{ cal/cm}^2 \cdot \text{day}$$

When this output of 26.6 cal/cm^2·day is compared to the input of 500 cal/cm^2·day, we find a photosynthetic efficiency of

$$(26.6 \text{ cal/cm}^2 \cdot \text{day})/(500 \text{ cal/cm}^2 \cdot \text{day}) = 0.053 = \mathbf{5.3\%}$$

Since 25% of sunlight has the correct wavelength and 60 to 70% of the incident light is absorbed by the plant's leaves, to have an overall efficiency of 5%

means that 29 to 33% of the light energy actually absorbed by the leaves is finally stored in biomass. This is quite consistent with our theoretical estimate of 35%. The actual net production of biomass in various locations is listed in Table 5.3.

Figure 5.1 shows the distribution of solar energy incident on the earth into various modes of absorption and scattering that take place in the atmosphere and on the surface of the earth. Of particular interest for the present purposes is the 40×10^{12} W of solar power that goes into photosynthesis. This is about 1/4000 of the total incident solar power. About half of the photosynthesis is thought to take place in the oceans, which cover 70% of the earth's surface. If we assume that the dry plant production is basically glucose with an energy content of 3744 cal/g, we can estimate the total number of grams of plant matter produced each year.

$$40 \times 10^{12} \text{ (J/sec)} \times (1 \text{ cal}/4.184 \text{ J}) \times (3.15 \times 10^7 \text{ sec/year}) \times (1 \text{ g}/3744 \text{ cal})$$
$$= 8 \times 10^{16} \text{ g/yr}$$

Another estimate of the earth's annual plant production is 320 grams per square meter per year averaged over the total earth's surface. Since the earth has a radius of 6.37×10^6 m, and the surface area of a sphere is given by $4\pi r^2$, the total production is

$$4\pi (6.37 \times 10^6)^2 \text{ m}^2 \times 320 \text{ g/m}^2 \cdot \text{yr} = 16 \times 10^{16} \text{ g/yr}.$$

The factor of two difference between these two results is characteristic of the uncertainties in information of this type.

Until about 1880, the main source of energy for heating, transportation, and industrial processes in the United States was wood (see Figure 1.1). Although the industrialized countries now rely primarily on fossil fuels for their energy, wood and agricultural and animal wastes are still important sources of energy in the developing countries. Would it now be possible for the developed countries to return to renewable biomass to meet their basic energy needs?

To meet the energy demands of a modern nation, wood and other vegetable matter can, of course, be burned directly to obtain heat energy. The vegetable

Table 5.3 Net Biomass Production for a Summer Day

Location	Plant Production (g/m² · day)	Solar Energy Conversion Efficiency (%)
Potential maximum	71	5
Polluted stream	55	4
Iowa cornfield	20	1.5
Pine forest	6	0.5
Wyoming prairie	0.3	0.02
Nevada desert	0.2	0.015

matter could be cultivated explicitly for such use as a fuel, or agricultural waste, municipal waste, sewage, algae, or seaweed from oceans or lakes could be used. Vegetable matter can be converted by well-established processes into liquid or gaseous fuels as substitutes for gasoline, oil, or natural gas. The most widely publicized example of this conversion process is *gasohol*—a mixture of 10% grain-based ethanol and 90% gasoline.

Before examining the details of the conversion processes, it is informative to obtain an overview of how much fuel energy could be provided by biomass each year in the United States and to compare this with the country's energy needs. One pound of dry plant material, which could be wood, cornstalks, or various other vegetable matter, will yield about 7500 Btu when burned directly. This is equivalent to 4300 cal/g, which is close to the amount of energy stored in glucose that we calculated earlier. The yield of any crop in g/m^2·yr is, of course, highly variable depending on such factors as soil, rainfall, type of crop, and agricultural practices. If we take as a practical average a value of 3300 g/m^2·yr (equal to 15 tons/acre·yr), which is in the middle of the range for corn in the United States (see Table 5.3), the annual yield for all of the U.S. land now harvested (350×10^6 acres) is

$$(15 \text{ tons/acre·yr}) \times (350 \times 10^6 \text{ acres}) \times (7500 \text{ Btu/lb}) \times (2000 \text{ lb/ton})$$
$$= 79 \times 10^{15} \text{ Btu}$$

As discussed in Chapter 1, the total energy used in the United States in 2003 was 98×10^{15} Btu. Although the comparison is far from exact, on the basis of the calculation shown here we could expect to furnish all our needs from cultivated crops.

In practical terms, wood is the logical biomass to consider as a fuel to generate electric power and provide direct heating. Wood comes from federally owned and state-owned timberlands and also from private land holdings and urban residues; waste wood comes from manufacturing as well as construction and demolition debris. Next to hydroelectric power, wood and other biomass material provided the largest amount of electric power from renewable energy in the United States in 2002, that is, 6650 MW$_e$ of installed generating capacity. In 2002 39×10^{12} watt-hours of electric energy was generated from 50×10^6 tons of wood. This was equivalent to about 200,000 barrels of oil per day.

Following the energy crisis of the early 1970s, the government made an effort to increase production of wood and residues as an energy feedstock. The U.S. Department of Energy formed the Short Rotation Woody Crops Program (SRWCP). This involved research into all aspects of hastening the growth of trees and in particular finding species capable of resprouting from the stump after harvesting. The program was reasonably successful, and more than 20 private tree plantations were established on marginal agricultural land just for feedstock wood. On these plantations, trees can be harvested every 3 to 10 years. In some regions, trees of eight inch trunk diameter can be grown in five to eight years. Four-year-old trees are shown in Figure 5.11. Plant growth rates of 4 to 10 dry tons/acre·yr were achieved, compared to average natural stand yields of 1 to 2 dry tons/acre·yr.

In the 1970s there was a rush back to the time-honored use of woodstoves for home heating. While these stoves were at first fueled largely by the use of firewood harvested in the traditional way, augmented by urban residues, more recently there has been a shift to the use of wood pellets. The pellets are a manufactured product made from finely ground wood pressed into cylindrical pellets about a quarter inch in diameter by three-quarter inch long. In 1999 there were over 600,000 pellet stoves in use in the United States, replacing the equivalent of over a million barrels of imported crude oil per year. The pellet stoves have the advantages of the possibility of automatic feeding of fuel in response to thermostat demand, improved efficiency, and improved emissions, features not available in traditional wood stoves. The pellets can be made of wood not normally suitable for use as firewood, but at some cost in energy for the manufacturing process. On the negative side, many pellet stoves are equipped with blowers and automatic feeders and cannot operate when the electricity is interrupted. Also, the cost of the pellets, on a Btu basis, can be higher than that of natural gas, propane, fuel oil, or traditional firewood.

Although there has been a large increase in the use of fuelwood since the early 1970s, at this point the future looks limited. Government programs to encourage increased use of biomass energy have been curtailed. However, there are now about 1000 wood-fired power plants in operation. Many of these do not provide power for a utility grid, but are owned by industries such as those in pulp and paper production to provide in-house steam, heat, and electric power. Factors now limiting growth are relatively low coal prices, the expiration of government tax credits for biomass electricity generation, and a limited supply of waste wood due to constraints on the logging industry. It appears that in the long term, however, biomass fuels will become more important because of their renewable nature and distinct environmental advantages over coal and other fossil fuels. Plans to make extensive increases in the growing of wood for fuel may bring into question the wisdom of diverting land from use for food crops. Harvesting of wood for fuel also affects the habitat of wildlife species.

5.6   Biomass: *Municipal Solid Waste*

Solid municipal waste is a growing problem in the United States and the rest of the world. Over the 41 years from 1960 to 2001 the generation of municipal solid waste (MSW) in the United States grew from 88 million tons per year to 229 million tons per year. Much of this increase was due to population growth, but the rate of solid waste generation per person also increased from 2.7 pounds per person per day in 1960 to 4.4 pounds per person per day in 2001. Landfill sites are rapidly being filled and the availability of new sites is limited, especially near cities. In addition to the problem of site availability, there are problems of groundwater contamination, odor, disease, toxic chemicals, and escaping methane. Every day, 350 thousand tons of waste are added to landfills in the United States.

The majority of this solid waste has its origin in photosynthesis: paper (36%), food wastes (11%), plant waste (12%), and wood (6%). Glass (6%), plastic (11%), and metals (8%) make up the bulk of the rest.

The pattern of MSW management has changed rather drastically since 1960 when 63% of the waste was landfilled and 31% was combusted without energy recovery. In 2001 the fraction going into landfills was slightly reduced to 56%, but 29% was recycled or composted, and 15% was incinerated with energy recovery. The average heat content of MSW is about 4300 Btu/lb (10,000 kJ/kg), about one-third that of good coal. While disposing of the waste is the primary goal, sufficient heat can be released by burning waste to make a worthwhile contribution to steam-generated electric power. The waste incinerators must be designed to dry the waste as well as to burn it. This is accomplished by injecting enough air into the firebox for complete burnout. The hot exhaust gases must be cleaned by electrostatic precipitators or bag houses to remove particulates, combined with wet or dry scrubbers to remove acid gases. The incinerators are designed differently depending on whether they are intended for mass burning of the normal mix of municipal wastes or for incinerating wastes that are separated prior to burning.

The waste-to-energy facilities are very capital-intensive, and many municipalities find it difficult to make the necessary investment. There has been a trend toward private rather than public ownership. In 2001 over 100,000 tons per day of waste-to-energy capacity was available in 97 different facilities. Most of the existing facilities generate electricity alone, and the rest generate steam or combined steam and electricity. An example of a plant which generates electricity is shown in Figure 5.12. While the energy derived from MSW is far from negligi-

Figure 5.12 The Otter Tail Power Plant in South Dakota. This plant burns refuse to generate electricity. Plants such as this can extract up to 700 kWh of electrical energy per ton of processed solid waste. (*Source*: ©Philip Shepherd/Dept. of Energy/National Renewable Energy Laboratory)

ble, the contribution of the recycling program to a reduction of the solid waste going into landfills is probably more significant.

5.7 *Biomass-Derived Liquid and Gaseous Fuels*

The famous Ford Model T automobile of 1908 was designed to run on ethanol or gasoline or any combination of the two. During the 1930s some 2000 service stations provided gasoline that contained 6 to 12% ethanol made from corn. Ethanol is one of the common forms of alcohol; it is an oxygenated hydrocarbon, C_2H_5OH. The conversion of biomass to liquid fuel has a long history in the United States. The renewable energy resources discussed so far have in a variety of ways provided electricity, heat, or steam. There is obviously also a need for a liquid fuel or combustible gas that can be easily transported to provide for space heating or transportation. In recent years this need has stimulated a renewed effort to convert biomass to a fuel useful for transportation. The effort focuses on helping to alleviate our growing dependence on imported oil and on reducing emissions from automobiles. The addition of ethanol to gasoline increases the oxygen content of the fuel, thus leading to more complete combustion and reduction of carbon monoxide emissions. In certain parts of the country prone to high carbon monoxide levels, the U.S. Environmental Protection Agency has mandated that these ethanol–gasoline blends be used during the winter months.

The particular blend of 10% ethanol to 90% gasoline is known as *gasohol*. Gasohol now receives a reduction of 5.2 cents per gallon on the current federal excise tax of 18.4 cents per gallon of gasoline. This amounts to a subsidy of more than 50 cents for each gallon of ethanol. Ethanol now has a per gallon cost about twice that of gasoline and has only about two-thirds the energy content per gallon.

The manufacture of ethanol serves as an example of converting biomass into a liquid fuel. Although other grains can be used, the principal crop utilized in the United States has been corn. After harvesting, the entire stalk and cobs are chopped up, ground, and mixed with water. The resulting material is then cooked to help convert the starches into sugars by enzymatic action. The sugars are then converted by fermentation into alcohol. Distillation removes the alcohol from the rest of the material; it then is blended, usually with unleaded gasoline, to make a product directly usable in unmodified auto engines. Figure 5.13 shows diagrammatically the process of conversion as well as the ways in which fossil fuel energy enters into the various stages.

Because there is usually a sizable surplus of corn grown in the United States, fuel tax exemptions have been provided by a number of states to encourage the use of corn to augment the domestic supply of gasoline. This is in addition to the federal subsidy mentioned earlier. It has been unclear whether gasohol would be economically viable if it enjoyed no subsidy but was left to a free market. Because of the considerable use of fossil fuels in agriculture, it is also unclear whether or not more Btu of fossil fuel are consumed in the process than are made avail-

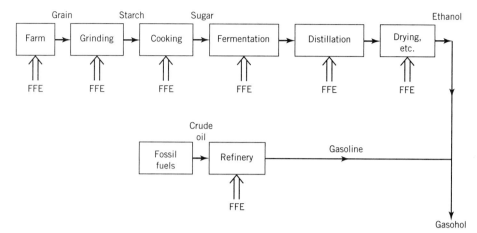

Figure 5.13 Diagram of the production of gasohol from grain such as corn. In addition to the ethanol product, the stalks and cobs as well as by-products of the fermentation and distillation have economic value. Fossil fuel energy (FFE) is required for almost every stage of the process, including steps not indicated such as transportation.

able in the ethanol. The answer to these questions is that ethanol is apparently close to the break-even point, but much depends on whether prime crops or agricultural wastes are used, and on the details of the conversion technique.

About 1.6 billion gallons of ethanol are currently produced annually for fuel in the United States. Ethanol production is the largest industrial market for corn. About 95% of ethanol is derived from corn, predominantly in the midwestern states.

Brazil, which has limited oil reserves, has made a serious effort to shift to alcohol as a transportation fuel. Sugarcane is widely grown in Brazil and can be converted to ethanol by the process shown in Figure 5.13. In the 1980s, nearly 90% of the new automobiles in Brazil were designed to operate on pure ethanol, but now, owing mainly to higher per-mile ethanol costs, no more than 20% of Brazil's cars run on straight ethanol. However, gasoline in Brazil typically has a 25% mixture of ethanol. Flex-fuel cars, which run on either gasoline or ethanol, or any combination of the two, are becoming increasingly popular in Brazil. They now represent 24% of new-car sales, and there are predictions that this number will grow to 100%.

In addition to producing a liquid fuel, biomass can be converted into a usable gaseous fuel, methane. Methane is the hydrocarbon, CH_4, that makes up typically 85% of the natural gas extracted from underground. The methane produced from biomass has about the same heating value as ordinary natural gas, which consists of methane plus several percent of other gases. Several processes are used to convert biomass to methane. The one that is most common is simply the fermentation of organic matter by the action of bacteria in the absence of oxygen (anaerobic fermentation). The organic material used can be crops, agricultural waste (either vegetable or animal), waste from lumber mills, waste from breweries, algae, sludge from sewage treatment plants, or waste from municipal disposal sites. In the presence of water and absence of oxygen such organic material will ferment naturally, and 60 to 80% of the carbon in the organic material is converted into methane and carbon dioxide. Figure 5.14 shows the process involved. After purification, the volatile gases are almost pure methane with a heating value of about 1000 Btu/ft^3. One pound of dry organic material will produce about 5 cubic feet of methane. This amounts to about 5000 Btu per pound of input material, not too much less than the earlier estimate of 7500 Btu per pound for dry biomass. We see here that the waste-to-energy conversion is quite efficient, if we can ignore the energy needed to operate the conversion facility.

The process of converting waste to methane has been in operation for some time at places such as sewage plants and breweries, and there has been good success in producing useful amounts of methane. In spite of the high conversion efficiency, in the range of 50 to 70%, the overall costs of operating the process suggest that it will be economical only if the organic material is waste that otherwise would be disposed of. It does not appear that growing crops for methane production will be economically competitive with natural gas as long as the domestic supply of natural gas holds up.

There is an excellent example of waste-to-energy conversion now operating at the municipal sewage plant in Boulder, Colorado. This system manufactures methane from sewage, then uses the methane to power an engine

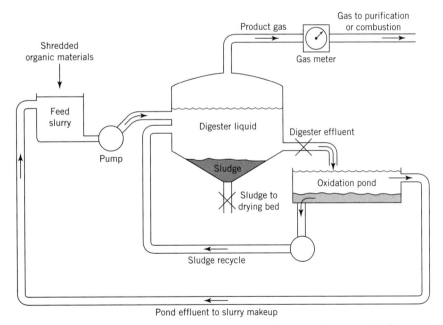

Figure 5.14 A unit for the continuous conversion of biomass by anaerobic fermentation into methane and other combustible gases. (*Source*: Solar Energy as a National Resource, NSF/NASA Solar Energy Panel, December 1972.)

which drives a generator. The generator produces 700 kW of electrical power, which is more than sufficient to run the sewer plant. The surplus electricity is sold to the local electric utility and provides an annual revenue of $93,000. The system was put into use in 1987 at a construction cost of $2.1 million. Considering both the avoided cost for purchase of electricity and revenue from the sale of electricity, the payback time for the installation is estimated at ten years.

Biodiesel fuel is produced from vegetable oils, recycled cooking greases or oils, and animal fats. It is a substitute for regular diesel fuel obtained from petroleum. All of the ingredients of biodiesel have their origin in plants, as even the animal fats are from animal life sustained by vegetation. Thus biodiesel is a renewable energy resource. Since vegetation growth removes carbon dioxide from the atmosphere in about the same amount as is returned when biodiesel is combusted, there is little effect on the global atmospheric carbon dioxide burden. Other tailpipe emissions, particularly particulates, are reduced by using biodiesel, but nitrogen oxide emissions are increased. The use of biodiesel potentially can reduce our dependence on fossil fuels.

The most commonly used biodiesel fuel is a blend of 20% biodiesel and 80% regular diesel, designated B20. The use of B100 requires special handling and engine modifications.

The various oils and fats in biodiesel cannot be used directly as a fuel. They have to be converted into long chain monoalkyl esters. These chemicals are also known as fatty acid methyl esters or FAME. The conversion process involves reactions with methanol in the presence of a catalyst. Glycerine is a by-product.

The main use of biodiesel at the present is as B20 in trucks and school buses. Biodiesel costs more per gallon than regular diesel, partly because the ingredients such as soybean oil and yellow grease have a higher cost, by about a factor of two or three, than petroleum. Government estimates predict that biodiesel use will increase to comply with the Energy Policy Act of 1992, reaching at least 6.5 million gallons in 2010.

In addition to its use in vehicles, biodiesel can be used in heating furnaces as a way of reducing emissions and saving on fossil fuels. The heating value in Btu per gallon of biodiesel is about 7% less than that of regular diesel.

The growing dependence of modern agriculture on fossil fuel energy is often overlooked. In many ways a modern farm uses land to convert oil, natural gas, and coal into food, and the energy input from the sun is just part of the picture. In the United States in 1900, one farmer was needed to feed five people. By 1974, 50 people were fed by one farmer, and the trend continues. The 10-fold increase has come about through increased use of machinery, electricity, irrigation, pesticides, herbicides, fertilizers, and improved seeds. All of these ingredients use energy, mostly from fossil fuels. The production and delivery of these ingredients also employ workers who could properly be counted within the agricultural enterprise along with the farmers.

One prime example of the use of fossil fuels (natural gas) in farming is that of the fertilizer anhydrous ammonia, a compound of nitrogen and hydrogen, which has become an essential ingredient in modern high yield crop production, and hence in the amount of biomass that can be grown for food and as an energy resource. Anhydrous ammonia has increased crop yield by a factor of two or three and it has become a major factor in our ability to help feed the world's population.

Anhydrous ammonia is manufactured by a steam re-forming process using hydrogen from natural gas and nitrogen from air at high temperature and pressure. The process was developed in 1909 in Germany by Fritz Haber (1868–1934). It is estimated that about 2% of North American natural gas production is used to produce ammonia fertilizer. Our ability to continue high-yield farming could be seriously reduced by depletion of the natural gas resource.

5.8 Geothermal Energy

There are only two nonsolar sources of renewable energy that have sufficient potential to warrant consideration at this time. These are geothermal energy, based on the naturally occurring heat from the interior of the earth; and tidal energy, based on the potential and kinetic energies of the earth–moon–sun system that is bound together by gravitational forces.

There is a continuous flow of heat energy from the hot interior of the earth to the cooler surface. As shown in Figure 5.1, the total heat flow is 32×10^{12} W, which is a vast amount of power compared to our needs. The density of this power over the earth's surface, however, is quite small, averaging only about 1/16 W/m^2. This amounts to only about 6 watts coming up through your 100 m^2 basement floor. There are, however, many deviations from the average where the heat flow is far greater, and volcanoes, geysers, fumaroles, and hot springs can occur there. The internal heat of the earth is mainly from the decay of long-lived radioactive nuclei that were incorporated into the earth when it was formed. This means that geothermal energy, like solar energy, is fundamentally nuclear energy. There is also some residual geothermal heat from impact and gravitational compression that occurred during the formation of the earth and from accretion of incoming meteorites. The temperature of the inner core at a depth of 6400 km is estimated to be 7200°F (4000°C). Closer to the earth's surface, at a depth of 80 to 100 km, there is molten rock with temperatures between 1200 and 2200°F (650 to 1200°C). It is this molten rock (magma) that penetrates upward into the earth's crust in certain areas such as Yellowstone National Park and produces a variety of remarkable thermal features.

For thousands of years people have used geothermal energy for bathing and space heating. It has also been used widely for aquaculture, greenhousing, and industrial process heat. The first use for electricity generation was in Italy in 1904. While the total world's geothermal energy reserve is vast, 10^8 QBtu, a million times more than we use in the United States each year, there are only a limited number of sites where practical, economical electricity generation can take place. The Geysers, a geothermal electric power plant, shown in Figure 5.15, is the notable example in the United States. Located north of San Francisco, California, on a 70 km^2 site, it had an installed generating capacity of 1967 MW$_e$ in 1989. The capac-

158 Chapter 5 Renewable Energy Sources II: Alternatives

Figure 5.15 The Geysers, a geothermal steam field in northern California, showing venting geothermal wells and gathering pipes. In this area, which covers 30 square miles, over 200 wells have been drilled; the deepest of these is more than 10,000 feet deep. (*Source*: Courtesy Pacific Gas & Electric Company)

ity is now less than that. Even though geothermal is often classified as renewable energy, experience at the Geysers illustrates the fact that at any particular site, the geothermal resource is finite, and the use of geothermal energy can be equivalent to mining a limited resource. It was reported in 1991 that "By 1988 generating capacity (at The Geysers) had more than doubled from its 1981 level, to 2043 megawatts, and the number of players had grown. . . . Steam delivery is dropping by 11% per year. . . . By 1987, power production peaked, then began nosing down. . . . It's probably too late to stop, much less reverse, the decline of The Geysers." The thermal energy at The Geysers is being withdrawn from the ground 80 times faster than it is being replenished by natural processes.

In 2005 the reported total geothermal capacity in the United States was 2800 MW_e for electric utilities and 5400 MW_e for nonutility energy producers.

The total geothermal resource is diverse in nature and adaptable to various applications, although it appears that in the coming decades electricity generation will predominate. In one study, the resource was divided into six categories:

Hot Water Reservoirs These are geothermally heated reservoirs of underground water, very large in magnitude in the United States, but not generally appropriate for electricity generation. Hot water has been used to produce electricity in some locations outside of the United States. The hot water may also be appropriate for space heating.

Natural Steam Reservoirs This is the type of geothermal resource found at The Geysers, where the naturally occurring steam is used to drive turbines. This is a highly desirable type of resource, but because it is very rare, it is unlikely that it will contribute significantly to our national energy budget.

Geopressured Reservoirs This is brine completely saturated with natural gas under considerable pressure because of the weight of the overlying land. It may be an important energy resource because of both heat energy and production of natural gas.

Normal Geothermal Gradient Even in dry rock the normal geothermal gradient (30°C/km) produces useful temperatures any place on the globe. Drill holes 20,000 feet deep are achievable, corresponding to a temperature of about 190°C above the surface temperature. This temperature is adequate for electricity generation. Although the normal geothermal gradient resource is enormous in magnitude, no technology has yet been developed to extract useful and economic energy from this resource. Attempts have been made near Los Alamos, New Mexico, and elsewhere to circulate a working fluid through dry rock, but no commercially promising results have yet been reported.

Hot Dry Rock This is technically the same as the normal geothermal gradient resource, but it occurs at more favorable locations where the geothermal gradient in the earth's crust is greater than 40°C/km. It is estimated that such conditions are found under 5% of the U.S. land area. As in the case of the previous category, no commercial successes have been reported in circulating a working fluid through a system of channels, natural or man-made, in hot dry rock.

Molten Magma No technology yet exists to exploit the high temperatures in molten magma found at volcanic locations. The high temperatures could be attractive for electricity generation. Possible sources in the United States are in Hawaii and Alaska.

Example 5.7

The Geysers geothermal site near San Francisco, California, covers an area of 70 km², and the thickness of the subsurface zone from which heat is recoverable is 2.0 km. In this zone the temperature is 240°C and the volumetric specific heat is 2.5 joules/cm³·°C.

(a) Calculate the heat energy content in joules (at temperature above the mean annual surface temperature of 15°C).
(b) For how many years can this site provide power for a 2000 MW$_e$ plant if 1.9% of the thermal energy can be converted to electricity? Do not consider any possible replenishment of the heat energy; the rate at which this happens is very slow compared to the rate at which the heat energy is withdrawn. These numbers are representative of The Geysers site.

Solution

(a) The heat energy content is given by the volume times the specific heat times the temperature difference.

The volume is given by the surface area times the thickness.

$$V = 70 \text{ km}^2 \times 2.0 \text{ km} = 140 \text{ km}^3$$

This volume in cubic kilometers then is converted to cubic centimeters to match the units for the specific heat. 1 km × 10^3 m/km × 100 cm/m = 10^5 cm. Therefore 1 km³ = $(10^5 \text{ cm})^3 = 10^{15}$ cm³.

$$V = 140 \text{ km}^3 \times 10^{15} \text{ cm}^3/\text{km}^3 = 1.4 \times 10^{17} \text{ cm}^3$$

The heat content, Q, is given by the volume times the volumetric specific heat times the temperature change:

$$Q = V \times (\text{v.s.h.}) \times \Delta T$$

where

$$\Delta T = 240°C - 15°C = 225°C$$

Then,

$$Q = (1.4 \times 10^{17} \text{ cm}^3) \times (2.5 \text{ J/cm}^3 \cdot °C) \times 225°C = \mathbf{7.9 \times 10^{19}} \text{ J}$$

(b) For each year of operation, the electrical energy produced will be

$$E = P \times t = 2000 \text{ MW}_e \times 1 \text{ yr} = 2000 \text{ MW}_e \cdot \text{yr}$$

To produce this much electrical energy at an overall efficiency of 1.9% requires that heat energy be extracted each year in the amount of

$$\frac{(2000 \text{ MW}_e \cdot \text{yr})}{0.019} = 105{,}000 \text{ MW}_e \cdot \text{yr} = 1.05 \times 10^{11} \text{ W} \cdot \text{yr}$$

In units of joules, this is

$$(1.05 \times 10^{11} \text{ W} \cdot \text{yr}) \times (3.15 \times 10^7 \text{ sec/yr}) \times (1 \text{ J/W} \cdot \text{sec})$$
$$= 3.31 \times 10^{18} \text{ J/yr}$$

To obtain the number of years for which a 2000 MW$_e$ plant can be operated before the heat reservoir is exhausted, divide the available energy by the amount withdrawn each year:

$$\frac{\text{available energy [from (a)]}}{\text{amount consumed per year}} = \frac{7.9 \times 10^{19} \text{ J}}{3.31 \times 10^{18} \text{ J/yr}} = \mathbf{23.9 \text{ yr}}$$

Of the geothermal resources listed, only hot water and natural steam reservoirs have been put to use. For these two categories the environmental effects are, therefore, fairly well known; for the other four categories, the probable effects are highly uncertain. Many of the hot water reservoirs, particularly those that are at higher temperatures and more saline, pose the potential for soil salination if the extracted water is not reinjected into the ground after its heat is extracted. There is also a risk of land subsidence and aquifer disruption when large amounts of water are extracted from the ground. The gaseous air pollutant hydrogen sulfide is liberated into the atmosphere by some hot water reservoirs as well as by natural steam reservoirs. Other possible environmental effects include induced seismicity if water is injected into dry rock formations or if explosive (possibly nuclear) fracturing techniques are used in normally impermeable rock formations.

Some idea of the size of the resource can be obtained from Table 5.4. The first two categories, hot water and natural steam, are characterized by proven technology and are, thus, fairly accurately represented in this table; the other four are speculative and much more uncertain, especially with regard to the numbers in the second column. If these numbers are compared to the 2003 national energy budget of 98 QBtu, it can be seen that although natural steam and molten magma may have local importance, they do not appreciably affect the nation's energy future. The other four resources represent extremely large amounts of thermal energy.

It has been demonstrated at The Geysers and elsewhere that geothermally produced electricity can be cost-competitive with electricity produced by other means. In 2005, the National Renewable Energy Laboratory (NREL) reported that the cost of geothermally generated electricity in the United States was approximately 4 to 6 cents per kWh, well within the range of other technologies.

In making estimates of the future of geothermal energy, it is apparent that the limiting factor is not the size of the resource, because it is so enormous, but rather other factors, such as the economics relative to other energy sources, the unknown technology of the future, and even population distribution. The largest

Table 5.4 Estimated U.S. Geothermal Resources[a]

Reservoir Type	Total Resource (QBtu)	Total Potentially Producible (QBtu)
Hot water	12,000	6,000
Natural steam	180	45
Geopressured	73,000	2,400
Normal gradient	1,250,000	12,500
Hot dry rock	160,000	1,600
Molten magma	3,500	35
Total	1,500,000	22,600

[a]To 6,000 meter depth, $T \geq 80°C$, national parks excluded.
Source: Adapted from CONAES, 1980.

resource, the normal geothermal gradient heat stored in the earth's crust, has never been successfully tapped for energy, except for small demonstration projects. It is a diffuse distribution of thermal energy, and the various attempts so far to circulate water or steam through drillholes into deep beds of rock have not been encouraging. The more concentrated geothermal resources often lie far from population centers; in the United States the attractive sites are almost exclusively in the western states, including Hawaii and Alaska.

When considered in its entirety, it is apparent that although of local importance, the geothermal option is not likely to rank among the most important contributors to the national energy budget in the near future. The long-term prospects may be more encouraging, depending on the success of technologies that are now only speculative. Outside the United States, the prospects for geothermal energy are not much different. In 2005 the worldwide installed geothermal generating capacity was estimated at 8000 MW_e. If all the geothermal power plants now operating or planned would produce at capacity, they would provide less than 1% of the worldwide electricity consumption. The country of Iceland is a notable exception; geothermal energy there is about one-third of the total energy consumed.

5.9 Tidal Energy

Tidal energy differs from all other energy sources in that the energy is fundamentally gravitational; it is extracted from the potential and kinetic energies of the earth–moon–sun system. The well-known ocean tides result from this interaction, producing variations in ocean water levels along the shores of all continents. On the U.S. coasts, the vertical tidal range varies from about 2 ft in Florida to 18 ft and more in Maine. As the water level fluctuates twice daily through this range, it alternately fills and empties natural basins along the shoreline, suggesting that the currents flowing in and out of these basins could be used to drive water turbines connected to generators. The technology employed is very similar to that of low-head hydropower.

Tidal power is complicated by the variability of the tides both in time and in range. A complete tidal cycle takes about 12.5 hours, so there are usually two cycles a day, and the times during the day of high and low tides are continuously changing. About once a month there are unusually high tides (spring tides), and about once a month there are unusually low tides (neap tides). Irregularities are also brought about by storms and other meteorological conditions.

To enhance the efficiency of capturing energy from natural tidal currents, damlike structures can be built across the mouths of natural tidal basins, with gates or channels to direct the flow through the turbine locations. The gates can be opened or closed in sequence with the tides, permitting water to flow only when there is sufficient head to power the turbines. The turbines are designed to work in either direction, so water flow both in and out of the basin produces electricity. Because of the intermittent nature of this flow, the duty factor of such

Tidal Energy   **163**

an installation is appreciably less than 100%. A tidal power station produces only about one-third as much electric energy as would a hydroelectric power plant of the same peak capacity operating continuously.

The three large tidal power stations now operating are outside of the United States. A 1 MW plant, on the White Sea in Russia, was completed in 1969, and a 240 MW plant was completed on the estuary of the Rance River near St. Malo, France, in 1967 (see Figure 5.16). The Rance plant has turbines that can also serve as pumps; thus, the installation can function as a pumped hydro storage facility to even out loads on a large electricity generating and distribution system. In this way water pumped up into the basin during times of low power demand increases the head on the turbines at other times. The average tidal range at Rance is 28 feet, with a maximum of 44 feet. The dam extends 2500 feet across the estuary, enclosing a basin of about 8.5 square miles in area. The project generates about 540 million kWh of electricity in one year. This is about 18% of the tidal energy available in the basin.

The third operating tidal power station is on the Annapolis River in Nova Scotia, Canada, about 100 miles west of Halifax. It began providing electricity in 1984 with a peak capacity of 18 MW_e and since then has been producing 30

Figure 5.16 The Rance River tidal power installation on the northern coast of Brittany, France. This 750 meter long dike impounds tides that reach a height of more than 13 meters. (*Source*: ©Michel Brigard/Phototheque EDF)

to 35 million kWh of electricity per year. This site is one of about 20 in the Bay of Fundy suitable for tidal power.

For the United States, the most attractive sites are in Cook Inlet and Bristol Bay in Alaska and in the Bay of Fundy region in the northeastern United States and southeastern Canada. If the latter site, probably the most favorable in the world, were fully developed, it would provide about 15,000 MW_e to the United States and a similar amount to Canada. The U.S. share would certainly be important in the New England states, but on a national scale it would provide at most a few percent of our electrical needs. Further development of the Bay of Fundy sites has been under study by the Americans and Canadians for many years, but no definite plans have been funded. The most favorable tidal-power sites represent only about 63,000 MW_e worldwide or about 50 times less than the world's potential hydroelectric capacity.

It appears that although tidal power offers the great attraction of having no fuel costs, the necessary large capital investments have hindered development of the few favorable sites. Extensive tidal power development also must face the problem of interference with other uses of the tidal basins, such as shipping and fishing. Marine life in the basins is also disturbed. From an engineering point of view, the lack of continuous electric output reduces the value of tidal power to the utilities.

Efforts are underway to exploit tidal energy in new ways and new places. One such program involves a plan to place up to 300 underwater turbines in the East River in New York City where tidal currents run in the range of 4 knots. This turbine farm would produce 10 MW of electricity at times of peak tidal flow. A prototype turbine was tested at the site in 2003, and plans are now pending for larger scale deployment.

An even more innovative program is being proposed by a company called Tidal Electric. Their initial proposal is to build a lagoon, or an impoundment, with walls several meters high on the shallow seabed 2 km off the coast of Wales, where the tidal range is large. As the lagoon fills and empties through apertures in the walls, turbines at these locations would generate 30 MW of electricity. Depending on the success of this first installation, larger such tidal power stations along the coast of the United Kingdom would follow, each generating 400 MW or more. In this way, many of the objections to disturbing natural waterways will be avoided, and the potential contribution of tidal energy to the world energy budget could be very much enlarged beyond estimates based on natural sites. Sites at other locations around the world are also being studied.

5.10 Wave Energy

In many coastal regions around the world, ocean waves represent an attractive source of energy to drive electric generators. The waves get their energy from wind, which in turn results from solar energy. The wave energy resource is considerable. The U.S. Department of Energy estimates a total power of 2 to 3 million MW in the waves breaking on the world's coastlines, and an energy density that can average 40 MW per kilometer of coastline in favorable locations. One source reports a wave power resource around the United Kingdom of 120,000 MW, or more than double the total electricity demand. Another source estimates the economically recoverable fraction of the U.K. wave energy resource to be about one-quarter of the U.K. electricity demand.

While the earliest attempts to harness wave power go back thousands of years, it is only in recent decades, following the energy scares of the 1970s, that truly large-scale efforts have been underway. There are many ingenious proposals now at the planning and prototype stage of development. None have yet been fully commercialized.

One example of a current project is at Port Kembla in Australia, scheduled for completion in 2005. In this shore-mounted installation, waves are focused by a parabolic wall onto a vertically oscillating water column. The oscillating water column in turn forces a high-speed airflow back and forth through a turbine connected to a generator. The plant will produce in excess of 500 MWh of electric energy per year. A notable feature of this plant is that it will also produce desalinated water, using a fraction of the power from the generator and ocean water as inputs. This application may become important at many places in the world.

Another interesting installation is the Pelamis (named for *Pelamis platurus*, the yellow-bellied sea snake) project in Scotland, now in the prototype stage, rated at 750 kW. This device is of the sea snake configuration, a large flexible linear assembly, 120 meters long and 3.5 meters in diameter, anchored offshore. The wave-induced flexing motion drives hydraulic pumps, delivering a flow of high pressure hydraulic fluid to hydraulic motors connected to generators.

It is too soon to say what the future will hold for wave energy, but as fossil fuels become more expensive, economic factors will certainly open new possibilities for this emerging energy technology.

5.11 Summary

By looking back to Table 4.1, and considering the contributions that the renewable energy technologies covered in this chapter are making to the national energy budget, it is apparent that we are a very long way from significantly relieving our dependence on fossil fuels. The largest contributors of the renewables, biomass and hydroelectric, each produce only a few percent of the total national energy consumption. All other renewables are substantially less than that.

The massive scale of our present system of electric power plants, and the extent to which this system is dependent on fossil fuels, are such that no alternative energy technology, no matter how attractive, is going to bring about major changes in the near future. It will take several years, at a minimum, for electric power generation systems using new technology to be designed, constructed, and put on-line. In the longer run, on the scale of decades, it seems possible that our dependence on fossil fuels can be markedly reduced.

Key Terms

Hydroelectricity
Hydrologic cycle
Pumped storage
Wind turbine
Coriolis effect
OTEC
Solar spectrum
Photosynthesis
Photon

Carbohydrates
Chlorophyll
Glucose
Gasohol
Sucrose
Ethanol
Methanol
Normal geothermal gradient
Neap and spring tides

Suggested Reading and References

1. *Home Wind Power*. U.S. Department of Energy, 1981. Charlotte, VT: Garden Way Publishing.
2. *Assessment of Solar Energy Technologies.* American Solar Energy Society, 1989.
3. Cohen, R. "Energy from the Ocean." *Philos. Trans. R. Soc. London,* **A307** (1982), pp. 405–437.
4. Goldin, A. *Oceans of Energy—Reservoir of Power for the Future.* New York: Harcourt, Brace, Jovanovich. 1980.
5. Golob, R., and Brus, E., *The Almanac of Renewable Energy.* New York: Henry Holt and Company, 1993.
6. Ross, D. *Energy from the Waves.* Oxford: Pergamon Press Ltd., 1979.
7. Blackburn, J. O. *The Renewable Energy Alternative.* Durham, NC: Duke University Press, 1987.
8. Kerr, R. A. "Geothermal Tragedy of the Commons," *Science,* **253** (1991), pp. 134–135.
9. *Proceedings, National Regulatory Conference on Renewable Energy.* Washington, D.C.: National Association of Regulatory Utility Commissioners, October 1993.
10. *Renewable Energy Annual—1995.* Energy Information Administration. Washington D.C.: U.S. Department of Energy.
11. Hayden, H. C. "Rosetta Stones for Energy Problems." *The Physics Teacher,* **19,** No. 6 (September 1981), pp. 374–383.
12. "Assessment of Geothermal Resources of the United States—1975." D. E. White and D. L. Williams, Eds. *Geological Survey Circular* 726.
13. Moretti, P. M., and Divone, L. V. "Modern Windmills." *Scientific American,* **254,** No. 6 (June 1986), pp. 110–118.
14. Smith, D. R. "The Wind Farms of the Altamont Pass Area." *Annual Review of Energy,* **12** (1987), pp. 145–183.

15. Britton, P. "How Canada Is Tapping the Tides of Power." *Popular Science*, **226,** No. 1 (January 1985), pp. 56–58.
16. Greenberg, D. A. "Modeling Tidal Power." *Scientific American*, **257,** No. 5 (November 1987), pp. 128–131.
17. Solar Energy Research Institute. "The Potential of Renewable Energy." March 1990. SERI/TP-260-3674.
18. Wasserman, H. "Inherit the Wind—Will Clean Energy Pass Us By?" *The Nation*, July 16, 1997.
19. Schwartz, R. D. "Population, Fossil Fuel, and Food." Forum on Physics and Society, American Physical Society, January 2002.
20. Pasqualetti, M. J. "Wind Power." *Environment* (September 2004).
21. Parfit, Michael; photographs by Leen, Sarah; After Oil, Powering the Future; National Geographic 208 2 (August 2005), pp. 2–31.

Questions and Problems

1. Calculate the pressure in pounds per square inch at the bottom of a static column of water 300 meters high.
2. (a) How much electric *energy* can be generated by the water in a lake 2000 meters wide by 8000 meters long by 100 meters deep if all the water falls through a vertical distance of 500 meters? Assume that the generator is 90% efficient. Express your answer in joules.
   (b) What would the electric *power* output be if the lake were drained over a period of one year? (1 year = 3.15×10^7 seconds.) Express your answer in megawatts.
   (c) How large a community would this serve at the typical rate of 1 MW_e per 1000 people?
   (d) At $0.05/kWh, what is the value in dollars of this electrical energy?
3. (a) Calculate the energy in joules made available when 1 kg of water falls 30 meters if 90% of the energy can be converted to a useful form.
   (b) Calculate the energy in joules made available when 1 kg of water is cooled by 2°C if 3% of this energy can be converted to a useful form.
   (c) Compare these two numbers. Does this say something about the relative amounts of water that must pass through a hydroelectric plant and an OTEC plant?
4. If a windmill produces 23 kW of electric power at a wind velocity of 10 miles per hour, how much power will it produce at a wind velocity of 15 miles per hour?
5. A windmill has a diameter of 2 meters. It converts wind energy to electrical energy at an efficiency of 60% *of the theoretical maximum* when connected to an electrical generator.
   (a) What is the electric power output at a wind velocity of (1) 10 mph? (2) 20 mph? (3) 30 mph?
   (b) How many 60 watt lightbulbs can be supplied with electricity under conditions of (1), (2), and (3)?
6. How much thermal energy in joules is made available by cooling 1 cubic meter of rock from 240°C to 100°C? The specific heat is 2.4 $J/cm^3 \cdot °C$.

168 Chapter 5 **Renewable Energy Sources II: Alternatives**

7. A geothermal-powered steam turbine operates between a steam temperature of 210°C and an environmental temperature of 25°C. What is its maximum (ideal) efficiency? What percentage of the total steam energy must be discharged as a waste heat?

8. Calculate the overall efficiency of an OTEC plant that operates with the ideal Carnot efficiency between the temperatures of 20°C and 5°C, but which uses two-thirds of the energy extracted to run pumps and make up other losses.

9. (a) Starting from the results of Examples 5.4 and 5.5, estimate the number of cubic meters of water that would flow each second through an OTEC plant large enough (1000 MW$_e$) to provide electricity for Miami. Use the approximation of an ideal heat engine as in the examples.
(b) If this water flows at a velocity of 4 m/s, what would be the necessary diameter of the pipes?

10. A tidal basin with an area of 14 square kilometers and a depth of 12 meters empties in 6 hours with the water passing through turbines.
(a) How many cubic meters per second must flow on average during this six-hour period?
(b) How many square meters of cross-sectional area must the turbine pipes have if the flow velocity is 7 m/s?

11. Estimate the number of Btu that would be generated annually in the United States if all the municipal waste were incinerated. Assume 1000 pounds per year per person of burnable waste at 4300 Btu/lb. How does this compare with the total energy consumption of the United States?

12. Estimate the forest area needed to supply fuel continuously for a 1000 MW$_e$ power plant. Note that this requires about three times as much thermal power as electric power.

Multiple Choice Questions

1. Fifteen kilograms of water 90 meters above a generator represents ___ joules of potential energy.
   a. 2102
   b. 60
   c. 980
   d. 9.8
   e. 29,160
   f. 1350
   g. 1380
   h. 13,230

2. If a 90-meter high waterfall has a flow rate of 15 kilograms per 0.1 second, what is the power in watts in the stream of water as it hits the bottom?
   a. 210.2
   b. 600
   c. 98
   d. 9.8
   e. 291,600
   f. 135
   g. 13,800
   h. 132,300

3. Hydroelectricity accounts for approximately ___ percent of the electrical energy consumption in the United States.
   a. 1
   b. 7
   c. 40
   d. 70

4. How much electric power could be obtained from a small hydroelectric station if the elevation change was 150 meters and if 10,000 kg of water passed through the turbines every second, with the overall efficiency being 85%?
   a. 1.3 MW$_e$
   b. 25 MW$_e$
   c. 12.5 MW$_e$
   d. 12.5 kW$_e$

5. A windmill system which produces 5 kW of electric power when the wind is blowing at 3 m/s, will produce ___ kilowatts when the wind is at 9 m/s.
   a. 1.67
   b. 15
   c. 45
   d. 135

6. A wind of 30 m/s produces ___ times as much power per square meter as does a wind of 10 m/s.
   a. 27
   b. 9
   c. 3
   d. 36

7. If the wind blows for one year at a steady 10 m/s, the energy we could get in this year per square meter of cross section through the use of a windmill would be about ___ kWh.
   a. 2200
   b. 600
   c. 200
   d. 60

8. A modern windmill can extract about ___% of the power in the wind.
   a. 1
   b. 10
   c. 20
   d. 40
   e. 60
   f. 65
   g. 70
   h. 90

9. Of the total kinetic energy content of a one square meter cross section of wind, one might expect to convert ___ to electricity with a modern windmill.
   a. 1%
   b. 40%
   c. 10%
   d. 99%

10. In the OTEC plants envisioned, the colder water is at a temperature of ___, and is drawn in from a depth of ___.
    a. 5°C, 1000 ft
    b. 5°C, 1000 m
    c. 5 K, 1000 ft
    d. 5 K, 1000 m
    e. 25°C, 1000 ft
    f. 25°F, 1000 m
    g. 20°C, 500 m
    h. 20 K, 5000 m

11. Ocean thermal gradient heat engines operate between temperatures of ──────
    a. 75°C and 45°C
    b. 15°C and −10°C
    c. 25°C and 5°C
    d. 50°C and 20°C

12. The following is true of geothermal energy:
    a. It will easily supply all the energy needs of the United States.
    b. Since it basically comes from the sun, it is renewable.
    c. It has its origins in the earth's radioactivity, and it can be exhausted locally.
    d. The Geysers in California has a capacity of 512 GW$_e$.

13. A geothermal-powered steam turbine operates between a steam temperature of 210°C and an environmental temperature of 25°C. What is its maximum (ideal) efficiency?
    a. 13%
    b. 18%
    c. 23%
    d. 28%
    e. 33%
    f. 38%
    g. 43%
    h. 48%

14. The normal geothermal gradient is ―――
    a. 3°C/km
    b. 30°C/km
    c. 3°C/m
    d. 30°C/m
    e. 3°C/1000 ft
    f. 1/16 W/cm^2
    g. 16 W/m
    h. 3 K/m

15. Tidal energy has its origin in ―――
    a. the phase of the moon.
    b. just the oceans sloshing back and forth.
    c. the decay of primordial radioactive nuclei within the earth.
    d. the kinetic and gravitational potential energy of the earth–moon–sun system.
    e. trade winds acting on the earth's surface.
    f. the hydrologic cycle.
    g. the Carnot dynamics of ideal heat engines.

CHAPTER 6

The Promise and Problems of Nuclear Energy

(*Source*: Courtesy Pacific Gas & Electric Company)

6.1 Introduction

The story of nuclear energy is complex; it involves science, engineering, economics, health and safety, psychology, and politics. It is not easy to set aside the strongly negative image formed by the relationship of nuclear energy to nuclear weapons and the serious accidents that have occurred with radioactive materials and nuclear facilities. The mere terms *nuclear energy*, *radioactivity*, *radiation*, *criticality*, and *meltdown* often have a frightening connotation, and they are not well understood by a large part of the public. These words sometimes find their way into the headlines, even though they often don't belong there.

Certainly there are problems with the nuclear power industry that do not stem only from public apprehension. No new orders have been placed for reactors by the public utilities in the United States since 1978, and there have been many cancellations of orders placed before that time. This is largely because of the unfavorable economics experienced with some power reactors. Legal liability, increasingly stringent regulatory procedures, radioactive waste storage, and siting problems have added to the negative side of the ledger, and therefore no new orders are presently being realistically discussed by the public utilities. This is not true in some other countries; both Japan and France, for example, are moving ahead with nuclear electric power programs.

There are certain positive features associated with nuclear power reactors that should not be overlooked in considering the possible future of nuclear power. Reactors are not affected by the looming shortage of fossil fuels, and they emit no CO_2, SO_2, CO, or particulates into the atmosphere. Uranium reactor fuel is now reasonably abundant and inexpensive, and a breeder reactor technology could extend the resource to many thousands of years. There now are 104 power reactors operating in the United States with a total capacity of 97,000 MW_e. They produce about 20% of the nation's electricity. In France, 78% of the electricity comes from nuclear reactors. The reactor safety record is as good or better than that of any other electricity-producing technology. The Three Mile Island incident was costly, but endangered few people. The reactor industry is now studying ways to produce more reliable, less expensive, and safer reactors, making full use of the experience gained during the first 25-year period of development from 1953 to 1978.

A second path to utilizing nuclear energy, by way of nuclear fusion reactions, has problems and benefits quite distinct from those of nuclear fission reactors. Nuclear fusion, the joining together of two light nuclei to form a heavier nucleus, releases energy, but it is yet to be shown that nuclear fusion can be a practical source of electrical energy. Intense efforts to develop fusion reactors have been underway for more than 40 years, and progress is being reported, but it is not seen as likely that commercial electricity generation from fusion reactors will be a reality within the next few decades. The technical challenges are daunting, and it is possible that fusion power will never make its way to the power grid. By contrast, it may be noted that the time from the first laboratory demonstration of nuclear fission in 1938 to the first demonstration of a simple reactor in 1942 was only four years, and fifteen years later, in 1957, a power reactor at Shippingport, Pennsylvania, was producing commercial electricity.

6.2 *A Short History of Nuclear Energy*

In the early 1930s, laboratory experiments established that the nucleus of the atom was made up of neutrons and protons. The neutron and the proton have very nearly the same mass (1.67×10^{-27} kg), about 1840 times heavier than the electron. The neutron has no electric charge, and the proton is positively charged. In the shorthand of designating nuclear properties, the atomic number, Z, is the number of protons in the nucleus, and the atomic mass number, A, is the sum of Z and N, the number of neutrons. In addition to merely investigating nuclear masses and atomic numbers, scientists in Europe, the United States, and a few other countries conducted intensive research to determine more detailed properties of atomic nuclei. The scientists Hahn, Strassman, Meitner, and Frisch discovered in 1938 and 1939 that the nuclei of uranium atoms did a remarkable thing. When bombarded with neutrons, these heavy nuclei *fissioned*, that is, they often split into two fragments, at the same time emitting more neutrons. The two fragments are lighter nuclei, each having a mass roughly half that of the uranium nucleus. These fragments are called *fission products*, and they carry off about 160 MeV of kinetic energy, divided between the two fragments, from the fission reaction.

The energy release in fission can be explained in terms of nuclear mass energy. If one adds together the masses of the two fission fragments and the emitted neutrons, the total will be appreciably less than the mass of the initial uranium nucleus, ^{235}U, along with that of the neutron which initiated the fission.

The difference in mass, Δm, is related to the energy release, ΔE, by the Einstein equation, $\Delta E = \Delta m c^2$. The magnitude of the energy that can be obtained from direct conversion of nuclear mass to energy is impressively large compared to the energy released from burning a fuel such as coal. Chemical reactions involve a few eV of energy per atom; nuclear fission reactions involve hundreds of MeV per atom, a factor of about 10^8 times larger. Pushed apart by Coulomb repulsion, the positively charged energetic fission fragments move away from the site of the fission reaction and come to stop in the surrounding material. In this way their kinetic energy very quickly becomes heat energy in the material. The fission process is shown in Figure 6.1.

After the first reports of the laboratory findings concerning fission, it didn't take long for physicists to realize that all the ingredients of an exponentially increasing chain reaction were at hand. Because each fission releases more than one neutron, the neutrons from one fission reaction could induce further fission

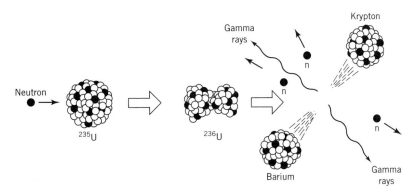

Figure 6.1 Three steps in the neutron-induced fission of ^{235}U. The combination of a neutron and ^{235}U forms ^{236}U in a highly excited state that promptly fissions into two lighter nuclei, emitting neutrons and gamma rays in the process.

reactions in more than one neighboring uranium nucleus. One becomes two, two becomes four, four becomes eight, and so forth, in a simple model. The process would then repeat itself at a growing rate and release a large amount of energy in a short time. The potential for enormous explosive power was obviously there. It was also apparent that if the rate of the chain reaction could be controlled by some means so that it was steady, rather than increasing, then one would have a source of heat energy appropriate to drive an electric generator. Because of the onset of World War II with the Nazi military occupation of much of Europe in the years immediately following the discovery of fission, the initial efforts to exploit this new phenomenon centered on producing a nuclear weapon rather than a nuclear power reactor. The first nuclear reactor was assembled at the University of Chicago in 1942 by Enrico Fermi and co-workers, and they successfully demonstrated a chain reaction, even though it was at a very low power level, about 200 watts.

The fear that Nazi Germany would develop a nuclear bomb spurred on intense efforts in the United States to get there first. The American project, given the code name the Manhattan Project, was centered at large laboratories in Los Alamos, New Mexico; Oak Ridge, Tennessee; and elsewhere. The project faced the difficult task of accumulating enough of the uranium isotope, ^{235}U, to form the critical mass[1] needed for a workable bomb. Only two isotopes of uranium are found in nature. They are $^{235}_{92}$U$_{143}$ and $^{238}_{92}$U$_{146}$. The left superscript gives the atomic mass, either 235 or 238, the left subscript gives the atomic number, 92, and the right subscript gives the number of neutrons, either 143 or 146, in the nucleus. The difficulty comes about because ^{235}U is the only isotope found in nature that is able to sustain a chain of fission reactions, and ^{235}U is only 0.7% abundant. That is, only 0.7% of uranium as it is mined from the earth is ^{235}U; the remaining 99.3% is ^{238}U, which is not useful for a nuclear fission weapon. Enormous engineering efforts were undertaken to process natural uranium so that it was enriched to the 90% ^{235}U necessary for a nuclear explosive. After some years of effort, the Manhattan Project did produce three nuclear bombs. Two of these were fueled with ^{239}Pu,[2] the other with ^{235}U. The first weapon was exploded as a test in July 1945 in New Mexico, and the other two were dropped on Hiroshima and Nagasaki in August of 1945 at the end of the war with Japan. It turned out that neither Germany nor Japan were close to having nuclear weapons during World War II, although their scientists were certainly aware of the fundamental principles of these weapons.

Following the war, efforts were made to utilize nuclear reactors for the generation of commercial electricity and also for the propulsion of submarines and other vessels for the U.S. Navy. The first production of electricity for the civilian market by a reactor was at Shippingport, Pennsylvania in 1957. That was a small power reactor largely financed by the federal government. Following this

---

[1] A critical mass is the minimum amount of a fissionable isotope needed to sustain a nuclear chain reaction. It is often reported as about 15 kilograms for ^{235}U, and about 5 kilograms for ^{239}Pu.
[2] The origin and uses of ^{239}Pu are discussed in Section 6.4 of this chapter.

demonstration, several much larger reactors having capacities of more than 100 MW$_e$³ were ordered by various public utilities. The Dresden I reactor station in Illinois was the first of these to go on-line in 1962. From 1953 to 1978, 253 nuclear power reactors were ordered in the United States, but 118 of those orders were canceled. The generating capacity of the 135 that were not canceled was 114 GW$_e$. In the United States, there have been no new orders placed for power reactors since 1978.

Up to about 1978 there was great optimism that an inexpensive, abundant, environmentally acceptable source of energy had at last been found. In order to understand some of the reasons why this optimism has not been fulfilled, a few more details on nuclear physics and engineering are needed.

6.3 Radioactivity

The word *radioactivity* has become apart of everyday conversation, but it has different meanings for different people, and understanding is often vague and limited. Radioactivity is an important part of any serious discussion of nuclear power because almost all of the fission products are radioactive, as are the nuclear fuels uranium and plutonium. It is also true, to some degree, that our food, the air, our natural surroundings, and even our bodies are radioactive.

For our purpose, radioactive refers to an atomic nucleus that is unstable. It can spontaneously decay, most commonly becoming a nucleus of another element, and emit an energetic electron or alpha particle in the process. Some types of radioactive nuclei decay very soon after they are formed; others have a low probability for decay and are likely to survive for billions of years after their formation.

An example of the decay of a radioactive nucleus is the beta decay of ^{137}Cs, a common fission product of exceptional concern because of its thirty-year half-life and the likelihood of its being taken up by living things. It decays as follows:

$$^{137}_{55}Cs_{82} \rightarrow \,^{137}_{56}Ba_{81} + \beta^- + \bar{\nu} \qquad T_{1/2} = 30 \text{ yr.}$$

Here we see that one of the neutrons in $^{137}_{55}Cs_{82}$ has become a proton in $^{137}_{56}Ba_{81}$, emitting an electron (also known as a beta particle) and an antineutrino in the process. These light particles carry off the majority of the decay energy. The energetic electron is responsible for much of the damage to living things that can be done by this radioactivity. Some beta-decaying radioactive nuclei emit positive rather than negative beta particles.

Another type of radioactive decay, important to nuclear reactors and their fuel, is alpha particle emission. The alpha particle is the nucleus of the ordinary helium atom, and it is stable against radioactive decay. It is common for the heav-

---

³This means 100 megawatts of electrical output. The thermal power of the reactor is about three times larger.

iest radioactive nuclei to decay by emitting alpha particles. The decay of plutonium-239 is an example of alpha decay:

$$^{239}_{94}Pu_{145} \rightarrow \,^{235}_{92}U_{143} + \,^{4}_{2}He_2 \qquad T_{1/2} = 24{,}000 \text{ yr.}$$

In this process, two of the protons and two of the neutrons originally in the plutonium combine to form the alpha particle. When heavy nuclei undergo alpha decay, most of the decay energy appears as kinetic energy of the alpha particle. In this typical example the alpha particle kinetic energy is about 5.2 MeV.

The half-life for radioactive decay is that characteristic time during which half of the nuclei in any given sample will undergo decay. This time is different from one nuclear species to another. For example, it is 1.3 billion years for potassium-40, 24,000 years for plutonium-239, 30 years for cesium-137, 12 seconds for oxygen-15, and 5730 years for carbon-14. The half-life is also, correspondingly, that time during which the rate of emission of radiation from a radioactive sample will decline by a factor of two. The process is continuous; the radiation or number of radioactive nuclei in a sample will decline by a factor of two in one half-life, a factor of four in two half-lives, a factor of eight in three half-lives, and so on.

There is a large variety of radioactive nuclei present in a collection of fission products. Their half-lives range from very short, a fraction of a second, to very long, in excess of a million years. Some of the fission product nuclei are not radioactive. Once formed, they remain as they are. All radioactive nuclei found naturally on earth must have half-lives at least comparable to the age of the earth or be formed by the decay of these long-lived radioactivities, except for a very few such as ^{14}C, with a half-life of 5730 years, that is formed continuously in our atmosphere by cosmic rays arriving from outer space.

Example 6.1

Estimate the number of years needed for the following radioisotopes to decay to one-thousandth of their original activity. [Note that $1/1000 \approx (1/2)^{10} = 1/1024$.]

$$^{239}Pu \; (T_{1/2} = 24{,}000 \text{ yr})$$

$$^{137}Cs \; (T_{1/2} = 30 \text{ yr})$$

$$^{3}H \; (T_{1/2} = 12.4 \text{ yr})$$

$$^{89}Sr \; (T_{1/2} = 50.5 \text{ days})$$

Solution

Each radioisotope will decay to one-thousandth of its initial activity in about 10 half-lives. For the listed radioisotopes this will require:

$$^{239}Pu: \textbf{240{,}000 yr}$$

$$^{137}Cs: \textbf{300 yr}$$

$$^{3}H: \textbf{124 yr}$$

$$^{89}Sr: \textbf{1.38 yr}$$

As a general rule the emissions from radioactive nuclei are classified as ionizing radiation, that is, they are capable of removing a bound electron from an atom or molecule by impact. If the atom or molecule happens to be a part of a biological system, the ionizing event may lead to the breaking of a molecular bond, with severe consequences, such as the disordering of genetic information or the changing of normal cells into cells that eventually become cancerous.

The human race has evolved in the presence of ionizing radiation that comes continuously from cosmic rays and from the natural radioactivity in the earth. Although such radiation certainly has the potential to be harmful, the effects have not been of sufficient magnitude to thwart the development of the human race or that of innumerable other animal and plant species. In fact, radiation certainly has caused some of the mutations needed for the evolutionary process to proceed as it has.

In the past 80 years, however, a number of different sources of ionizing radiation have been introduced by man. We are now at the point where the average exposure of people in the United States to man-made radiation is at roughly the same level as the exposure to natural sources. The nuclear power reactor industry could potentially add to our exposure to ionizing radiation, and an important part of the design and operation of a power reactor is devoted to the containment of this radiation.

6.4 Nuclear Reactors

In a nuclear power plant, the reactor is the source of heat energy which boils water to make the steam used to drive a turbine connected to an electric generator. Rather than burning coal, the fissioning of uranium is just another way of producing heat energy. The turbine and generator are essentially the same as those in a coal-burning power plant. Both types of power plants are subject to the Carnot efficiencies discussed in Chapter 3.

The heat from nuclear fission is generated in the reactor core where the ^{235}U fissions occur. The uranium fuel in the core, when freshly loaded, is about 3% ^{235}U and 97% ^{238}U, but the chain reaction takes place only in the ^{235}U. The reason for this is given by Figure 6.2, which shows the probability of fission taking place with neutrons bombarding ^{235}U and ^{238}U. It is clear that neutrons of very low kinetic energy (10^{-2} eV) have about a thousand times more probability of fissioning ^{235}U than do more energetic neutrons (10^6 eV) on either ^{235}U or ^{238}U. At first glance, this is a problem because the neutrons coming directly from the fission reaction have an average kinetic energy of about 2 MeV, corresponding to a very low fission probability, far too low to sustain a chain reaction in such a dilute assembly of ^{235}U.

Thus the first step in designing a reactor is to find a way to slow down the fission neutrons, decreasing their kinetic energy to the very low value where fission probability is large. This is done by having the fission neutrons bounce around in some material (the moderator) such as water, or graphite, until they lose most of their kinetic energy by many successive elastic scatterings off the

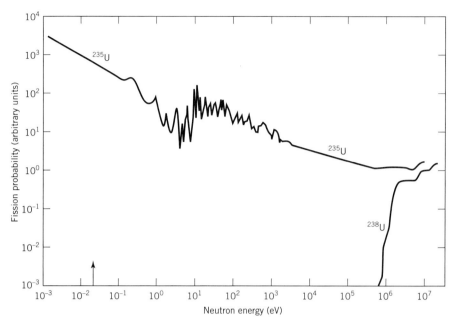

Figure 6.2 The fission probability for ^{235}U and ^{238}U as a function of neutron kinetic energy. The arrow at 0.025 eV indicates the kinetic energy of thermal neutrons. For ^{238}U the fission probability becomes appreciable only above about 1 MeV neutron kinetic energy.

moderator. Eventually the neutrons will have their kinetic energy reduced to where it is the same as the thermal energy of the moderator. This occurs at a kinetic energy of 0.025 eV, low enough for a high ^{235}U fission probability. Neutrons of this low energy are known as *thermal neutrons*.

Of the 2.5 neutrons that on average come from the fissioning of ^{235}U, not all will survive to cause other fission events. Some of the fission neutrons may be captured by ^{235}U or ^{238}U with the emission of gamma rays rather than resulting in another fission. This type of neutron capture can also take place in other materials in the reactor, such as the fuel cladding or structural parts. Neutrons can also escape from the reactor core into the surrounding shielding or containment vessel.

It is important to note that the capture of a neutron by ^{238}U results in the formation of ^{239}Pu, another important fissionable nuclear fuel. This isotope is not found in nature; it must be manufactured in a reactor. The sequence starts with a neutron being captured by ^{238}U to become ^{239}U, which then undergoes beta decay ($T_{1/2}$ = 24 min) to ^{239}Np, which then beta decays ($T_{1/2}$ = 2.3 days) to ^{239}Pu. The end result is ^{239}Pu that lives long enough ($T_{1/2}$ = 24,000 yr) to play an important role in reactors and in nuclear weapons. The probability of fission with thermal neutrons is even greater for ^{239}Pu than for ^{235}U. Thus, as ^{239}Pu builds up in the reactor core, it can fission and contribute to the power of the reactor. About one-third of the energy of the entire reactor fuel cycle comes from ^{239}Pu, and by the end of the fuel cycle 60% of the fissionings are due to ^{239}Pu.

A reactor must be carefully designed so that, given the losses of neutrons noted in the previous paragraphs, exactly one neutron, on average, will be available from each fission reaction to be moderated and finally find its way to induce a second fission. This is necessary for operation at a constant power level. The geometrical arrangement of the reactor is clearly important. In addition, there is a certain critical size below which the reaction will not be self-sustaining because the neutron losses through the surface of the reactor are too great.

To control the power level of a reactor, a set of control rods made of a material, often a boron compound, that readily absorbs neutrons is inserted into the core of the reactor. If the control rods are fully inserted, the reactor will be shut down. If they are fully extracted, the power level will rise, perhaps to a dangerous level. By carefully regulating the positions of the control rods, the reactor operators can make the reactor operate at a steady power level.

In addition to the prompt neutrons that instantaneously come from a fission event, there are also delayed neutrons that have their origin in the decay of some of the fission products. A few of the fission products decay by emitting a neutron rather than the more common radioactive emission of an electron. These neutron-emitting nuclei are formed by beta decay from fission product nuclei with half-lives that range from a fraction of a second to many seconds. Of all the neutrons produced in a reactor, only about 0.5% of them are delayed. The presence of these delayed neutrons is an essential point for reactor operation because they do not permit the neutron population to be changed instantaneously. Were it not for this tiny fraction of delayed neutrons, the control rods could not be moved in and out with sufficient speed to control the reactor power level.

When a reactor is to be shut down, the control rods are inserted and the neutrons are sufficiently absorbed so that the reactor goes subcritical (will not sustain a chain reaction) and power is no longer generated by fission. Because of the radioactive fission products built up in the fuel rods, however, an appreciable amount of heat will continue to be generated. This residual heat production is an important concern for reactor safety; it is discussed in Section 6.8 of this chapter.

6.5 *The Boiling Water Reactor*

The boiling water reactor (BWR) is used here to illustrate how a power reactor is constructed and operated. The other main type, the pressurized water reactor (PWR), also plays a major role in power plants, and it is used in naval propulsion systems. In the PWR, the water in the pressure vessel surrounding the reactor core is maintained at such high pressure that it remains a liquid. The hot pressurized water then is pumped through a heat exchanger where it boils water to steam to power a turbine and produce electricity. It is the PWR design that has been used so successfully in submarines. In this application, the PWR can provide power continuously for more than 15 years without refueling because the initial fuel loading is enriched far beyond the 3% common in commercial power reactors.

Table 6.1 lists many of the specifications for a 1220-MW$_e$ BWR. A diagram of the power plant is shown in Figure 6.3. The reactor heats water as it flows through passages in the core, causing the water to boil, under 71 atmospheres of pressure, as it reaches the top of the core. The steam is then piped to the turbines for the generation of electrical power. Figure 6.4 shows the pressure vessel and core of the reactor in more detail. The 46,000 fuel rods are about 12 feet long and only a half inch in diameter. The fuel rods are made of zirconium alloy tubes filled with pellets of uranium oxide. The uranium, when first put into the reactor, is enriched to 2.8% ^{235}U; it is removed when the enrichment has

Table 6.1 Characteristics of a Boiling Water Reactor

Plant electrical output	1220 MW$_e$
Plant efficiency	34%
Core diameter	193 in.
Core (or fuel rod) active length	150 in.
Core power density	54 kW per liter
Cladding material for fuel rods	Zircaloy-2
Fuel material	UO$_2$
Fuel pellet size	0.4 in. dia. $\times$ 0.4 in. long
Number of fuel rods	46,376
Control rod type	"Cruciform" control rods inserted from bottom
Number of control rods	177
Amount of fuel (UO$_2$)	342,000 lb
Coolant material	Water
Coolant pressure	1040 lb/in^2
Coolant temperature	551°F (288°C)
Fresh fuel enrichment	2.8% ^{235}U
Spent fuel assay	0.8% ^{235}U, 0.6% ^{239,241}Pu
Refueling sequence	About 1/4 of fuel per year
Vessel wall thickness	6 in.
Vessel wall material	Manganese–molybdenum–nickel steel
Vessel diameter	20 ft
Vessel height	70 ft

Source: Adapted from General Electric specifications.

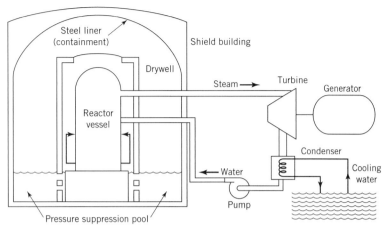

Figure 6.3 A diagram of a boiling water reactor power plant. Steam is produced in the reactor vessel and flows at high pressure to the turbine. After the steam is condensed to water at the low-pressure side of the turbine, the water is recirculated through a pump to the reactor core. The components of the reactor containment building are discussed in the text. (*Source*: A. V. Nero, Jr., *A Guidebook to Nuclear Reactors*, Berkeley: University of California Press, 1979.)

been reduced to 0.8%, which takes a year or more of steady operation. There are about 170 tons of uranium oxide in the reactor. The control rods are inserted into the core from the bottom. The water circulates between the fuel rods, serving both as the neutron moderator and as the coolant. Figure 6.5 shows fuel being loaded into a reactor core.

The reactor design includes a number of safety features. The core is surrounded by a pressure vessel, steel liner, and containment building. There is a system to add boron to the coolant water if the control rods fail to function, thus increasing the neutron absorption and shutting down the reactor promptly. In the event that feedwater fails to enter the reactor vessel, auxiliary systems are provided to maintain a normal water level in the reactor.

6.6 *Fuel Cycle*

The uranium fuel in either a BWR or PWR is in the chemical form of uranium oxide. It is a fine powder which is compacted and then sintered into pellets that are placed into zircalloy tubes (the fuel rods), which are then loaded into the reactor core. The fuel is typically left in the reactor core for three years, with one-third of the fuel rods being taken out and replaced every year. Initially most of the energy comes from fissioning ^{235}U, but as the reactor continues in operation, ^{239}Pu is formed following neutron capture on ^{238}U, as explained earlier. The ^{239}Pu fissions with thermal neutrons just as ^{235}U does, contributing a sizable

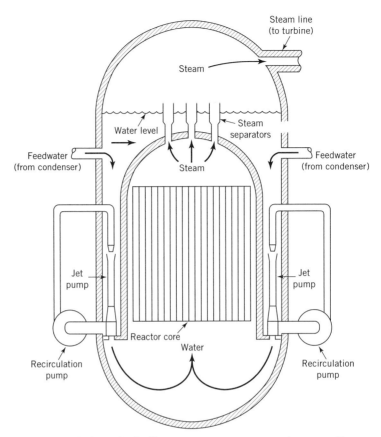

Figure 6.4 A detailed view of a boiling-water reactor core and surrounding components. (Based on WASH-1250.)

amount to the total power of the reactor toward the end of the fuel cycle. If fuel having an initial enrichment of 3.2% $^{235}UO_2$, 96.8% $^{238}UO_2$ is removed after three years of operation, it will have by weight 0.85% ^{235}U, 0.55% ^{239}Pu, 0.15% ^{241}Pu, and 3.8% fission products, along with some ^{236}U, ^{240}Pu, and ^{242}Pu. Of course, the bulk of the fuel rod content will still be $^{238}UO_2$.

The fuel rods removed from the reactor at the end of a fuel cycle are hot, both thermally and radioactively. The rods are immediately immersed in a water-filled cooling pool near the reactor. It was originally thought that after 150 days or so, the rods would be transferred to another site where the uranium and plutonium would be extracted for reuse. For various reasons, this has not generally happened in the United States, and the spent fuel rods have stayed in the cooling pools for 20 years or more. In the 1970s some efforts were made to reprocess the spent fuel from commercial power plants. However, mainly because of technical difficulties, the efforts were not successful. In 1977 the Carter administration decided to defer indefinitely the reprocessing of spent fuel from

Figure 6.5 Loading the first fuel bundles into the core of the nuclear power reactor at the Duane Arnold Energy Center near Palo, Iowa. The 550 MW$_e$ reactor uses about 19 tons of nuclear fuel per year. (*Source*: Courtesy U.S. Department of Energy)

power reactors. This decision was, to a large extent, based on fear of proliferation of nuclear weapons, due to the large amount of ^{239}Pu becoming available and possibly being diverted for nuclear weapons manufacture. The plans now are to dispose of the fuel rod contents, without reprocessing, in deep geologic repositories. Yucca Mountain in Nevada was selected as the site for this in 1987, but as of 2005 that facility is not yet prepared to receive spent fuel rods. The nuclear waste problem will be addressed in more detail later in this chapter.

6.7 *Uranium Resources*

Uranium ore found in the earth is mostly in the form of an oxide, U_3O_8, known as *yellowcake*. It varies widely in concentration from one ore deposit to another, leading to a considerable variation in the recovery costs of useful uranium. The more one is willing to pay, the greater the size of the resource. For this reason, uranium resources are classified according to price range. In the least expensive classification, up to $130 per kilogram of U, there is estimated to be 1.7×10^6 tonnes[4] of uranium in the United States and another 5.4×10^6 tonnes in the other major uranium-producing countries. In the second category of $130/kg to $260/kg there is thought to be an additional 1.3×10^6 tonnes in the United States and 12.2×10^6 tonnes in the other countries.

[4]The tonne is the metric ton, equal to 1000 kilograms. It is also equal to 1.102 ton, or 2205 pounds. One ton is the same as 2000 pounds.

In 2003 the domestic price of uranium was $28 per kilogram ($11 per pound of U_3O_8). At this price, the uranium accounts for only about 1% of the cost of electricity from a nuclear power plant. It thus appears that a uranium cost of up to $260/kg could be borne without seriously affecting the cost of electricity. At this price, the domestic resource of uranium is the sum of the two numbers given above, or 3.0×10^6 tonnes.

The uranium burnup in either a BWR or PWR is about 1.1 tonne of ^{235}U per $GW_e \cdot yr$, corresponding to roughly 200 tonnes of mined uranium per $GW_e \cdot yr$. The capacity of the 104 operating reactors in the United States is 97 GW_e. From this we estimate a resource lifetime of

$$\frac{(3.0 \times 10^6 \text{ tonnes U})}{(200 \text{ tonnes U/GW}_e \cdot \text{yr} \times 97 \text{ GW}_e)} = 155 \text{ years}.$$

This is for continuous operation of the 104 reactors. If all the electricity in the United States were to be provided by nuclear reactors, at today's rate of electricity consumption, then the lifetime of the economically recoverable resource would be about five times smaller, or only about 30 years.

Although it is clear that there is a significant domestic uranium resource for nuclear power, the available energy is far less than what is available from coal, and ^{235}U-fueled reactors offer no long-term solution.

Another type of reactor, the breeder, offers enormous potential for extending the lifetime of the uranium resource. The breeder can be said to produce more fuel than it consumes. This can happen when some of the neutrons emitted by fissioning ^{235}U or ^{239}Pu convert ^{238}U, which will not sustain a chain reaction, to ^{239}Pu, which does sustain a chain reaction and is a good reactor fuel. Existing ^{235}U would be used as the primary reactor fuel in the early stages of a breeder program. Once sufficient ^{239}Pu has been produced, it alone will be the reactor fuel which sustains and expands the program to as many reactors as are desired. Since natural uranium has about 140 times as much ^{238}U as ^{235}U, such a process has the obvious potential to extend the lifetime of the uranium resource from 30 years (at today's rate of electricity consumption) to 4200 years. This is a very large increase in the resource lifetime, and it is probably even greater than that. Because of the very large amount of energy derivable from a kilogram of uranium ore when the ^{238}U is utilized in a breeder reactor program, much lower grades of ore become economically attractive. It has been estimated that it would take as long as 50,000 years (at today's rate of electricity consumption) to deplete the uranium resources in the United States with a full-scale breeder reactor program.

The operation of a breeder reactor technology depends strongly on the number of neutrons emitted by each fission. The breeder technology requires that, on average, one neutron from each fission proceeds to induce another fission, thus maintaining the power level, and that at least one more emitted neutron beyond this be available for capture by ^{238}U to form ^{239}Pu. It turns out that this condition is met with ^{239}Pu fuel for fast neutrons (having kinetic energy in the keV to MeV range) but not for the thermal neutrons which sustain the chain reaction in ^{235}U. For this reason, a plutonium-fueled breeder reactor must have much less neutron moderation than a normal thermal reactor. The power densities in a breeder reactor are very high and a coolant with minimal neutron moderation and superior heat-transfer properties is needed. Liquid metals such as molten sodium have been used for this purpose, and these reactors are called liquid metal fast breeder reactors (LMFBR).

The United States has a fast breeder project at Clinch River, Tennessee, but that program and the related integral fast breeder program at Argonne National Laboratory in Illinois have been terminated. There are now only four operable liquid metal fast breeder reactors worldwide. France, Japan, Kazakhstan, and Russia each have one. In addition, three are planned in Russia, one is in suspended status in France, and one is permanently shut down in Germany. Many of these reactors have had a troubled operating experience, mainly because of sodium leaks.

There are several reasons behind the failure of the United States to pursue liquid metal fast breeder reactors more vigorously. Because uranium is now relatively abundant, and its price so low, there is little incentive for obtaining ^{239}Pu via the breeding process. Inherent in a breeder program is the reprocessing of fuel rods, with its many problems, and also the matter of security of the abundant ^{239}Pu. The threat of nuclear weapons proliferation and terrorist diversion of plutonium is very real. In addition, there are safety concerns associated with the high power density in the breeder reactor core and the dangers of liquid sodium leaks.

6.8 Environmental and Safety Aspects of Nuclear Energy

As we have seen, nuclear energy is the only fundamentally new source of energy in more than 100 years, and yet it has not been universally adopted by public utilities in the United States or in many other countries. There are now no new orders for nuclear power reactors in the United States, and some operating power reactors are being permanently shut down before their scheduled retirement time. It is important to sort out the reasons for the collapse of what once was a vigorous nuclear power reactor construction industry, and to make a judgment of the role that nuclear power should play in the United States.

In its normal operation, a nuclear reactor produces some emissions into the environment, but they are not of the kind produced by coal-burning plants. Although great care is taken to confine the radioactive fission products within the fuel rods, there can be some escape of the radioactive gaseous fission products krypton and xenon by diffusion and possibly through microscopic cracks in the fuel cladding. Radioactive tritium, ^{3}H, is also released in some amount in the discharged water. Various measures such as filters and holdup systems have been incorporated into reactor designs to ensure that the radioactivities released do not exceed the stringent limits designated by the Nuclear Regulatory Commission (NRC).

A nuclear power reactor cannot explode in the way that a nuclear weapon does. The distribution and enrichment of the fuel make that impossible. However, serious accidents can occur with nuclear reactors. It is possible for a chain reaction to get out of control to the extent that the core overheats to the point of melting and destruction. This type of occurrence is a criticality accident, such as was experienced at Chernobyl. Such an event is highly unlikely with the BWR or PWR types of water-moderated reactors widely used in the United States because of built-in negative feedback and shutdown systems.

A second type of accident can occur even after reactor shutdown. Because of the large buildup of radioactive fission products during an extended period of operation, considerable heat energy is generated in the fuel rods by radioactive decay. Immediately after reactor shutdown, after the fission reactions have ceased, about 7% of the normal thermal power generation still remains in the fuel rods due to this radioactive decay. At one hour after shutdown, this residual power level will be down to 1% of the normal operating reactor power level. This level of power generation is an important safety issue; it may be sufficient to melt the core and destroy the reactor if adequate cooling water is not provided to carry away the heat. If the cooling system fails, there can be a loss-of-cooling accident (LOCA). This possibility is of great concern to reactor designers, and elaborate safety systems are included to prevent such an occurrence. One of the important safety features of a power reactor is some secondary means of carrying away the decay heat after reactor shutdown in the event that there is a loss of the primary coolant system. This system is the emergency core cooling system (ECCS).

The accident in 1979 at Three Mile Island in Pennsylvania was a loss-of-cooling accident (in a reactor which first went into operation in 1978), and very severe damage was suffered by the core; however, there was no major release of radioactive material to the environment. At about the time of the Three Mile Island accident, a popular film, *The China Syndrome*, was released. The film fired the public imagination. It dealt with a fictional event in which there was imminent danger that a reactor core would melt down and continue to burn its way through the bottom of the container vessel into the earth down to the water level, at which point there would be a steam explosion and massive release of radioactivity. This movie influenced much of the public attitude toward nuclear power.

A power reactor is a highly complex array of pumps, valves, interlocks, and controls. A basic safety question asks, what is the probability of some of these devices malfunctioning so that the emergency core cooling system does not protect the core from overheating? It is also true that the people at the controls can make the wrong decisions at the time of a crisis.

To assess the probability of a loss-of-coolant accident (LOCA), a diagram called an *event tree* is constructed. The event tree allows one to follow the probable consequences of some initiating event. Probabilities are assigned to each component of the tree, such as a pump or valve, either to function or not function, and in this way various possible paths to a LOCA are evaluated. With this systematic method, there is no need to try to estimate in a single step the reliability of a complete complex system. The probabilities of failure of each component can be reasonably estimated on the basis of experience with individual pumps, valves, and so forth. It is not a matter of guessing, but of looking back at the record of how such components have performed in the past, in many applications.

Several studies of this type have been carried out under the auspices of the former Atomic Energy Commission (AEC) or the Nuclear Regulatory Commission (NRC). The results of the studies are often cast in terms of probability. The Rasmussen Report (designated WASH-1400) completed in 1975 was the most comprehensive of these studies and included the effects of natural radioactivity on people and the dangers of earthquakes as standards against which to compare the threat of reactor accidents. The report estimated that a LOCA has a probability of about 1 in 2000 per reactor year. The probability of significant release of radioactivity was stated to be 1 in 2 million per reactor year. With 100 operating reactors, the chance of an accident killing 100 people was said to be about the same as the probability of 100 people being killed by the impact of a meteor.

While these reported risks were comforting to the reactor community, the report did receive some criticism, and after the Three Mile Island accident, many questioned the findings of the report as well as the general safety of reactors. The NRC did require some changes in both procedures and equipment for operating reactors and those under construction.

Example 6.2

Calculate the combined hypothetical probability that if a plane crashes, it will be on a day that is rainy ($P_1 = 0.12$), that it will be a Thursday ($P_2 = 1/7$), and that it will be in October ($P_3 = 1/12$).

Solution

For compounding individual, causally unrelated probabilities, it is correct to compute the product of the individual probabilities. Thus, the probability of a given plane crash occurring on a rainy Thursday in October would be:

$$P_T = P_1 \times P_2 \times P_3$$
$$= 0.12 \times 1/7 \times 1/12$$
$$= \mathbf{1.4 \times 10^{-3}\text{, or roughly 1 in 700}}$$

A more recent (1990) NRC study entitled "Severe Accident Risks: An Assessment for Five U.S. Nuclear Power Plants" (known as NUREG-1150) is more relevant than earlier reports to safety issues for existing PWR and BWR in their present state. The study was carried out for five specific reactors representative of those now operating. Advanced analysis techniques were used, and the newer safety improvements required by the NRC were, of course, included. Table 6.2

summarizes some of the results of the study. The table indicates the probabilities per reactor year for core damage and large releases of radioactivity for internal events (failures initiated inside the reactor) and external events (seismic, fires, etc.) for the five reactors. Example 6.3 shows that for all the 104 reactors now operating in the United States, over a 30 year lifetime, there is about a 1% chance of a large release due to internal events.

Example 6.3

If the average probability of a large release of radioactivity for the five reactors evaluated in NUREG-1150 is 4×10^{-6}/RY, what is the probability that there will be a release during the 30-year operating lifetime of the 104 power reactors now operating in the United States?

Solution

$$(4 \times 10^{-6}/\text{RY}) \times (104 \text{ reactors}) \times (30 \text{ years}) = \mathbf{0.013 = 1.3\%}$$

If all of the present level of electric power in the United States was provided by nuclear reactors similar to those in the NUREG study, there would be a high probability of a large release about every 400 years. The NUREG study

Table 6.2 Estimated Probabilities of Reactor Accidents for Reactors Studied in NUREG-1150[a]

	Reactor Studied				
	Surry 1	Zion 1	Sequoyah 1	Peach Bottom 2	Grand Gulf 1
Type	PWR	PWR	PWR	BWR	BWR
State	VA	IL	TN	PA	MI
MW$_e$	781	1040	1148	1100	1142
Operational	1972	1973	1981	1974	1985
Core damage probability (10^{-6}/RY):					
Internal events	40	60	57	4.5	4
Seismic events	54			15	
Fires	11		20		
Large release probability (10^{-6}/RY):					
Internal events	4	6	7	2	1

[a]Probability is in units of 10^{-6} per reactor-year of operation.
Source: Adapted from D. Bodansky, *Nuclear Energy—Principles, Practices, and Prospects*. Woodbury, New York: American Institute of Physics, 1996.

reports for two reactors the probabilities for core damage due to external events, namely, earthquakes and fires (see Table 6.2).

The study calculated the health effects of reactor accidents both for early fatalities and for long-term latent effects, such as cancer. They found that for each of the reactors studied, there is less than one chance in a million per reactor year for a single early fatality. The risk to the populations surrounding the reactors due to latent fatal cancers ranged from 10^{-3}/RY to 2×10^{-2}/RY. The risk of such a fatal latent cancer to any single individual living near a reactor is only 1×10^{-8}/RY.

The overall risks are small by any standard. They are particularly small in comparison to alternative ways of furnishing electric power, such as coal, or oil, or even solar energy.

There has been one major release of radioactivity from an operating power reactor, which occurred during the accident in 1986 at the Chernobyl reactor in the Ukraine. Although that reactor is far different from the power reactors in the United States, it can be used as an example of what can go wrong.

6.9 The Chernobyl Disaster

On April 26, 1986, an explosion and fire destroyed a reactor of the Chernobyl power plant in the Ukraine. This was clearly the most disastrous accident in the history of nuclear reactors, and it is important to understand its causes and consequences for nuclear development in the United States and elsewhere. The 1000 MW$_e$ reactor was a boiling-water graphite-moderated reactor, unlike any power reactor in the United States. The RMBK reactor, as reactors of that type are called, is designed to produce ^{239}Pu for nuclear weapons, as well as electric power. In order to produce ^{239}Pu with a minimum content of ^{240}Pu, which is unwanted for nuclear weapons, it is important to have a graphite-moderated reactor where the fuel rods can be taken out after only a few weeks of operation. This is done without shutting down the reactor. Removing the fuel rods during operation requires a large open space above the reactor, which means not having a steel or reinforced concrete containment vessel of the type common to power reactors in the United States. The RMBK reactor is also particularly unstable and hazardous in that as the water in the reactor is boiled and turned to steam, there is reduced neutron absorption. This causes the power level to increase and induce more water to boil. This positive feedback condition is compensated at high power levels by a negative temperature coefficient, but at low power levels it is a cause of instability and concern. In principle, the positive feedback can be controlled by insertion of control rods, but this takes time. As discussed earlier, the delayed neutrons normally provide the necessary safety factor of time for control rod insertion. The design of the RMBK reactor, however, permits the reactivity (basically the number of neutrons) to rise to the point where the reactor is prompt critical. This means that the reactor is critical without the contribution of the delayed neutrons; hence the safety factor of time no longer exists. Once a reactor goes into a condition of prompt criticality, the power level can very rapidly rise to the point of meltdown and destruction.

On the night of April 25, 1986, the operators of Unit 4 had planned to conduct an experiment at low power levels near 200 MW before a shutdown for routine maintenance. In order to perform the measurements, certain safety systems, such as the emergency cooling system, were shut off, in violation of safety rules. To control the reactor at this low power level, the operators had to withdraw the control rods and turn off the generator that powered the cooling water pumps. At this point the reactor went out of control; the thermal power surged, the fuel started to disintegrate, the cooling channels ruptured, and an explosion ripped open the reactor, exposing the core and starting many fires. The graphite moderator began to burn.

Some 50 megacuries[5] of radioactivity, including a large fraction of the cesium and iodine in the fuel rods, were released to the environment. Some of it rose high into the atmosphere and was carried by winds to great distances. The radioactivity was deposited most heavily within a radius of 30 kilometers, and 135,000 people in that area were evacuated. Significant levels of radioactive fallout were experienced by much of Europe. It has been estimated that about 47,000 eventual excess cancer deaths in Europe and Asia may occur in the next 50 years due to the radioactivity. That number of additional cancers will be extremely hard to detect against the background of about 500 million cancer deaths normally expected in the same population group over that period. In addition, 200 plant personnel and firefighters experienced acute radiation sickness, and 31 of those individuals died from acute radiation exposure.

The causes of this tragic accident can be laid at least partly to design flaws in the RMBK reactor and to an insufficient containment system. There were also obvious errors made by the operators. While power reactors in the United States do not share the particular design flaws of the RMBK reactors, and presumably have better containment vessels, the Chernobyl disaster presented a sobering demonstration of what might be expected from catastrophic reactor failure. It has been a serious blow to public confidence in nuclear power.

6.10 Nuclear Weapons

Although a nuclear reactor cannot explode like a bomb, there is, unfortunately, a relationship between nuclear weapons and nuclear reactors, even those designed to produce electric power. Two essential facts enter into a discussion of the relationship between reactors and bombs. First, any fissionable material can be used for a bomb, whether it is ^{233}U, ^{235}U, or ^{239}Pu, but it must be highly enriched, to 90% or better. Second, the amount of fissionable material needed for

---

[5]A curie of any radioactive isotope is the amount which has a radioactive decay rate equivalent to that of one gram of radium, or 3.7×10^{10} decays per second. It is also equivalent to 3.7×10^{10} becquerels, as the bequerel is defined as one decay per second.

a critical mass (the minimum size assembly that will sustain a chain reaction, and explode) depends strongly on the shape of the material and on the surroundings, but it is about 20 kg for ^{235}U and 10 kg for ^{239}Pu.

Since normal light-water reactor fuel is only enriched to about 3% ^{235}U, it cannot be used for bomb manufacture without further enrichment to about 90% ^{235}U. During World War II, the United States tried a variety of schemes to enrich uranium. The process that proved most effective and economical is gaseous diffusion. This is a complex and expensive process, and it would be extremely difficult for any country or group of people to set up and operate a gaseous diffusion plant to enrich either natural uranium or uranium reactor fuel without its being detected. A second method of separation is through the use of centrifuges; that is what was under development in Iraq during the 1980s and early 1990s. Also a new process of uranium enrichment through the use of lasers is being explored, which may eventually prove feasible. That process may prove to be far simpler and less expensive than the others.

It is a far more likely scenario for a nation to construct a uranium-fueled reactor and run it either for military purposes or for electric power generation. Fissionable ^{239}Pu could then be separated out from the spent fuel rods, where it amounts to about 0.6% of the fuel rod loading. One of the weapons that the United States exploded in Japan was made in this way, as were the weapons subsequently made in Great Britain, the Soviet Union, France, China, and India. Except for India, all of these countries have special reactors for plutonium production. In 1998, Pakistan also demonstrated a nuclear capability. Apparently, every country that has ever tried to explode a nuclear weapon has succeeded.

To control the spread of nuclear weapons, a Nuclear Non-Proliferation Treaty was signed in 1957–1958 by more than 100 countries; the primary aim of this treaty was to control the reprocessing of spent fuel rods by the International Atomic Energy Agency. Under the treaty, the United States, for example, has control over the reprocessing of fuel rods manufactured from uranium produced or enriched in the United States. Many of the nations that are not party to the treaty have the scientific and industrial capability to produce nuclear weapons. Countries such as Israel, South Africa, Brazil, Spain, and Argentina are probably in such a category. Thus, it is to be expected that as more and more countries such as Iran and North Korea construct nuclear reactors, some will divert sufficient plutonium to make nuclear weapons. It would be surprising to many if Israel and South Africa have not already done so.

The connection between the United States embarking on a vigorous nuclear power reactor program and the proliferation of nuclear weapons is somewhat indirect, and it is also complicated. One of the complications is the fact, mentioned earlier, that the continuation of a 500-reactor power program beyond roughly the year 2030 depends on the development and use of breeder reactors, which will use ^{239}Pu as their primary fuel and produce more in the process. This means that all spent fuel rods will be reprocessed to remove the ^{239}Pu and, of necessity, requires that large amounts of plutonium be transported and stored. The possibility of some of this plutonium being clandestinely diverted to nuclear weapons manufacture by those other than the U.S. government would be increased. It was this consideration that led President Carter in 1977 to discontinue funding for the Clinch River breeder reactor and to suspend the reprocessing of spent fuel rods in the United States.

Arguments can be made that other countries, such as France, because of their lack of fossil fuels, must be firmly committed to nuclear power, and what the United States does unilaterally will not have major influence on worldwide nuclear proliferation. It is also argued that with new simpler techniques for uranium enrichment based on atomic excitation by lasers, or through the use of new ultracentrifuges, nuclear weapons will become more widely available whether or not we have more nuclear power reactors. In presenting this argument, we recall that nuclear weapons may be based either on enriched uranium derived from ordinary uranium ore or on plutonium from reprocessed fuel rods, with equivalent results.

On the other hand, many in the United States and elsewhere feel that since both nuclear reactors and nuclear weapons originated here, this nation has a moral obligation to set an example for the rest of the world by renouncing entirely the use of nuclear energy. Such a step, it is argued, would be a moral deterrent to any nation further developing a nuclear program. A less drastic step would be for all nations to follow the Carter initiatives and to forego reprocessing of all fuel rods and, of course, the use of breeder reactors. Under such a program, fuel would be used only on a once-through basis. The spent fuel rods would eventually be put into permanent storage after removal from the reactor. Obviously this option precludes the long-term use of nuclear energy, as we would be dependent on our ^{235}U resources from relatively high-grade ores. This option would give the world about 50 years of nuclear fission energy, and might permit a transition to renewable energy sources or perhaps to nuclear fusion energy.

The weapons aspect of a nuclear reactor policy is the most difficult one to assess and about which to draw absolute judgments. The questions raised involve national security as viewed by many different nations and other international policy matters that are difficult to fold into an optimization calculation. The fact that the United States, along with the former Soviet Union and other countries, has already stockpiled sufficient nuclear weapons to wipe out essentially all civilization as we know it is a concern that to many completely dwarfs the problem of further proliferation.

6.11 The Storage of High-Level Radioactive Waste

A direct legacy of producing electric energy with nuclear reactors is the accumulation of radioactive waste. In the years following the first development and use of nuclear reactors, the U.S. Atomic Energy Commission appeared to recognize the radioactive waste as a problem but considered it a quite solvable problem, and postponed pursuing any real long-term solution. With the passage of time, the amount of radioactive waste has increased, but solutions to the problem have not come about. There is no easy solution as clean as simply sending the waste on a rocket into the sun, or putting it forever at the bottom of the deepest oceans. The only solution accepted or near acceptance in the United States or anywhere else is deep geologic disposal.

Disposal in this case means to keep the radioactive material out of the biosphere for a hundred thousand years or more. This time is so great, and beyond any human experience, that it is difficult for the public to imagine the requirements on the design of such disposal schemes. It is also challenging for the engineers and geologists to ensure that the radioactive waste will be safely isolated from the effects of earthquakes, changes in ground water level, and other natural phenomena for time periods on such a scale.

High-level waste is produced by reactors operated both for commercial and military purposes. The waste in the spent fuel from commercial power reactors is by far the largest in terms of activity. Much of the military waste comes from reactors operated for plutonium production for nuclear weapons.

As discussed earlier, when the spent fuel rods are first removed from a reactor they are placed in a nearby water pool to dissipate the heat from radioactive decay and to provide a shield to absorb some of the emitted radiation. It was originally intended that the spent fuel rods would only be in the water pool for about 150 days, after which they would be removed either to long-term storage or for reprocessing. In fact, since there is no long-term storage facility, and reprocessing is not allowed, the spent fuel now stays in the water tanks; some has been there for over 20 years. A small fraction of the fuel rods have been transferred from the water pools to dry storage casks at the reactor site where they are cooled by natural air flow. It appears that local on-site storage of the fuel rods cannot continue much longer, as the storage facilities are being filled to capacity. As of 1999, 40,000 tonnes of spent fuel rods were awaiting disposal at 72 U.S. power plant sites. Each year a nuclear reactor adds 20 to 30 tonnes to the accumulated total, which amounts to about 2000 tonnes added each year for all U.S. power reactors. It is expected that by 2010 there will be 60,000 tonnes accumulated and 80,000 tonnes by 2020. A 1000 MW_e reactor which has operated for a year or more has a total radioactivity of about 70 megacuries (MCi) in its spent fuel at one year after it is shut down or after the fuel is removed from the reactor. At ten years the 70 MCi will have decayed to 14 MCi, at the end of 100 years it will be 1.4 MCi, and after 100,000 years the radioactivity will be down to 2000 Ci.

In 1982 a Nuclear Waste Policy Act was enacted that called for the disposal of spent nuclear fuel in a repository in a deep geologic formation unlikely to be disturbed for thousands of years. The Department of Energy was to develop such a repository and a fee (one dollar per 1000 kWh) was assessed on nuclear generated electricity to be paid to the Nuclear Waste Fund to cover the cost.

By 1987 a site was selected for the deep geological depository in welded tuff at Yucca Mountain, Nevada, near the nuclear weapons test site. *Tuff* is the soft rocklike residue of material from ancient volcanoes. The site was selected because of its remoteness from population, climate, depth of the water table, and nearness to the nuclear weapons test site. The spent fuel assemblies would be put in cylindrical steel cannisters about 5 meters long by 2 meters in diameter. A stabilizing material would be packed into the cannisters to provide support and heat transfer. The cannisters would be placed in tunnels 200 to 500 meters below the surface and 300 meters above the water table. The cannisters would initially be retrievable and tunnels would be open for monitoring. This would continue for an unspecified time until it was decided the repository was safe and secure. The tunnels would then be back-filled and the site decommissioned. Figure 6.6 shows a Nevada test facility. The capacity of the Yucca Mountain repository would be limited to 70,000 tonnes of spent fuel. It is expected that this amount will be achieved well before 2030. Seven thousand tonnes of capacity is reserved for reprocessed high-level waste from the weapons program. A new repository would be needed for any waste beyond 70,000 tonnes.

The suitability of the Yucca Mountain site for storage of high-level radioactive waste was studied for about 20 years and then received the approval of the Department of Energy. In 2002 both the president and the U.S. Congress approved of the project. This paved the way for the Department of Energy to seek licensing from the NRC. The State of Nevada officially disapproved of the decision and challenged the constitutionality of the procedure. In 2004 the courts turned down the challenges except with regard to the 10,000 year compliance period, stating that it should be extended. So the project has not yet had final approval from all parties and the waste continues to pile up. There is still some hope that waste storage can begin in 2010. It is estimated that it would take 24 years to ship all the waste to Yucca Mountain. There would be about 550 shipments by train and 100 shipments by truck each year. Although some individu-

Figure 6.6 A spent reactor fuel storage test facility in Nevada. The storage test assesses the effects of heat and radiation on the behavior of granitic rock. (*Source*: Courtesy U.S. Department of Energy)

als have regarded the nuclear waste problem as easily solved, delays in implementing solutions may spell a very limited future for nuclear energy—our one really new source of energy.

6.12 *The Cost of Nuclear Power*

Establishing a real cost at the present time for nuclear power is difficult because of a number of complicating factors. The initial capital cost for many of the reactors in the United States was relatively high owing to the customized basis depending on the needs of the individual utility and its particular site requirements. Thus, there has been little standardization of design and no economy of scale. Each reactor has had to have separate approval by the Nuclear Regulatory Commission (NRC) and frequently changes were required in the mechanical and electrical details during construction. The required changes were costly and added many months to the construction time. The time it took to construct a nuclear power plant in the United States increased from 60 months in 1961–1966 to about 150 months in 1985–1990. In Japan over this entire time period the reactor construction time has stayed under five years.

Since there have been essentially no new reactors built in the United States in the last 20 years it is difficult to quote a meaningful capital cost, but it would appear to be about $1,800 to $2,000 per kW_e of capacity. Money must also be set aside for decommissioning cost (about 9–15% of the initial capital cost) and for costs associated with storage and disposal of the spent fuel (about 10% of the overall cost of the power output assuming that the fuel is not reprocessed).

As mentioned previously, there is a levy of 0.1 cent/kWh to fund the $18 billion spent fuel program. The cost of the actual reactor fuel (U_3O_8) is small, about 0.35 cents/kWh.

In summary, the cost of nuclear electric energy has been estimated, assuming a 10% amortization rate on the capital, to be $51/MWh, or $0.051/kWh. This is more than for coal-powered plants at $37/MWh or gas turbine combined cycle plants at $35/MWh. On a comparative basis, solar electric is $202/MWh, solar thermal is $158/MWh, and wind-generated electricity is $55/MW.

Recent experience with nuclear reactors in the United States has been quite encouraging. The capacity factor of the nuclear plants has increased from an average of 68% in 1990 to 90% in 2002, and the down time for refueling has been reduced from 100 days to 42 days.

With the first phase of nuclear reactor development in the United States at an end, there are now efforts being made to design reactors for the future that are safer and less costly than those of the first generation. One design by Westinghouse Electric and the U.S. Department of Energy is a 600 MW_e light-water reactor that can be standardized and receive NRC approval prior to construction. These smaller reactors are designed to have passive stability, simplified controls, greater ruggedness, ease of operation, and a higher level of standardization. The designation *"passive stability"* means that the reactor has the inherent property of having its reaction rate decrease as the temperature of the coolant or fuel increases above the normal operating point, without relying on action by the operator or external control devices. A reactor design designated as AP-600 is an example of the new approach. It has all the features listed here, but also has the number of valves, pumps, heat exchangers, ducting, and control cables reduced by 35 to 80% compared to a conventional reactor of the same capacity.

Another effort is devoted to development of large evolutionary reactors. These reactors are of the general class of PWR that are over 1000 MW_e capacity and have been in common use for many years. The goal is to incorporate in an evolutionary way various changes and improvements based on the years of operating experience. A study at Oak Ridge National Laboratory in Tennessee made cost projections for the two new classes of reactors, including costs for decommissioning. The smaller designs, such as the AP-600, were projected to deliver electricity at a total cost of 4.0 cents per kilowatt-hour, while the large evolutionary reactor was projected to produce electricity at 3.8 cents per kilowatt-hour. These costs are comparable to that of electricity from natural gas–powered plants or from coal-fired plants. Such optimistic projections have been proved wrong before for nuclear energy.

Society faces important decisions concerning whether thermal energy from nuclear reactors should to a large extent replace that from burning fossil fuels. With nuclear reactors, the construction costs dominate; with fossil fuel plants, the fuel is the important cost. What is often missing in such comparisons is the economic value assigned to environmental impacts such as the emission of CO_2, SO_2, and particulates in the case of coal, and the chance of radioactive releases in the case of nuclear.

6.13 *Nuclear Fusion as an Energy Source*

We have already dealt with the phenomenon in which energy is released when a nucleus fissions. In this process, looked at in more detail, what has happened is that a nucleus, ^{235}U, for example, has had its constituents rearranged mainly into two smaller nuclei—the fission products. In order for energy to have been given off, it is necessary for the final state (mainly the fission product nuclei) to have lower total mass than the initial state (mainly the ^{235}U). Einstein made that clear. It is also true that the fission product nuclei have their nucleons bound together more tightly than they were in the uranium nucleus. In nature, it is always true that whenever we have a system moving from less to more binding, energy is given off. We see this in burning any carbon-based fuel. The carbon and oxygen become carbon dioxide molecules which bind the carbon and oxygen atoms together in a lower total energy state than they had in the original fuel and air. Anything that moves to a state of lower energy gives off its excess energy in doing so, and it also loses mass. Now it's time to look at another nuclear process, *fusion*. In fusion, two separate nuclei with relatively loose binding, such as two isotopes of hydrogen: deuterium (^2H, a neutron and proton loosely bound together) and tritium (^3H, a proton and two neutrons bound together), can combine to form helium, in which the nucleons are very tightly bound together. This fusing, or fusion, will be true to the laws of nature and release energy as mass is lost. The helium nucleus has less mass than the sum of the masses of the two hydrogen nuclei.

Examples of nuclear fusion reactions in which two hydrogen nuclei combine to form a helium nucleus and give off energy in the process are as follows:

The D–T reaction:

$$^2_1H_1 + {^3_1}H_2 \rightarrow {^4_2}He_2 + n + 17.6 \text{ MeV}$$

The two D–D reactions:

$$^2_1H_1 + {^2_1}H_1 \rightarrow {^3_2}He_1 + n + 3.2 \text{ MeV}$$

$$^2_1H_1 + {^2_1}H_1 \rightarrow {^3_1}H_2 + {^1_1}H_0 + 4.0 \text{ MeV}$$

The D stands for the deuteron, $^2_1H_1$, and the T stands for tritium, $^3_1H_2$.

These particular fusion reactions are important for at least two reasons. First, they are the basis for much of the research effort that is going into the attempt to develop a revolutionary new source of energy for the production of electricity. Second, nuclear weapons of the thermonuclear type, known as hydrogen bombs, gain their energy release through the D–T reaction.

It is not a simple matter to make fusion reactions occur. One cannot merely put deuterium gas into a bottle and expect that D–D reactions will happen. The reason is that the deuterium nuclei, the deuterons, have a positive electric charge and thus repel each other. In order for them to interact in a way that will produce fusion, the deuterons must somehow be brought nearly into contact, and only then will they be drawn together by an attractive nuclear force that only acts over very short distances. One way to force the deuterons together closely enough so that they will fuse is to raise their temperature so high that their thermal velocities can overcome the electric repulsion. If two deuterons are approaching each other at very high speed, they can come together in spite of the force that tries to keep them apart. We know that fusion reactions do occur in the core of the sun, producing solar energy, where the temperature is 20 million K.

The predominant energy-producing reactions in the solar core are not those we have discussed so far, but start with another fusion reaction, the fusing of two hydrogen nuclei (protons) to form a deuteron:

$$^1_1H_0 + {^1_1}H_0 \rightarrow {^2_1}H_1 + \beta^+ + \nu + \text{energy}. \tag{6.1}$$

In this reaction, a proton becomes a neutron in the newly formed deuterium nucleus. A positron (β^+) and a neutrino (ν) are also emitted. This reaction would happen at an impossibly slow rate under conditions obtainable on earth, but because of the extremely high density along with the high temperature in the solar interior, the reaction rate there is appreciable.

Once reaction (6.1) is accomplished, there will be both protons and deuterons moving about in the solar interior. This leads to a second reaction:

$$^1_1H_0 + {^2_1}H_1 \rightarrow {^3_2}He_1 + \text{energy}, \tag{6.2}$$

in which a mass-3 helium nucleus is formed. At this point there is a way to make helium-3 nuclei which can then fuse together to form helium-4 (also known as an alpha particle) and protons:

$$_2^3\text{He}_1 + {}_2^3\text{He}_1 \rightarrow {}_2^4\text{He}_2 + 2\,{}_1^1\text{H}_0. \tag{6.3}$$

If we now let reactions (6.1) and (6.2) each occur twice, and reaction (6.3) once, and sum up the constituents, the net effect is

$$4\,{}_1^1\text{H}_0 \rightarrow {}_2^4\text{He}_2 + 2\,\beta^+ + 2\,\nu + \text{energy}.$$

Here we see that four protons (the ${}_1^1\text{H}_0$) have effectively joined together to form an alpha particle (the ${}_2^4\text{He}_2$) plus some lighter products, and have released energy in doing so. Hydrogen has been consumed to form helium. One might think of the protons as the fuel and the helium as the ashes. This is commonly known as the *proton–proton cycle*. A second cycle of thermonuclear reactions in the sun is based on carbon as a catalyst. Through a series of reactions starting with protons on carbon, four protons are again fused into an alpha particle. Both the proton–proton cycle and the carbon-catalyzed cycle provide about 25 MeV of energy, but the proton–proton cycle is dominant in the sun, furnishing all but 1 or 2% of the total energy production.

Each day our sun burns 5.3×10^{16} kg of hydrogen into helium. In this process, 0.76% of the mass of the protons is converted into energy, meaning that the amount of mass lost is

$$(0.76 \times 10^{-2}) \times (5.3 \times 10^{16}\text{ kg}) = 4.0 \times 10^{14}\text{ kg}.$$

The equation $E = mc^2$ gives the rate of solar energy production by the proton–proton cycle:

$$E = (4.0 \times 10^{14}\text{ kg/day}) \times (3 \times 10^8\text{ m/s})^2 = 3.6 \times 10^{31}\text{ J/day}.$$

6.14 Controlled Thermonuclear Reactions

It is impossible to duplicate the conditions of the interiors of the sun and other stars on earth. However, it may be possible to create conditions where fusion reactions can take place between two deuterons or between a deuteron and a triton. The minimum temperature needed for the D–D reaction is 100×10^6 K, and for the D–T reaction 40×10^6 K. Although these are extremely high temperatures, they correspond to average kinetic energies for the hydrogen nuclei of only about 5 keV, very modest by laboratory standards. In spite of enormous effort over the past 50 years, in many countries, it has not yet been possible to use fusion for practical energy production. The necessary conditions of temperature and particle density have not been achieved for a sufficiently long time. Research in this field is driven by recognition of the vast available store of fuel to feed the fusion reactions, and the fact that there are no radioactive reaction products.

In normal water, such as in the world's oceans, there is naturally one deuteron for every 5000 atoms of hydrogen. A simple calculation shows that a mere 1% of the deuterium in the world's oceans has an energy equivalent through the D–D fusion reaction of 500,000 times the Q_∞ for all the world's fossil fuels. If controlled fusion reactors could be built for the D–D reaction, the world's energy needs would be met for millions of years.

It would also be possible to consider the D–T reaction for fusion energy production. This reaction can proceed at a lower temperature than D–D, and it is the focus of current efforts in controlled fusion. The D–T reaction does not hold the promise of providing energy for the really long term, however. The tritium, T, is radioactive, with a half-life of only 12 years. It is not available naturally in appreciable quantity, so it must be manufactured. Tritium is now produced in a fission reactor by inserting lithium into the reactor and letting the intense neutron flux interact with lithium nuclei by the reaction

$$\text{n} + {}^6\text{Li} \rightarrow \alpha + \text{T}.$$

If a D–T fusion reactor could be brought into operation, the emerging neutrons from the D–T reaction could be used to produce tritium from lithium by a process similar to that in which neutrons from fission are now used. This is shown in Figure 6.7. The number of lithium-6 atoms in the world is far less than the number of deuterium atoms, and one atom of lithium is used for every atom of tritium produced by the processes just mentioned. Thus, the world's resources of lithium would be entirely consumed by the D–T reaction, which uses D and

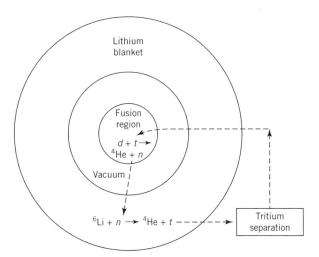

Figure 6.7 A D–T fusion reactor. Deuterium and tritium react in the high-temperature fusion region producing 14 MeV neutrons that enter a surrounding blanket of liquid lithium. The neutrons deposit their kinetic energy in the lithium and induce reactions that provide tritium for the fusion reaction. Heat energy is then removed from the lithium blanket by a circulating coolant and used to drive a steam turbine.

T in equal number, long before the deuterium was exhausted. It has been estimated that all of the lithium-6 in the United States would be consumed in a few hundred years by a D–T fusion power program that produced all of the nation's electricity. After that, lithium from the world's oceans could possibly extend a D–T fusion power program to several thousands of years.

6.15 *A Fusion Reactor*

The main laboratory efforts at present are focused on producing a plasma at sufficient temperature and particle density, held together long enough, that the fusion reaction is a net producer of energy. At the temperature encountered in a fusion reactor, the kinetic energies of the particles are so high that the atoms of deuterium and tritium have been stripped of their electrons. The resulting electrically neutral mixture of dissociated electrons and nuclei at high temperature is known as a *plasma*. It has been demonstrated that for the D–T reaction to be a useful producer of energy, the numerical product of particle density and the time duration of particle confinement at this density must exceed a certain minimum value. This finding is known as the *Lawson criterion*, which specifies that $n\tau > 10^{14}$ sec/cm^3, where n is the particle density in number per cubic centimeter and τ is the confinement time in seconds. This criterion could be met by having 10^{14} particles/cm^3 held together for 1 second, 10^{15} particles/cm^3 confined for 0.1 second, or by any other combination of density and time giving a product equal to or greater than 10^{14} sec/cm^3. For comparison, normal air has a density of about 10^{19} molecules/cm^3, several orders of magnitude greater than the density anticipated in fusion reactors. Of course, the necessary high temperature must also be achieved in addition to the Lawson criterion.

If and when the fusion reaction can be demonstrated as a useful energy producer on a laboratory scale, a later step will be the construction of a fusion reactor system for producing electricity for our distribution system. A diagram of such a system is shown in Figure 6.8. The fusion plasma region shown in the figure is the region where the plasma is confined by magnetic fields or by some other means. The high temperature region must be isolated from contact with container walls, not so much because the container would be destroyed if brought up to the temperature of the hot plasma but because the plasma would be cooled on contact and the reaction extinguished. The energy density in the fusion plasma is not as high as might be supposed; it is one or two orders of magnitude less than in fission reactors and inadequate to destroy materials.

A common approach to confinement and isolation from container walls is to use magnetic fields. Charged particles, as in a plasma, cannot move across strong magnetic fields. If the fields are properly shaped, the plasma can be confined within the desired region. One such confinement scheme, shown in Figures 6.9 and 6.10 is the Tokamak (from the Russian *to*—toroidal, *ka*—chamber, *mak*—magnetic), which was initially developed in the Soviet Union and is currently the most promising magnetic confinement device.

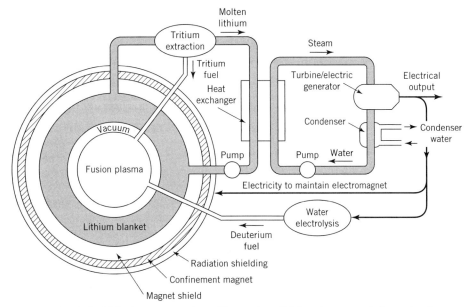

Figure 6.8 A D–T fusion power plant. The molten lithium is pumped through a heat exchanger that produces high-pressure steam to drive a turbine. The fusion plasma and lithium blanket regions are in the shape of a torus, a cross section of which is shown.

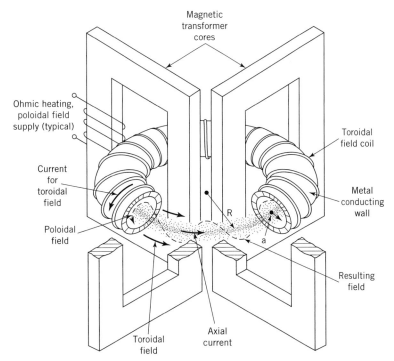

Figure 6.9 A Tokamak magnetic confinement system. A magnetic field confines the plasma within a torus.

Figure 6.10 The Tokamak fusion test reactor at the Princeton University Plasma Physics Laboratory. (*Source*: Courtesy U.S. Department of Energy)

Progress on achieving the necessary temperature, confinement time, and ion density has been slow because the plasma physics is highly complex and the experiments are expensive. Unexpected instabilities and turbulence in the plasma have been encountered. Because of the size and cost of Tokamaks, a joint effort of the United States, Russia, Europe, China, South Korea, and Japan has been formed. It is called the International Thermonuclear Experimental Reactor (ITER).

It is proposed to build a $5 billion 16 meter diameter superconducting Tokamak capable of producing 400 MW of fusion power. It would be the world's first controlled self-sustaining fusion burn. About 80 MW from energetic alpha particles will remain in the plasma for self-heating and 40 MW of external power will be used for heating and control.

Progress on the ITER was held up for some time because of an impasse in choosing a site. Europe, China, and Russia insisted on a site in southern France, while the United States, Japan, and South Korea backed the choice of a site in northern Japan. In June of 2005 it was decided that ITER would be built at the Cadarache site in southern France.

Other proposed confinement schemes, classified as inertial confinement, include fusion induced by either laser beams or beams of energetic heavy ions. The laser method is illustrated in Figure 6.11. The heavy-ion method is similar, but beams of particles from accelerators rather than laser beams converge on a D–T pellet. These inertial schemes are still in early development stages.

Fusion reactors will produce neither the problematic emissions now experienced from fossil fuel–burning power plants nor the long-lived fission products

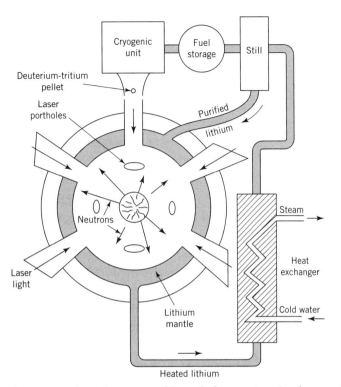

Figure 6.11 A cutaway view of a proposed laser fusion reactor. A microscopic deuterium–tritium pellet is injected into the reaction chamber several times per second. Precisely timed and sharply focused pulsed laser beams of high intensity converge in the chamber, thereby creating the requisite conditions of temperature and density to initiate fusion. As in the magnetic confinement schemes, the resulting energetic neutrons are absorbed in a surrounding mantle of molten lithium, which is circulated to remove the heat energy. The tritium produced in the lithium mantle is separated out and incorporated into new D–T pellets.

and transuranic elements resulting from fission reactors. They will, however, contain a large inventory of radioactive tritium and, during operation, the reactors' structural components will become radioactive to varying degrees. Neither of these factors, however, is considered to present anything near the hazards associated with fission reactors. Studies that have considered lithium fires, failure of magnets, and other possibilities have concluded that there would be no severe damage outside the plant itself and little potential for damage to the surrounding community. Such findings indicate further possible economies in the use of waste heat and in power distribution, as the reactors could be located within cities, near the point of need for both electricity and heat.

At this time, it is not certain that we shall ever achieve controlled fusion. The optimists see progress over the past decades, but others point out that there still remain at least four barriers. These are proof of scientific feasibility of the basic reactions, economics of electric power generation, materials damage to the reactor due to the intense neutron flux, and materials availability for reactor construction. The last point arises primarily with regard to the metals beryllium, manganese, chromium, nickel, tin, niobium, and vanadium, some of which are needed for superconducting magnet construction. Reliance on foreign markets is necessary for these materials, and it is possible that not even all the world's resources would be adequate for large-scale fusion power generation. The other problems are being addressed with intense engineering and scientific effort, and several ideas are emerging. The overall prospects for the fusion power program have not been improving appreciably with the passage of time, and, because of this, the U.S. Department of Energy has actually reduced its level of funding for the program. If one asks researchers in fusion power when it will be a reality for practical electric power generation, the standard answer seems to be *50 years*, whether the question was asked in the 1950s, 1960s, 1970s, 1980s, 1990s, or is asked now.

Key Terms

Radioactivity	Delayed neutrons
Half-life	Boiling water reactor
Ionizing radiation	Pressurized water reactor
Criticality	Curie
Meltdown	Bequerel
Three Mile Island	Yellowcake
Proton	Enrichment
Neutron	Breeder reactor
Fission	Event tree
Fusion	*China Syndrome*
Fission product	Loss-of-coolant accident
Manhattan Project	Chernobyl
Alpha particle	Critical mass
Gamma ray	Passive stability
Neutrino	D–T reaction
Thermal neutrons	Proton–proton cycle
Moderator	Tritium
Chain reaction	Lawson criterion
Control rods	Tokamak

Suggested Reading and References

1. Choppin, G. R., and Rydberg, J. *Nuclear Chemistry—Theory and Applications.* New York: Pergamon Press, 1980.

2. Devins, D. W. *Energy, Its Physical Impact on the Environment.* New York: John Wiley, 1982.
3. Inglis, David R. *Nuclear Energy—Its Physics and Its Social Challenge.* Reading, MA: Addison-Wesley, 1973.
4. Krenz, Jerrold H. *Energy—Conversion and Utilization*, second edition. Boston: Allyn and Bacon, 1984.
5. Nero, Anthony V., Jr. *A Guidebook to Nuclear Reactors.* Berkeley: University of California Press, 1979.
6. "Energy for the Planet Earth." *Scientific American*, **263,** No. 3 (September 1990), pp. 54–163.
7. Bethe, H. A. "The Necessity of Fission Power." *Scientific American*, **234,** No. 1 (January 1976), pp. 21–31.
8. Cohen, B. L. "The Disposal of Radioactive Wastes from Fission Reactors." *Scientific American*, **236,** No. 6 (June 1977), pp. 21–31.
9. Inhaber, H. "Risk Evaluation." *Science*, **203,** No. 4384 (February 4, 1979), pp. 718–723.
10. Lewis, H. W. "The Safety of Fission Reactors." *Scientific American*, **242,** No. 3 (March 1980), pp. 53–65.
11. Ross, M. H., and Williams, R. H. *Our Energy: Regaining Control.* New York: McGraw–Hill, 1981.
12. *Reactor Safety Study: An Assessment of Accident Risks in U. S. Commercial Nuclear Power Plants.* Washington, D.C.: U.S. Nuclear Regulatory Commission (WASH-1400), 1975.
13. "Report to the American Physical Society by the Study Group on Nuclear Fuel Cycles and Waste Management." *Reviews of Modern Physics*, **50,** No. 1, Part 2 (January 1987), pp. S1–S183.
14. Wilson, Richard. "A Visit to Chernobyl." *Science*, **236,** No. 4809 (1987), pp. 1636–1640.
15. Upton, Arthur C. "Health Effects of Low-Level Ionizing Radiation." *Physics Today*, **44** (August 1991), p. 34.
16. Cohen, B. L. "The Nuclear Reactor Accident at Chernobyl, USSR." *American Journal of Physics*, **55,** No. 12 (December 1987), pp. 1076–1083.
17. Bodansky, David. *Nuclear Energy—Principles, Practices and Prospects*, second edition. Springer, 2004. New York, N.Y.
18. *Physics Today*, **50** (June 1997). "Special Issue: Radioactive Waste."
19. *Physics Today*, **58** (February 2005). "Integrated Simulation of Fusion Plasmas."

Questions and Problems

1. (a) Calculate the number of joules that can be obtained from the fissioning of 1 kg of ^{235}U, assuming 198 MeV average energy release per fission.

    (b) How much energy can be obtained from the ^{235}U in one kg of natural uranium?

2. Calculate how much energy, in joules, can be obtained from burning 1 kg of coal.
3. Compare the total energy in the U.S. resources of coal and ^{235}U.
4. In the following reaction for the spontaneous fission of ^{238}U, apply an appropriate conservation law to determine the number of neutrons emitted.

$$^{238}_{92}U_{146} \rightarrow {}^{140}_{55}Cs_{85} + {}^{92}_{37}Rb_{55} + (?)n$$

5. Explain how ^{239}Pu builds up in the fuel rods of a reactor such as the BWR or PWR with time, and what role it plays in the operation of the reactor.
6. Explain how the buildup of ^{239}Pu in a nuclear reactor is related to the manufacture of nuclear weapons.
7. Estimate the number of years that a reactor such as Zion 1 would have to run for the probability of a large release of radioactivity from internal events to be greater than 50%, according to NUREG-1150 (Table 6.2).
8. Now imagine that all the electric energy in the United States was furnished by 700 Zion 1 nuclear reactors. According to NUREG-1150 (Table 6.2), how many years would it take for a large release from internal events to happen from at least one of the 700 reactors?
9. It has been estimated that in the year 2010, the United States will need 10^{12} W of electric power produced by one thousand 1000 MW$_e$ power plants. These would most likely be powered either by coal or by nuclear reactors.

   (a) If the energy is provided by ^{235}U-burning reactors, how many tonnes of ^{235}U will be needed annually? Assume an efficiency of 33%.

   (b) How many tonnes of uranium must be mined annually?

   (c) If the energy is provided by coal, how many tonnes of coal must be mined each year? Again, assume an efficiency of 33%.

   (d) If the coal is obtained from land that provides 170,000 tonnes per acre (as would be true for a 100-ft thick seam), how many square miles of land would be mined annually? One square mile equals 640 acres.

10. Compare the probabilities of a serious release of radioactivity as presented by the Rasmussen Report (WASH-1400) and the NUREG-1150 report.
11. (a) How would the high-level radioactive waste problem be changed if the United States went from a complete dependence on thermal LWR reactors to fast breeder reactors?

    (b) What changes would be brought about in the radioactive waste problem if all of the electric energy in the United States were produced by fusion reactors?

12. Discuss briefly the reasons for the sharp curtailment of new reactor orders in the United States in the 1970s.
13. You buy a lottery ticket that gives you a 10% chance of winning a prize and you also go to a bingo game where there is a 5% chance of winning a prize. What is the probability that you will win both the lottery and bingo?
14. It is estimated that the Three Mile Island accident released into the environment 20 curies of ^{131}I, which has a half-life of 8 days. How many curies remained after 48 days?
15. The Rasmussen report estimates the probability of having a loss-of-cooling accident followed by failure of the emergency core cooling system followed by a serious

release of radioactivity as one chance in two million reactor years. If we have 200 reactors operating in the United States, how frequently should we expect such accidents?

16. Describe briefly the major differences between the reactor that was involved in the accident at Chernobyl and a typical U.S. light-water reactor.

17. For a new nuclear power plant, the designers estimate the following probabilities: The probability of a loss-of-coolant accident is 10^{-3} per reactor year. The probability that the emergency core cooling system will work is 80%. What is the probability per reactor year that the reactor will have a loss-of-coolant accident and be saved by the ECC system?

18. In 1999 the world produced 24×10^{11} kWh of electricity with nuclear reactors. If a 1000 MW$_e$ nuclear power plant, operating continuously, can produce enough plutonium for 20 nuclear bombs each year, how many bombs could be produced from the power plant plutonium made in 1999, assuming continuously operating reactors?

Multiple Choice Questions

1. The mass of your pencil is 10 grams. The equivalent mass energy is _____
   a. 2.5×10^{14} kWh
   b. 0.83×10^{3} kWh
   c. 2.5×10^{8} kWh
   d. 0.83×10^{6} kWh

2. The total mass energy of 200 grams of carbon is _____
   a. 1.8×10^{16} joules
   b. 9×10^{16} joules
   c. 1.8×10^{16} Btu
   d. 3.6×10^{16} calories

3. The following reaction takes place: $^{9}_{4}Be_5 + ^{4}_{2}He_2 \rightarrow ^{12}_{6}C_6 + X$
   What is X?
   a. a neutron
   b. a proton
   c. an alpha particle
   d. a deuteron

4. In the radioactive decay of $^{14}_{6}C_8$ to $^{14}_{7}N_7$, the following particles are emitted _____ :
   a. an electron (negative) and an antineutrino
   b. an electron (positive) and a neutrino
   c. only an electron (negative)
   d. an alpha particle

5. The half-life of ^{137}Cs is 30 years. In how many years will a 1000 Ci source be down to 62.5 Ci?
   a. 150
   b. 60
   c. 90
   d. 120

6. The reason that thermal neutrons (0.025 eV) are used in conventional light-water reactors such as the BWR is _____
   a. the probability of a thermal neutron causing a fission reaction with ^{235}U is very high
   b. the probability of a thermal neutron causing a fission reaction with ^{238}U is very high

Chapter 6 The Promise and Problems of Nuclear Energy

   c. there is no choice since the neutrons are going to be slowed down in the coolant anyway
   d. the neutrons that come from fission are thermal to begin with, so there is no choice

7. In a typical fission of a uranium nucleus, the number of neutrons emitted is about _____
   a. zero to 1
   b. 2 to 3
   c. 10 to 12
   d. 235

8. In a typical 1000 MW$_e$ nuclear power plant, about how many uranium nuclei are fissioning per second?
   a. 10^3
   b. 10^{10}
   c. 10^{15}
   d. 10^{20}
   e. 10^{25}
   f. 10^{30}
   g. 10^{35}
   h. 10^{40}

9. The three isotopes known to be good fission reactor fuels are _____
   a. ^{12}C, ^{90}Sr, ^{239}Pu
   b. ^{233}U, ^{235}U, ^{239}Pu
   c. ^{90}Sr, ^{137}Cs, ^{232}Th
   d. ^{231}U, ^{237}U, ^{239}Pu

10. The only naturally occurring nucleus that fissions with slow neutrons is _____
    a. ^{232}Th
    b. ^{233}U
    c. ^{235}U
    d. ^{239}Pu

11. After the control rods are inserted into a reactor and it is shut down, a great deal of energy release continues in the core. The source of this energy is largely _____
    a. the heat capacity of the mass of the core
    b. the radioactivity of the fission products
    c. the steam returning from the turbines
    d. the control rods heating up from absorbing the neutrons

12. The fuel rods in a commercial thermal light water reactor such as the BWR have in them initially _____
    a. natural UO_2 pellets
    b. UO_2 pellets with the ^{235}U enriched to about 3%
    c. UO_2 pellets with 100% ^{235}U
    d. UO_2 pellets with 100% ^{238}U

13. About how many 1000-MW$_e$ BWR can be fueled for their 30 year lifetime with the economically recoverable uranium resources of the United States?
    a. 5
    b. 50
    c. 500
    d. 5000

14. Most of the energy liberated in nuclear fission is in the kinetic energy of _____
    a. neutrons
    b. protons
    c. gamma rays
    d. fission fragments

15. In order for uranium ore to be useful for the American type of fission reactors, it must first be enriched to _____

a. about 0.7% uranium-235
b. 3% or greater uranium-235
c. 3% or greater uranium-238
d. at least 90% uranium-235
e. at least 99% uranium-238
f. at least 99% plutonium-239
g. at least 0.7% uranium-238
h. at least 0.7% thorum-232

16. The main reason for going to plutonium breeder reactors is to _____
    a. reduce thermal pollution
    b. extend the useful lifetime of our uranium
    c. reduce the amount of plutonium produced
    d. reduce the threat of nuclear weapons proliferation

17. Eventually, the breeder reactor will have to be used if we are to have energy from nuclear fission because _____
    a. the BWR and PWR are too dangerous
    b. we will need to use ^{235}U instead of ^{233}U
    c. we will need to use ^{238}U instead of just ^{235}U
    d. there is not enough ^{238}U in our uranium resources

18. The solar proton–proton cycle may be approximately summarized as _____
    a. $4\,^1_1H_1 \rightarrow \,^2_2He_2 + 2\,\beta^+ + 2\,\nu + energy$
    b. $4\,^1_1H_0 \rightarrow \,^4_2He_2 + 2\,\beta^+ + 2\,\nu + energy$
    c. $4\,^4_2He_2 \rightarrow \,^1_1H_0 + 2\,\beta^+ + 2\,\nu + energy$
    d. $6\,^1_1H_0 \rightarrow \,^4_2He_2 + 2\,\beta^+ + 2\,\nu + energy$

19. High temperatures are required in fusion reactors because _____
    a. the two combining particles are both positively charged
    b. the two combining particles have opposite charges
    c. the high temperature increases the density of the plasma
    d. the high temperature reduces the pressure of the plasma

20. The fuel for the D–T reaction may be considered as _____
    a. deuterium and lithium-7
    b. deuterium alone
    c. deuterium and uranium-238
    d. deuterium and lithium-6

21. The D–T reaction gives off most of its energy as _____
    a. kinetic energy of neutrons
    b. kinetic energy of alpha particles
    c. gravitational potential energy
    d. ultraviolet radiation

22. In the year 3000, the only sources of energy that can be expected to be of major importance in the United States (excluding smaller contributors such as geothermal, tidal, and hydroelectric) are _____
    a. D–D fusion reactors and thermal ^{235}U reactors
    b. D–D fusion reactors, solar energy, and fast breeder reactors
    c. coal, D–T fusion reactors, and solar energy
    d. oil shale, fast breeder reactors, and solar energy

CHAPTER 7

Energy Conservation

(*Source*: Courtesy Owens-Corning/Projections)

7.1 A Penny Saved Is a Penny Earned

The old adage about pennies is just as true for barrels of oil or gallons of gasoline. When one less barrel of oil is used to heat your home, not only does the oil not have to be taken from the earth but since it is not burned there are no emissions to foul the environment and you save money in the bargain. It is better to save a barrel of oil than to find a new one.

One side of energy conservation is energy efficiency. How can the same goal be achieved with a lowered expenditure of energy? The addition of insulation to the outside walls of a house means that the same inside temperature can be maintained with less fuel to the furnace. Use of fluorescent, rather than incandescent, lightbulbs provides the same level of lighting and saves energy and money.

An example of a second approach to energy conservation would be to lower the thermostat setting so that the house is cooler. Some temperature reduction can often be tolerated without undue discomfort, but carrying this approach too far may mean that saving energy carries with it a sacrifice of comfort on the part of the home occupants. This element of sacrifice may appear in many situations. Examples are choosing to walk or ride a bicycle rather than driving a car, or hanging the laundry out to dry rather than using a clothes dryer. There can be significant compromise, such as when one chooses a light, fuel-efficient car over a heavier vehicle knowing that there may be some loss of safety in a collision.

Do not confuse what we call *energy conservation* in this chapter and The Principle of Energy Conservation discussed earlier. They are quite different. In the discussion of The Principle of Energy Conservation, it was stated that the amount of energy in a closed system can never change. The energy can be transformed from one form to another, but energy can never be created or destroyed. It is apparent from Figure 1.1 that the energy consumed in the United States each year increased approximately exponentially from about 1850 to 1975. The average growth rate over this 125 year period was close to 3% per year, but there were times of significantly slower and faster growth. In the early 1970s, several analysts made projections of energy use into the future based on a recently observed growth trend of 4.3% a year. These projections indicated that the United States would be using energy at the rate of about 160 QBtu per year by the year 2000. We have seen that by 2003 we were using only 98 QBtu/yr, far short of the common projections. Why were the projections wrong? After the oil embargo in 1973, the cost of energy rose very steeply. The increased fuel costs, coupled with federally sponsored energy conservation programs and help from public utilities and various community organizations, have provided the incentives, education, and means for a significant drive toward a reduction in the use of energy. If the energy savings had come along with an economic depression or reduction in the national industrial output, that would not be good news, but this was not the case. In 1973 the ratio of energy consumption to gross domestic product was 27,000 Btu/$. Since 1973 this ratio has come down steadily and in 1990 it was 20,000 Btu/$, and by 2004 it had fallen to 8,400 Btu/$. While this is a remarkable achievement, and in part reflects the changing nature of the country's industrial output, the ratio is still higher in the United States than in many other developed countries.

In discussing energy conservation, the main attention will be given to household use. The subject, about which a great deal of information is available, is of interest to many. Some more general consideration will be given to industry and agriculture. Energy conservation in transportation will be covered in Chapter 8. Residential uses of energy now amount to about 22% of the total energy consumed in the nation, commercial uses are another 18%, industry uses 33%, and transportation 27%.

It is important to recognize that, in addition to the largely technical measures we discuss for implementing energy conservation, certain policy acts of government are crucial. These are put forth at all levels of government from local to federal. The federal Energy Policy Act of 1992 forms the statutory basis of federal energy management activities. It is comprehensive and has had wide-reaching impact in setting mandatory standards. It covers issues ranging from buildings to transportation to fuels and numerous other elements of energy conservation. A more recent and equally comprehensive Energy Policy Act of 2003 was intended to update the Act of 1992 and it moved through various committees and subcommittees and through the House of Representatives but was never approved by the Senate, where it got bogged down by hundreds of amendments. It languished and never received approval and implementation.

In 2005 the Energy Policy Act of 2005 was passed by the House and the Senate and it was signed into law in August of 2005 by the President. The Energy Policy Act of 2005 contains numerous provisions intended to reduce the national energy consumption. Among other provisions, starting in January of 2006 it will give tax credits for hybrid and diesel autos which demonstrate superior energy efficiency. It will give credits on the cost of buying and installing energy-efficient items in the home. It will give tax credits for the purchase and installation of solar energy devices, including water heaters. The Act is broad-ranging and goes far beyond dealing with consumer items. The entire energy industry is affected by its enactment. The details of its implementation remain to be worked out.

In 1992 the U.S. Environmental Protection Agency established an important federally sponsored cooperative program between government and industry. The voluntary labeling program known as Energy Star was initially designed to identify and promote energy-efficient products to reduce greenhouse gas emissions. The labels are now seen on major appliances, lighting, office equipment, and other devices. This program has since been expanded to cover homes and commercial buildings. By advocating and implementing widespread knowledge and implementation of energy efficiency measures, the Energy Star program is playing an important role in managing the environmental effects of energy use.

At the state level, legislation has been passed that enforces standards for various energy-consuming products. Exit signs, ice makers, lighting and swimming pool pumps are among the items included in this legislation. This legislation has, among its intentions, a reduction of peak electric demand, thus reducing the need for new power-generating facilities. The states of California, Colorado, Connecticut, Maryland, and Rhode Island now have enforceable energy-efficiency standards for a variety of products.

7.2 *Space Heating*

As can be seen in Table 7.1, the largest amount of household energy use in the United States is for space heating, and natural gas is the predominant source of this energy. About 47% of the residential energy use in the United States is for space heating. Prior to the 1940s, coal and wood were important sources of home heat.

Table 7.1 Annual Household Energy Uses and Sources, 2001, in QBtu

	Source			
Use	Natural Gas	Electricity	Fuel Oil	LPG[a]
Space heating	3.32	0.39	0.58	0.28
Water heating	1.15	0.36	0.13	0.05
Appliances	0.37	1.67		0.05
Air-conditioning		0.62		
Refrigerators		0.53		
Lighting		0.31		

[a]LPG stands for liquefied petroleum gas, most commonly propane.
Source: Energy Information Administration, Washington, D.C.: U.S. Department of Energy, 2004.

If we could build an ideal house that had perfect thermal insulation in the walls, windows, roof, and floors and had no leakage of air into or out of the house, and no radiated energy coming in or going out of the windows, no energy source would be needed to keep the inside of the house at a constant temperature. Of course it's impossible to build such a house, and eliminating fresh air would make it impossible to live in. Any attempt to approach this ideal is limited in effectiveness by the cost and inconvenience of unreasonable amounts of insulation. Nevertheless, many well-designed houses have been built that require no separate heating system. They are heated adequately by existing internal sources of heat energy such as lights and cooking stoves.

The paths of heat loss for a typical home are shown in Figure 7.1, where losses by thermal conduction are shown through the windows, walls, and so forth. The figure also shows the amount of heat energy lost by air exchange through the cracks found in any typical home. As a general rule, 30 to 40% of the total heat loss is by this convective process.

7.2.1 *Thermal Insulation*

Let's first look at how insulation reduces heat loss by conduction. The number of Btu/hr conducted outward through a surface such as a wall is given by:

$$Q/t = kA(T_i - T_o)/l,$$

where k is the thermal conductivity of the wall material in units of

$$(\text{Btu} \cdot \text{in.})/(\text{hr} \cdot \text{ft}^2 \cdot °\text{F}).$$

A is the area of the wall surface in square feet, l is the thickness of the wall in inches, and $(T_i - T_o)$ is the temperature difference in degrees Fahrenheit between the inside and outside.

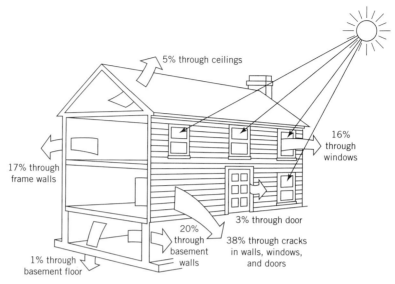

Figure 7.1 Heat losses from a typical conventionally insulated house. Solar energy entering through the windows can be a significant source of heat energy.

To facilitate calculation of the amount of heat lost per heating season, a quantity known as the number of *degree days* has been tabulated for various locations. This number is taken as a measure of the severity of the winter, but it does not consider wind or sunshine or several other weather factors. It is only a measure of the average outside air temperature. For each day of the heating season, the number of degree days accumulated is given by the average temperature difference between inside and outside, with the inside temperature taken to be a constant 65°F. For an entire heating season, the total number of degree days is obtained by summing these degree days for the individual days of the heating season. Table 7.2 lists the heating season degree days for a number of locations in the United States.

Example 7.1

(a) Calculate the number of degree days accumulated in one day in which the average outside temperature is 17°F.
(b) Calculate the number of degree days accumulated during a 150-day heating season in which the average outside temperature is 17°F.

Solution

(a)

$$\text{Degree days} = 1 \text{ day} \times (65°F - T_o)$$
$$= 1 \text{ day} \times (65°F - 17°F) = \textbf{48 degree days}$$

Table 7.2 Heating Season Degree Days

Location	Degree Days
Barrow, Alaska	19,990
Fairbanks, Alaska	14,160
Park Rapids, Minnesota	9,550
Portland, Maine	7,680
Madison, Wisconsin	7,300
Chicago, Illinois	6,310
Reno, Nevada	6,040
Denver, Colorado	5,670
New York, New York	5,050
Newark, New Jersey	4,810
Washington, D.C.	4,330
San Francisco, California	3,070
Birmingham, Alabama	2,780
Los Angeles, California	2,020
Tucson, Arizona	1,780
Miami, Florida	170

(b)

Degree days per heating season = 150 days $\times$ (65°F $-$ T_o)

$\qquad\qquad\qquad\qquad\qquad\quad$ = 150 days $\times$ (65°F $-$ 17°F) = **7200 degree days**

The thermal conductivity, k, is a property of a material without reference to its dimensions. In practice, one often knows both the material and the dimensions of a wall, for instance. In this case it is convenient to use the specification R-value, which incorporates both the wall thickness and the material. The R-value is often labeled on insulation and other building materials and it represents the thermal resistance of a material. For any slab of material it is given by

$$R = \frac{l}{k},$$

where l is the thickness of the slab and k is its thermal conductivity. This number can be obtained from standard reference books. The units of R are commonly

$$\frac{\text{hr} \cdot {}^\circ\text{F} \cdot \text{ft}^2}{\text{Btu}}.$$

For a wall or any other surface built up of a layered series of material each having thickness $l_1, l_2, l_3, \ldots$, and conductivity $k_1, k_2, k_3, \ldots$, the total R-value for the combination is given by the sum of the individual R-values, or

$$R_T = R_1 + R_2 + R_3 + \ldots$$

$$= \frac{l_1}{k_1} + \frac{l_2}{k_2} + \frac{l_3}{k_3} + \ldots.$$

Once the total R-value is determined, then the conductive heat loss in Btu is given as

$$Q = \frac{A}{R_T}(\Delta T)(t),$$

where A is the area of the surface, ΔT is the temperature difference, and t is the time in hours.

For a period of time characterized by a known number of degree days, the heat loss in Btu is given by

$$Q = 24\frac{A}{R_T} \times (\text{degree days})$$

where the number 24 accounts for the number of hours in a day.

Table 7.3 lists some typical R-values for materials used in building construction. In some cases it is given in terms of R-value per inch of thickness. In addition to the R-values of the actual materials, there are insulation properties in the static air layers formed on the surfaces of the materials, such as on the inside and outside surfaces of a wall, window, or whatever other surface is being considered. When materials are in contact, the surface air layer effect is not taken into consideration because the air layer must be several millimeters thick to be effective. Typical R-values for these air layers are indicated in Table 7.3. However, they are not exact, because the R-value of the outside layer varies considerably depending on factors such as wind velocity, for example. The thermal insulation properties of glass windows are almost entirely due to these air layers on the surface of the glass. The effective R-value for a single-pane window is obtained by adding the inner and outer air layer R-values to that of the glass alone to obtain $0.68 + 0.03 + 0.17 = 0.88$, using the units of Table 7.3. For a double-pane window, one would add together one outer layer, three inner layers, and two glass layers to get an R-value of 2.27. Two of the inner layers are on the glass surfaces facing one another. In practical applications one can often use the approximation that a single-pane window has $R = 1$, a double-pane window, $R = 2$.

Example 7.2

(a) Calculate the heat loss in Btu for one heating season having 6000 degree days for an ordinary insulated 2 × 4 stud wall with 3/4 inch of insulating sheating on the outside with an R-value of 2.06, fiberglass insulation 3 1/2

Table 7.3 *R*-Values for Building Materials

Material	R (hr·°F·ft^2/Btu)
Plywood, 3/4 inch	0.94
Insulating sheating, 3/4 inch	2.06
Fiberglass, per inch (battens or loose)	3.70
Mineral wool, per inch (battens or loose)	3.70
Polystyrene board, per inch	5.00
Polyurethane board, per inch	6.25
Urea foam, per inch	5.25
Gypsum board, 1/2 inch	0.45
Poured concrete, per inch	0.08
Brick, common, per inch	0.20
Stone, per inch	0.08
Concrete block, sand and gravel, 12 inch	1.28
Concrete block, cinder, 12 inch	1.89
Outside air layer	0.17
Inside air layer	0.68
Glass, 1/8 inch	0.03

inches thick, and a 1/2 inch gypsum board on the inside. The wall is 20 feet long and 8 feet high. For this problem, ignore the conduction directly through the studs; consider the entire wall to be of insulated construction.

(b) Repeat the calculation for the same wall without the fiberglass insulation.

(c) Calculate the savings of fuel cost in dollars for one heating season resulting from the addition of the fiberglass insulation. Use a furnace efficiency of 75% and a natural gas cost of $13.00 per million Btu.

Solution

(a) First calculate the total *R*-value.

Outside air layer	0.17
Insulating sheating	2.06
Fiberglass (3.5 in. × 3.70/in.)	12.95
Gypsum board, 1/2 inch	0.45
Inside air layer	0.68
Total *R*-value	16.31

Then

$$Q \text{ (Btu)} = \frac{24 \text{ (hr)} \times A \text{ (ft}^2) \times \text{(degree days)}}{R \text{ (hr} \cdot \text{ft}^2 \cdot °F/\text{Btu)}}$$

$$= \frac{(24)(20 \times 8)(6000)}{16.31} = \mathbf{1.41 \times 10^6 \text{ Btu}}$$

(b) With no fiberglass insulation, the R-value of the wall is 3.36. Then

$$Q \text{ (Btu)} = \frac{24 \text{ (hr)} \times A \text{ (ft}^2) \times \text{(degree days)}}{R \text{ (hr} \cdot \text{ft}^2 \cdot °F/\text{Btu)}}$$

$$= \frac{(24)(20 \times 8)(6000)}{3.36} = \mathbf{6.86 \times 10^6 \text{ Btu,}}$$

a conductive heat loss about five times larger than with the insulation.

(c) The value of the natural gas saved in one season will be

$$\text{Savings (\$)} = \frac{\text{heat energy saving in Btu} \times \text{cost/Btu}}{\text{efficiency of furnace}}$$

$$= \frac{(6.86 - 1.41) \times 10^6 \text{ Btu} \times \$13.00/10^6 \text{ Btu}}{0.75} = \mathbf{\$94.47}$$

Up until about 30 years ago, homes in many areas of the United States were commonly built without insulation in wall spaces. It was a feature that the prospective home buyer could not see and for the most part did not care about. Fuel was considered too cheap to save.

The cost of the fiberglass insulation alone in Example 7.2 is about $0.25/ft^2, or $40 for the wall in 2005. Thus, the payback time in fuel savings is less than a year, and for new construction the insulation is obviously an excellent investment. Even if the installation charges doubled the cost, it would still be a good investment. The installation of insulation is not always so favorable economically, especially in existing buildings. In some cases access to the place where the insulation is to be installed is extremely difficult and may require removing inner or outer wall surfaces. For certain types of construction, insulation can be blown into building cavities. In many houses the most easily accessible area to insulate is the attic, where the insulation can often be simply laid between the joists with little specialized labor required. The amount of insulation needed is determined by considering both climate severity and insulation cost. These considerations are included in local building codes which specify R-values for ceilings, walls, and floors, as well as whether single-, double-, or even triple-pane windows are required. Some recommended R-values are given in Table 7.4.

Window design can have a considerable effect on heat retention. A double-paned window loses only about half the heat energy that is lost by a single-paned window. It is best if the two glass sheets are separated by a centimeter or more because the boundary air layers are typically a few millimeters thick. In addition

Table 7.4 Recommended R-Values (hr·ft^2·°F/Btu)

Degree Days	Ceilings	Walls	Floors
Above 7000	38	17	19
3500–7000	30	12	11
2500–3500	22	11	0
Under 2500	19	11	0

to going from single- to double-paned windows, thermal insulation can be provided by draperies or other window coverings that can be closed, especially at night when there is no solar energy to be gained.

7.2.2 Air Infiltration

In a typical house, infiltration of outside air is sufficient to account for a complete change of air in a house about once per hour. This effect can account for a third of the total heat loss. The cold air comes in, the warm air goes out. Much of this air exchange occurs around doors and windows, or through leaky siding and other cracks in the structure. Fireplace chimneys and furnace vents can also provide a path for the escape of warm air.

Much of this leakage can be reduced by simple and inexpensive caulking and weather-stripping, by adding automatic flue dampers, and by closure of fireplace openings and chimneys. Flue dampers close off the chimney when the furnace and water heater are not running, and open when the furnace or water heater is running. They must be perfectly reliable to avoid the hazard of carbon monoxide accumulation in the living space. By these simple means, it is possible to reduce the air exchange rate to about once every five hours and the associated heat loss to about 10% of the total for the house.

A house that is sealed too tightly presents possible health hazards from radon, carbon monoxide, and nitrogen dioxide, as well as from building materials made with formaldehyde.

Radon is a radioactive gas that is a daughter activity of naturally occurring uranium. Because all soil contains uranium to some degree, radon gas can diffuse from the soil under a house into the living space, where it can accumulate at concentrations much higher than in the outside air. Based on the experience of uranium miners, it is known that inhalation of radon can be associated with lung cancer.

Carbon monoxide gas can be formed whenever any fuel is burned. If the combustion takes place in a properly operating heating stove, fireplace, furnace, or water heater, the carbon monoxide will be vented to the outside. On the other hand, the use of natural gas in cooking stoves or ovens can lead to the accumu-

220 Chapter 7 **Energy Conservation**

lation of carbon monoxide in the house. The hemoglobin in blood has a great affinity for carbon monoxide that effectively blocks the uptake of oxygen by the blood flowing through the lungs, leading to illness or even death.

These fumes, and others such as from urea-formaldehyde foam insulation, have not normally been major problems because air infiltration has been sufficient to keep their concentrations down. However, if a house is sealed too tightly without thought given to the possible presence of these gases, significant health problems could be encountered.

A device called an *air-to-air heat exchanger* can relieve the problem of unhealthful indoor concentrations of harmful gases without excessive waste of energy. As shown in Figure 7.2, the heat exchanger warms the incoming air, stripping the heat from the outgoing air to do it. The heat energy is transferred from warm air to cold air by conduction through a thin barrier between the outgoing and incoming air streams, bringing the incoming air up to the house temperature before admitting it to the living space. If the barriers are somewhat porous to moisture, as would be true for a paper barrier, the air's humidity can also be transferred to the incoming stream of fresh air.

This transfer of humidity is an important consideration for energy conservation. The outside air in winter is usually much drier than inside air. If outside air at 40°F and a relative humidity of 60% is brought into a house where the temperature is 70°F, its relative humidity will then be only 20%. This is usually considered to be too dry for comfort and health, possibly requiring water to be

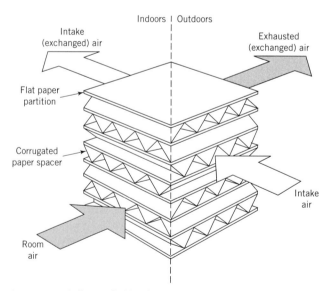

Figure 7.2 A commercially available air-to-air heat exchanger. The air flows are both driven by an electric fan. Both heat energy and moisture are interchanged through the paper partitions in this device to effect a maximum energy savings. (*Source*: From Mitsubishi Corporation.)

evaporated in a humidifier to bring the relative humidity up to a proper level. Under ordinary conditions, the humidified air then exfiltrates from the house, requiring that replacement air be continuously humidified. Of course, the evaporation of water requires energy (970 Btu/lb of water or 540 cal/g of water). This energy must be provided by the humidifier or the furnace. For example, if a house has a volume of 15,000 ft^3 and exchanges air once per hour with the outside, as is typical, it will require 3800 Btu/hr, 24 hours/day, to increase the relative humidity from 20 to 40% and maintain it at that level.

7.2.3 *Furnaces, Stoves, and Fireplaces*

Most living and working spaces are maintained at a comfortable temperature in winter by a furnace that burns natural gas or fuel oil. The furnace heats air or water that is then circulated around the building to maintain a reasonably uniform temperature throughout. The rated efficiency of furnaces is usually in the range of 60 to 90%, but for older units it may be as low as 50%. The energy not used for heating the air or water, and thereby wasted, escapes mainly up the chimney. A diagram of a furnace is shown in Figure 7.3. The electric energy used to operate the blowers or pumps must be taken into account when calculating the efficiency, but it is usually not an important factor.

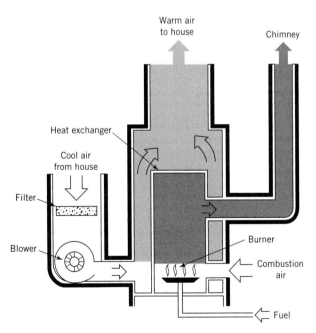

Figure 7.3 A hot-air furnace that can be fired by natural gas, propane, or fuel oil. Air circulation is forced by the electric blower.

Electric resistance heating is sometimes used; it has an efficiency of essentially 100%, measured from meter-to-heater, and is usually simple and inexpensive to install, and it requires no chimney. The drawback is the cost of operation, since in most areas of the country electric energy costs the consumer at least twice as much as energy from natural gas or oil. This should not be surprising because the power plant can produce electric energy from fossil fuel energy only at about 35% efficiency. Some of the electric energy is also lost in the transmission lines. This can be significant over long distances.

Another approach to using electric energy for heating involves electrically powered heat pumps, as discussed in Chapter 3. In the parts of the country where winter temperatures are not too low, the coefficient of performance of heat pumps can be large enough to overcome the energy cost disadvantage of electric resistance heating. Under good conditions, a heat pump can transfer several times more energy to the inside air than is consumed in operation. The same device also functions as an air conditioner when cooling is needed, thus helping to justify the capital investment, which is larger than for resistance heating.

Probably the single most important factor in conserving energy going into space heating is the setting of the thermostat. For many years, up until widespread recognition of the need for energy conservation, Americans regarded 72°F as the normal indoor temperature for living and working spaces. Because of rising fuel bills, 65 to 68°F is now more typical. There has been a widespread belief that one should not bother to set the thermostat back too much at night, or when a building is unoccupied, as it will require more energy to bring the house back up to temperature than was saved during the setback. There is no thermodynamic justification for such a statement. The fact is, of course, that the lower the inside temperature, the lower the rate of fuel usage. The main considerations concerning thermostat setback are possible discomfort during the time it takes the house to rewarm, and not having water lines freeze. The first of these can be countered by having a thermostat programmed to return to a normal daytime setting in advance of the time the greater warmth is wanted. For a properly insulated building, freezing of pipes should not be a problem down to an inside temperature of 55°F or so.

Newer building codes now require that air for combustion be provided directly to the furnace from the outside. Without this provision, the combustion air is often taken from the living space, and goes up the chimney. This creates a slight vacuum in the house, causing cold air to infiltrate into the living space from the outside to replace the warm air that went up the chimney.

Recovery of heat energy that would otherwise go up the stack is the essential basis of newer furnace designs having efficiencies of up to 95%. A conventional older furnace may have a stack temperature of 300°F, or higher, to provide sufficient draft and avoid condensation problems in the flue. Some of the newer designs include recovery of the heat energy in the water vapor formed during combustion by cooling the stack temperature, and thus forcing the vapor to condense to a liquid. This happens at any temperature below the boiling point of water, 212°F. It is possible to have stack temperatures as low as 100°F in these designs, eliminating the need for a conventional chimney. It is not unusual now to see plastic pipe used to conduct the cool exhaust gases to the outside. The water from the condensation goes down a floor drain. An electric blower provides the draft for combustion.

By federal regulation, all furnaces sold in the United States after 1991 are required to have an Annual Fuel Utilization Efficiency (AFUE) of at least 78%. Short of buying a new furnace, there can be less complicated and inexpensive ways of improving the efficiency of a heating system. Insulating the ducts or water pipes going through an unheated crawl space would be one example. Providing external combustion air has already been mentioned. Being sure that the pilot light is not set too high could be another energy-saving step because it reduces the amount of gas burned in the pilot flame and also reduces the draft of air continuously going up the flue even when the furnace is not running. Replacement of the pilot flame by an electronic ignition is even more effective.

The use of wood-burning stoves and fireplaces is not an energy conservation measure unless they are used to heat small areas of a house while the main furnace thermostat is turned down. The main effect of wood burning is to transfer the energy source from oil or natural gas to wood. If one has access to a supply of firewood at little cost, then money can be saved over the use of more expensive fossil fuels, but more total fuel energy will probably be used to heat the same area because of the lower efficiency of wood-burning stoves or fireplaces. Modern stoves have efficiencies of 40 to 65%.

Open fireplaces often have negative efficiency; they remove more heat from a house than they provide. Heated indoor air is drawn into the fireplace opening and then out through the flue. The effect is that while one may feel warm near the fireplace, the other parts of the house become colder. This unwanted effect can be minimized by using doors over the fireplace opening, but sufficient air must be admitted for combustion. The doors become hot from the fire, thus adding heat to the room. Home fireplaces are sometimes constructed with convective passages around the back of the firebox so that cool air is drawn in near the floor, heated, and then circulated back into the room. Such a fireplace, with convective circulation and doors across the opening, is likely to have a positive efficiency.

Firewood provides heat energy at the rate of about 12 to 30 million Btu per cord. A cord is a stack having a volume of 128 ft^3. Most commonly it is a pile 4 feet high, 8 feet long, and 4 feet from front to back. Wood of higher density, such as oak, has more heat per cord than wood of lower density such as cedar or fir. When the heat content for dry wood is considered on the basis of weight rather than volume, almost all wood is the same—about 8600 Btu/lb.

7.2.4 Solar and Other Sources of Heat Energy

In considering the areas for which energy conservation measures might be a good investment for a given building, a good first step is often an *energy audit*. This could be a process of adding up all the space heating energy losses for a year and then trying to balance those losses against all known sources of heat energy. The main energy input, fuel for the furnace, can be obtained from the statements sent out at the end of the month by the gas utility or the fuel oil provider. Other heat sources are electric appliances, lighting, people, pets, solar energy, and possibly wood burning. These miscellaneous sources typically add up to about 7% of the 1 million Btu total heat energy needed for a home on a winter's day. The wood-burning component can be estimated from information in the preceding section. The solar energy input can be important even if deliberate solar collection measures have not been incorporated into the building.

A reasonable estimate of the solar energy input can be calculated from the information given in Chapter 4. Although all windows receive some input during daylight hours from diffuse light, the direct sunlight through south-facing windows is usually the major component. The insolation values given in Table 4.2 for vertical south-facing windows must be multiplied by the fraction of time the sky is clear during the heating season. For many areas this fraction will be in the range of 60 to 75%. In addition, the transmission of sunlight through the glass window must be taken into account (about 86% transmission per pane), as well as the fraction of the transmitted light that is converted to heat. The conversion efficiency depends on the nature of the material the sunlight encounters in the room and hence varies widely, but it generally ranges from about 60 to 90%. Whatever solar energy is not reflected back out through the windows is converted to heat inside the building.

Example 7.3

Estimate the solar energy contribution for one heating season to the heating of a home at a latitude of 40° N that has 300 ft^2 of double-paned south-facing windows. Assume clear skies 75% of the time, an absorption efficiency of 90%, and clear-sky insolation from Table 4.2.

Solution

Estimate the heating season to be about six months long, starting in late October. From Table 4.2 the clear-sky solar energy on a vertical south facing surface from October 21 to April 21 can be estimated to be about 1650 Btu/ft^2 per day for the 180 days in this period. Multiplying these two numbers together gives about 0.3×10^6 Btu/ft^2 for the entire heating season. Next multiply this by 0.75 for the clear-sky fraction, by $(0.86)^2$ for transmission through the two panes, by 0.90 for absorption, and by the window area.

0.30×10^6 Btu/ft$^2 \times (0.75) \times (0.86)^2 \times (0.90) \times 300$ ft^2
$= \mathbf{45 \times 10^6}$ **Btu per heating season**

The total heat energy needed per year for a reasonably well-insulated single-family house in a region with 6000 degree days is about 90×10^6 Btu. In the preceding example, solar energy provides about half of this. Without the benefit of nighttime draperies or other insulating window covers, the heat losses through the same windows would be about 25×10^6 Btu per season. In general, south-facing windows can be an important component of a well-designed, energy-efficient house. Figure 7.4 shows an example of a room warmed by south-facing windows.

7.2.5 Standards for Home Heating

How many Btu should be needed to heat a house? The answer obviously depends on the climate and on the size of the house. One method of treating the matter of house size is to compute the heat energy needed relative to the number of square feet of floor area. Figure 7.5 contains data for various homes on the basis of the number of Btu per square foot (in units of 10^6 Btu/1000 ft^2) as a function of the degree days for the heating season. The data, compiled by a group at Lawrence Berkeley Laboratory, show the results for a group of typical houses and some building standards. For a particular house, a calculation shows how the space heating requirements compare with those shown on the plot. This

Figure 7.4 Proper window orientation is important for maximum use of solar energy for space heating. With sufficient heat storage, a room can be comfortably warmed even after the sun is no longer shining. A wood-burning stove can shift the fuel supply from fossil fuels to wood and offer a savings in heating expense if a supply of firewood is available. (*Source*: ©pamm Mcfadden/Dept. of Energy/National Renewable Energy Laboratory)

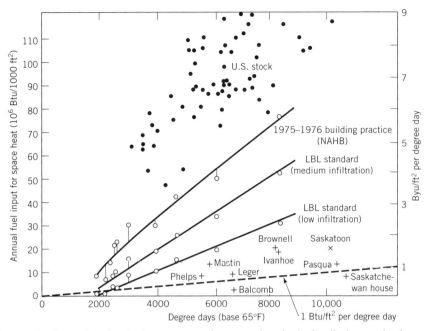

Figure 7.5 Space heating fuel energy requirements for single-family homes in the United States for one year in terms of 10^6 Btu/1000 ft^2 as a function of degree days per heating season. The black dots show typical requirements in various parts of the country. Also shown are standards recommended by the National Association of Home Builders. The LBL standards refer to building performance standards put forth by Lawrence Berkeley Laboratory with 0.6 air changes per hour (medium infiltration) and the lower line with 0.2 air changes per hour. The individual homes at the bottom of the plot illustrate what can be accomplished in specially designed, well-insulated homes in various locations. It appears that 1 Btu/ft$^2 \cdot$ degree day is an achievable goal. (*Source*: Lawrence Berkeley Laboratory Report LBL-11900.)

figure illustrates the dramatic difference between the energy efficiency of our customary building practices and what can be achieved in the way of energy conservation with improved building practices.

7.3 *Water Heaters, Home Appliances, and Lighting*

7.3.1 *Water Heating*

Water heaters use about 10 to 20% of the energy consumed in a typical home. In most cases, natural gas or electricity is used to heat a tank of 30 to 50 gallons of water to a temperature in the range of 120 to 140°F. The water is maintained at this temperature and continually loses heat through its thermal insulation. Gas water heaters commonly have an uninsulated flue going up through the center of the tank and connected to the chimney. A constant draft of air, induced by the pilot light and the water's warmth, flows up through this flue and carries off heat from the water. The pipes from the water heater to the plumbing also conduct heat away from the unit. Because of these losses, as well as the losses through the insulated surfaces, a steady expenditure of energy is required to maintain the water temperature, even if hot water is not being used. Energy conservation steps that can be easily implemented include the following:

(a) Lowering of the heater's thermostat to 120°F reduces the losses from those experienced at higher temperatures and still provides water at a sufficiently high temperature for household use.

(b) Insulation may be added to the tank. This is especially effective with an electric water heater where there is no flue but less effective for standard gas water heater. Insulating links can also be added to the connections of the pipes to the heater, and the hot water pipes can also be insulated.

(c) Reduction of the amount of hot water used obviously saves energy. This becomes automatic with the use of flow-restricting shower heads and faucets.

(d) Electronic igniters can eliminate the waste of energy due to the pilot light.

(e) A flue damper can be installed. For a fail-safe arrangement, this damper must be designed to open whenever the flame comes on.

7.3.2 *Appliances*

The electric energy used in one year by various appliances is listed in Table 7.5. Aside from electric water heaters in those homes that have them, the main consumers of electric energy are refrigerators, clothes dryers, and air conditioners. Some of our widely scorned symbols of energy waste such as electric carving knives and toothbrushes use a negligible amount of electric energy. The column marked CO_2 emissions will be considered when the greenhouse effect is discussed in Chapter 10.

For a number of years it was the practice of refrigerator manufacturers to minimize insulation thickness in order to provide a larger interior volume for given exterior dimensions. This trend finally led to the use of so little insulation that the outside walls would condense moisture from the air. To prevent this problem, electric heaters were put in the exterior walls, thus increasing the thermal gradient across the insulation, increasing the rate of heat leakage into the interior, and causing the refrigerator to run more frequently. This is a remarkable example of total disregard for energy conservation, an attitude that is now changing in response to a greater awareness of the need for energy efficiency and in response to government-directed energy efficiency requirements.

The example of refrigerators is interesting as it illustrates what can be achieved by pushing for increased efficiency. In 1972 it took about 1700 kWh/yr to operate an average American refrigerator, at an annual cost of about $130. Since then, there has been a series of standards put in place by the federal gov-

Table 7.5 Electric Energy Consumption in the Home

Use	kWh/yr	$/yr[a]	CO_2 Emissions, lb/yr[b]
Air conditioner (room)	1,070	89	2,140
Air conditioner (central)	3,230	268	6,460
Coffee maker	100	8	200
Clothes dryer	895	74	1,790
Clothes washer (incl. hot water)	1,080	90	2,160
Dishwasher (incl. hot water)	935	77	1,870
Electric blanket	93	8	186
Furnace fan	600	50	1,200
Home computer	116	10	232
Lighting	844	70	1,688
Microwave oven	167	14	334
Kitchen range	674	56	1,348
Refrigerator	1116	93	2,232
Stereo and radio	75	6	150
Television	151	13	302
Television, turned off	33	3	66
Toaster oven	50	4	100
Vacuum cleaner	25	2	50
VCR	40	3	80
Waterbed	960	80	1,920
Water heater	5,300	440	10,600
Water heater (standby losses)	795	66	1,590

[a]Based on $0.083/kWh.
[b]Based on 2 lb CO_2/kWh.
Sources: Adapted partially from *Homemade Money—How to Save Energy and Dollars in Your Home*. Richard Heede and staff of the Rocky Mountain Institute. Amherst, NH: Brick House Publishing Company, 1995; U.S. Department of Energy, Office of Energy Efficiency and Renewable Energy, 2005.

ernment and the state of California that have lowered the energy demand of the standard refrigerator. In 2001, the energy standard for a new 22 cubic foot refrigerator, larger than what is now in the average American household, was down to 535 kWh/yr, and the typical new refrigerator used only 500 kWh/yr. The operating cost was down to about $40/yr, a savings of nearly $100/yr from 30 years earlier. This is a considerable savings over the lifetime of the average refrigerator. If an older refrigerator is being replaced, within a few years the savings in electricity expense will exceed the cost of the new refrigerator.

Clothes dryers consume about 15% of the electric energy used in a household. Gas-fired dryers cost less to operate, but the energy consumption is about the same. Hanging clothes outside to dry is a very real energy conservation step. If clothes are hung inside to dry during the heating season, the energy comes from the fuel for the furnace rather than from a dryer. Venting the hot air from an electric dryer into the interior of a house can conserve energy but may present other problems with lint and excess moisture.

Air-conditioning draws a large and growing fraction of our electric energy as more and more people move to the sunshine states. About 76% (81 million) of U.S. households now have air conditioners. More than 80% of new homes have central air conditioners. It's clear that the cost of energy is not seriously limiting the use of air-conditioning. A sixth of the nation's electricity consumption now goes to air-conditioning, residential and commercial. At peak demand periods in the summer months, as much as 40% of our electrical power goes to air-conditioning.

As with heating systems, there is a wide range of efficiencies for air conditioners now on the market. Central air conditioners are rated by their seasonal energy efficiency ratio (SEER), which is the seasonal cooling provided in Btu divided by the energy input in watt-hours for the average U.S. climate. In this context, it is interesting to note that 3.4 Btu is the energy equivalent of 1 Wh. Some years ago, residential central air conditioners were typically rated at a SEER of 6 or 8. The national standard now requires a minimum SEER of 10, and this will be raised to 13 starting in January 2006. Residential central air conditioners are now available with a SEER as high as 19.5. The rating of room air conditioners does not consider their performance over the entire cooling season. The minimum standard for the room air conditioner energy efficiency ratio (EER), again in Btu/Wh, is about 9.7 for units of the most common type, but units with ratings of 11 and higher are available. The cost of the energy to operate air conditioners is sufficient to motivate consumers to pay serious attention to the house design, and to good thermal insulation and to limiting air exchange with the outdoors. All factors that need consideration for the heating of interior spaces should be weighed for space cooling.

The federal government now requires that an energy guide be posted for any new appliance being sold. As shown in Figure 7.6, the guide lists the cost in dollars to run the appliance for one year, based on the stated rates for electricity and natural gas. Such information can be important to a consumer weighing the possible higher initial cost of an appliance that consumes less energy against an appliance having a lower initial cost but higher energy consumption.

It is commonly assumed that when you click off the television with the remote control it stops using energy. This is usually not true, as some small amount of electrical power continues to be used to maintain the memory of certain settings, to keep the digital clock operating, and to keep the remote control alert. Some sets use electric power to keep the tube warm and lessen the time for a picture to appear when the set is turned on. Other household items such as computers, cable TV boxes, DVDs, microwave ovens, the furnace thermostat, the doorbell, and video games also draw power even when turned off. The small

Refrigerator-Freezer
With Automatic Defrost
With Side-Mounted Freezer
With Through-The-Door Ice Service

LG
Model LRSC26930SW
Capacity: 25.5 Cubic Feet

Compare the Energy Use of this Refrigerator with Others Before You Buy.

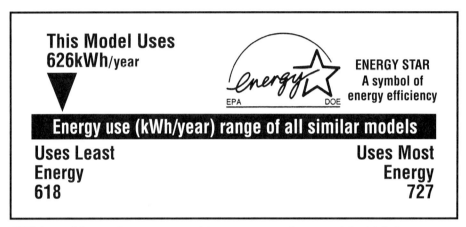

kWh/year (kilowatt-hours per year) is a measure of energy (electricity) use. Your utility company uses it to compute your bill. Only models with 25.5 to 26.4 cubic feet and the above features are used in this scale.

Refrigerators using more energy cost more to operate. This model's estimated yearly operating cost is:

Based on a 2001 U.S. Government national average cost of 8.29¢ per kWh for electricity. Your actual operating cost will vary depending on your local utility rates and your use of the product.

Important: Removal of this label before consumer purchase is a violation of Federal law (42 U.S.C. 6302).

3850JZ1114B

Figure 7.6 An energy guide of the type attached to all new appliances. This information can be important to a consumer choosing a new appliance.

transformers that plug into wall sockets to provide 12-volt power for radios, tape recorders, electric toothbrushes, battery-powered tools, and so forth also draw power continuously. It is estimated that a typical home uses about 50 watts of electric power drawn continuously for these purposes even with all the power switches in the off position. This is referred to as electricity "leakage." Although this level of power is trivial to the individual homeowner, it adds up when totaled for all the households in the nation and all the hours in the year. One estimate is that such leaks in the United States consume all the electricity generated by four large power plants, and that this leaked electricity has a market value of about 3 billion dollars per year. The only way one can be sure that an appliance is really off is to unplug it from the wall socket.

7.3.3 Lighting

About 20 to 25% of the electric energy generated in the United States goes to provide lighting. Thus sizable savings can be accomplished with energy conservation measures for lighting. Standards for light levels in public school classrooms have increased from 20 lumens (lm) per square foot in 1952 to 60 or more lumens per square foot today. Although 60 lm/ft^2 is a reasonable level for reading, many modern office buildings now have lighting levels in the range of 80 to 100 lm/ft^2, including corridors and stairways. To make matters worse, some buildings have master light switches that turn on whole floors at a time. It is not unusual to see an entire multistory office building glowing with light late at night when only a few custodians are present.

Table 7.6 lists the efficacy of several types of lighting devices in lumens per watt. A comparison of incandescent and fluorescent lights is of particular inter-

Table 7.6 Typical Characteristics of Selected Lamps

Light Source	Efficacy (lumens/watt)	Rated Life (hours)
Standard incandescent	5–18	750–1000
Tungsten–halogen	18–24	2000–3000
Tubular fluorescent	65–100	20,000
Compact fluorescent	40–70	10,000
High-pressure sodium	45–110	20,000
Low-pressure sodium	80–160	20,000
White LED (today)	30	20,000
White LED (future)	150–200	100,000
Sunlight	92	
Open gas flame	0.2	
Candle	0.15	

est for the home. A fluorescent light produces about five times more lumens per watt than does an incandescent light, resulting in a 15-watt fluorescent bulb producing the same light output as a 75-watt incandescent bulb. The energy saving alone makes the fluorescent light attractive, but in the past some people objected to the quality of light produced by the fluorescent lamps. This problem has been addressed by the manufacturers, and compact fluorescent lamps with excellent light quality and the common screw base are now commonly available. Some examples are shown in Figure 7.7. In addition to the savings in energy, the fluorescent lamps last about 10,000 hours, about ten times longer than conventional lightbulbs. This is an important factor for commercial and institutional use where there is an appreciable labor cost associated with replacing lightbulbs. The lower energy cost and the longer life more than make up for the greater initial cost of the fluorescent units. (See problem 17 at the end of this chapter.) In addition to the measurable savings to the consumer, it must be recognized that the use of a lightbulb with higher efficiency reduces the environmental impacts and resource depletion associated with energy production. This may be the most important consideration.

The tungsten–halogen lamp, a type of incandescent lamp commonly called a halogen lamp, is finding widespread use. It is 10 to 50% more efficient than an ordinary incandescent lamp, has excellent color quality, is compact, and can be dimmed. It is substantially more expensive than a standard incandescent lamp of the same output.

The light-emitting diode (LED) is a good-news emerging lighting technology. We often see LEDs now, in different colors, in exit signs, as arrays in traffic signal lights, as well as in vehicle taillights and turn indicators. We even find

Figure 7.7 Examples of compact fluorescent bulbs designed to replace ordinary incandescent lightbulbs, with a large savings in energy consumption. (*Source*: ©Paul Kevin Picone/PiCorp. Used with permission by OSRAM SYLVANIA, Inc.)

them in flashlights. In these uses they are inexpensive, durable, and have good brightness. Their use in general lighting applications is not yet common. Their efficacy, as seen in Table 7.6 for the white lighting needed for general illumination, is still not the equal of fluorescent lights, but laboratory developments are producing higher efficacy and lower costs at a remarkable pace. The numbers shown for future characteristics are reasonable estimates put forth by the Sandia National Laboratory of what might be expected within a decade or so. An efficacy at the level of the estimated future 150 to 200 lumens per watt and the very long lifetime will be revolutionary. The 20% of our electricity that now goes for lighting will come down, costs of maintenance of lighting systems will be reduced, and even the wiring requirements for lighting circuits will be lessened.

Many minor steps can be implemented to conserve the electric energy used in lighting. In addition to choosing light fixtures of high efficiency as noted earlier, lighting design can focus on providing light at the site of a task rather than throughout a large room or large area. Switches can be provided so that lights can be controlled individually; too often many lights are ganged together on one switch. Corridors and stairways can have their light levels at a much lower level than in work areas. Lights can be on automatic controllers that turn them off when no one is in the room, or users can be counseled to turn off unneeded lights. In many buildings, such as large retail outlets, and for many university classrooms, little attention is paid to admitting outside light through windows or skylights. The lights burn brightly even on the sunniest of days, leaving the impression that electricity is too cheap to save. Daylighting measures designed into a building can make a significant reduction in the use of electric lighting. Taken together these small measures can add up to considerable savings in electric energy. Furthermore, in the warm season, use of excessive energy for indoor lighting increases the load on the air conditioner system, causing even more use of electric energy.

7.3.4 *The Energy-Conserving House*

It is difficult to assess accurately the effects on the nation's energy consumption that could be brought about by implementing all or some of the energy conservation measures for the home that have been discussed so far. Houses differ, climates differ, personal habits differ, and, of course, there is the question of how much capital investment can be made or should be made to effect the changes.

A study was carried out for a 1200 ft^2 northern California house to determine, for one explicit case, what the energy savings would be for various levels of investment in the types of energy conservation measures discussed in this chapter. Figure 7.8 illustrates the various steps taken and their cost, as well as the effects on energy consumption. The largest effect was for space heating, which went from 120 million Btu per year to 35 million Btu per year, for an expenditure of $1600, starting with a house having *R*-11 in the ceiling. From this and other experiences, it appears to be relatively easy to reduce space heating requirements by about two-thirds in an average home using straightforward and not unduly expensive measures. Figure 7.8 also shows other conservation meas-

234 Chapter 7 **Energy Conservation**

ures taken to reduce the total energy (gas plus electricity) used in the home each year from about 235 to 100 million Btu with an investment of $2700.

Buildings and appliances now consume about 37% of the nation's 98 QBtu annual energy budget, amounting to 36 QBtu/yr. If the 1/2.35 ratio for the house of Figure 7.8 were applied across the nation, the 36 QBtu could be reduced to

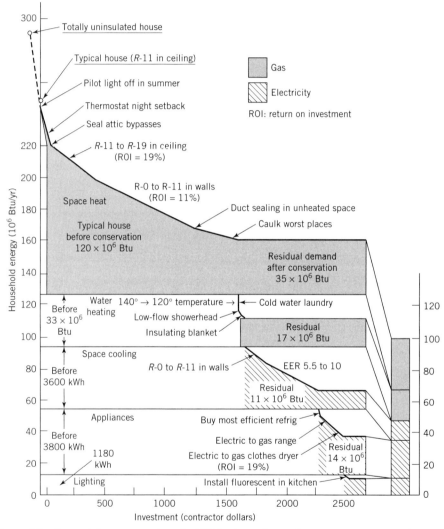

Figure 7.8 Potential energy savings in 10^6 Btu/year for various energy conservation measures for a northern California house (1200 ft^2, 3000 degree days) in terms of the investment required. The shaded areas correspond to energy demand in the form of natural gas, and the lined areas represent electrical energy demand. If all the recommended measures were adopted, the total energy demand per year would be reduced (compared to a totally uninsulated house) by a factor of three for an investment of $2700. (*Source*: Lawrence Berkeley Laboratory Report LBL-11650.)

15 QBtu at a cost of about $150 billion. Although this is far too simple to take as the basis for a national energy conservation plan, it does demonstrate that the potential for energy conservation in the home and in commercial buildings is very real and very important.

7.4 Energy Conservation in Industry and Agriculture

The industrial and agricultural sector of the energy economy is so large and diverse that it is difficult to discuss energy conservation measures briefly in explicit terms. The construction, mining, manufacturing, and agricultural enterprises each consume energy in a variety of ways. However, conservation measures can be divided into some areas which affect all types of activities.

7.4.1 Housekeeping

Housekeeping involves the same kind of considerations and changes that were discussed for homeowners, since the problems of space heating, water heating, appliances, and lighting are also commonly present in industry. Because of the scale, the savings may be impressively large and may even justify employing full-time staff to implement energy conservation measures. In one sense the problems here are more amenable to solution than in the home because larger units are involved and more technical help is available. This will aid in identifying the problem areas, finding solutions, and assessing the results. On the other hand, the average worker in a factory may not be as enthusiastic about shutting off lights and so forth as the individual homeowner who is directly paying the utility bills. Programs of education and persuasion may be needed in an industrial setting.

7.4.2 Waste Heat Recovery and Cogeneration

As discussed in Chapter 3, it is possible, with various heat engines and processes that use heat energy, to recover some of the heat that would otherwise be rejected to the environment. The recovered heat energy is often at a fairly low temperature, or, equivalently, of low grade, but still suitable for space heating, water heating, greenhouses, water preheating, or various other uses. There is a growing trend toward electricity generation on-site because of the recognized value of the heat energy recoverable from the generating facility. This trend is seen in industries, universities, and other large institutions.

7.4.3 Process Changes

In many industrial processes having large energy inputs, such as the smelting of metals (particularly aluminum), uranium enrichment, and the refining of chemicals and fuels, there is opportunity to make fundamental changes in the process that will lead to a reduction in energy use per unit of output. Some of these changes are being forced on industry by increasing energy costs, questions about long-term prospects for energy availability, and market price competition from newer technologies. The capital investment necessary for implementing newer, less energy-intensive technology is often enormous, but so is the potential for energy savings. The ongoing modernization of industry carries with it the opportunity to reduce energy use by introducing better technology as it becomes available.

7.4.4 Recycling

Our throwaway society built up over the past few decades is gradually being replaced by one that emphasizes recycling. Recycling saves basic resources such as aluminum, and at the same time it reduces the problem of overloaded landfills now being encountered by our cities and towns. It is an important step toward energy conservation, and it reduces pollution of our environment. However, we still continue to throw out most materials after their first use—about two-thirds of all aluminum, three-quarters of the steel and paper, and an even higher fraction of plastic.

Recycling can be effective both for industry and for individual households. Table 7.7 compares the energy content of containers for their first use and for use after recycling. From this table it would appear that it takes only a third as much energy to form a new beverage can from recycled aluminum cans as it would to make it from new aluminum stock, but other references indicate that the advantage of recycling is even greater. They conclude that it takes only one-twentieth as much energy to make a can from recycled aluminum as it would to make it from the original bauxite aluminum ore. Of course it would take still

Table 7.7 Energy Consumption Per Use for 12-Ounce Beverage Containers

Type	Btu
Aluminum can used once	7100
Steel can used once	6000
Recycled steel can	4000
Glass beer bottle, used once	3900
Recycled aluminum can	2500
Recycled glass beer bottle	2400
Refillable glass bottle, used ten times	600

Source: John E. Young, *State of the World, 1991*, p. 39, New York: W. W. Norton and Company, Inc, New York, 1991.

less energy to reuse an existing container rather than making a new container from recycled aluminum. The bottle deposit now required by some states encourages the reuse of containers. Owing to aluminum recycling programs, a significant fraction of aluminum beverage cans is put back into the recycling stream. More than 50% of the aluminum in new beverage cans is recycled material. About 35% of the material in new glass bottles is from recycled glass, about 28% of the material in new steel cans has been recycled, and about 35% of our new paper products are made from recycled material, and these fractions are steadily increasing.

There are significant differences in recycling rates from city to city. Most cities now have voluntary programs for aluminum, glass bottles, newsprint, and some plastics. As an example of how well recycling programs are succeeding, the state of Virginia reports an overall recycling rate in 2003 of 30% for all recyclable materials, principally paper, metal, plastic, glass, and wood. Some localities within the state had recycling rates as high as 45%, others were under 5%. Across the nation, recycling rates have barely increased over the past decade. In 1995, the national rate was 26%, and now it is hardly above 30%. Some of the reason for this apparent leveling off may be attributed to a fading commitment on the part of the public, or it might be because the amount of trash we each now discard has grown substantially over recent decades.

It takes only about a third as much energy to produce steel products from scrap as it takes to make them from iron ore. In recent years there has been a move in the steel industry to replace open hearth, coal-fueled furnaces with electric arc furnaces. These new furnaces turn out steel of high quality and use far less energy than the older furnaces. They can also be economically located near major sources of scrap steel rather than near the coal fields. About a third of the steel now produced in the United States is made from scrap using electric arc furnaces.

There are numerous other examples of successful recycling in industry. These programs have generally been effective in saving energy, reducing the waste stream, and reducing harmful emissions while increasing the companies' profits. Notable examples include the recycling of storage batteries, tires, glass, wood chips, and sawdust. What was once waste is now a resource.

7.4.5 *New Developments*

One of the most fundamental changes in the world with respect to energy conservation was the development of the transistor and eventually the microprocessor. Prior to the transistor age, which began in the 1950s, all electronics from radios to televisions to computers made use of devices called vacuum tubes. The vacuum tubes were bulky and consumed a lot of energy. They operated with electron beams drawn off hot filaments powered by electricity, consuming considerable energy in the process. The computing power we now have in a modest desktop PC would have required rack after rack of vacuum tubes a few decades ago, enough to occupy a good-sized room. The vacuum tubes frequently burned out, requiring replacement much as one now replaces a lightbulb, and they generated so much heat that large air-conditioning systems were needed to remove the excess heat.

The electronics we have today require much less energy both in their manufacture and in their operation. The need for energy to power a cooling system is also vastly reduced. Now some 50 years after the introduction of the transistor to the market, vacuum tubes have virtually disappeared.

In addition to the reduction in energy use for operation of electronics of all sorts, introduction of modern electronics into the control of automotive ignition and fuel injection has significantly reduced fuel consumption in transportation. Similar examples of energy conservation by microprocessor-controlled industrial processes are prevalent.

A second important example of energy savings by new technology is the use of optical fibers to replace copper cables in our communications systems. The optical fibers are superior in many ways. They provide faster data transmission, less interference, and a much lower cost for the same capacity. The demand for copper has been greatly reduced, thereby diminishing the environmental impact and energy demand of mining operations. Similar positive effects have been achieved by using microwave links rather than cable connections for long-distance telephone service.

7.4.6 Help from Public Utilities

Until recently, the profits realized by public utilities were directly dependent on how much electricity or natural gas they sold to their customers. This meant that the less efficient the use of energy by the customers, the greater the revenue to the utility. In most cases, the rates charged by the utilities are set by a state public utility commission. In 1989, new regulations were negotiated with some public utility commissions that adjusted rates so that the utilities would not lose profits by promoting efficient use of energy by consumers. One advantage, for example, to an electric utility was that by providing less energy to existing customers, the utility would not have to finance construction of new power plants to provide mandated service to new customers. In addition, the improved efficiency of energy use reduces fuel consumption by customers and by power plants and this reduces environmental pollution. From the customers' point of view, their more efficient energy use means smaller utility bills. Everybody wins.

The utilities have found a number of ways to sell the idea of efficiency rather than continuing in the old practice of selling kilowatt hours of electricity or Btu of natural gas. For example, Niagara Mohawk Power Corporation in New York provided participating households with low-flow shower heads, compact fluorescent lightbulbs, and insulation for electric water heaters and hot water pipes. A rather complicated scheme was developed to calculate the effect of these changes on the utility's charges. For a typical home, the difference between the annual loss of revenue from a household and the utility's cost for the service ($72/yr lost revenue minus $40/yr cost = $32/yr) was charged to the customer every year for eight years. This included a profit of $5/yr for the utility. In addition, each household was charged $6/yr for eight years for the equipment for a total added charge of $38/yr, substantially less than the $72/yr saved in energy cost. For the first eight years, each household thus saves ($72 − $38)/yr = $34/yr for a total savings of $272 over the eight year period, and after that the household continues to benefit from the improvements.

These ideas have been implemented in different ways by many utilities, using various marketing schemes. Some have simply given compact fluorescent lightbulbs to their customers, some have leased them, and some have sold them at below-market costs. Some utilities have given rebates to customers who purchase efficient appliances including air conditioners. Some have paid customers for every kilowatt-hour saved, regardless of how it was accomplished. They all have had the goal of avoiding the need to develop expensive new sources of energy to meet the demands in their service areas. The general effect of these programs has been to turn around the usual relationship between sales and profits, to the benefit of the utilities, the customers, the environment, and the conservation of natural resources.

Key Terms

Principle of Energy Conservation
Degree days
Thermal conductivity
R-value
Radon
Nitrogen dioxide

Carbon monoxide
Air-to-air heat exchanger
Energy audit
Energy guide
Recycling

Suggested Reading and References

1. Hickock, Floyd. *Your Energy Efficient Home.* Englewood Cliffs, N.J.: Prentice–Hall, 1979.
2. Lovins, Amory B. *Soft Energy Paths.* New York: Harper and Row, 1977.
3. Ross, M. H., and Williams, R. H. *Our Energy, Regaining Control.* New York: McGraw–Hill, 1981.
4. Rosenfeld, Arthur H., and Hafemeister, David. "Energy-Efficient Buildings." *Scientific American*, **258**, No. 4 (April 1988), pp. 78–85.
5. Ross, M. "Energy and Transportation in the United States." *Annual Review of Energy* **14** (1989), p. 131.
6. *Energy: Production, Consumption and Consequences.* Washington D.C.: National Academy Press, 1990.
7. Wilson, A., and Morrill, J. *Consumer Guide for Home Energy Savings.* Washington, D.C.: American Council for an Energy-Efficient Economy, 1993.

8. "Energy for Planet Earth." *Scientific American*, **263,** No. 3 (September 1990).

9. *Energy Efficiency and the Environment—Forging the Link.* Edited by Edward Vine, Drury Crawley, and Paul Antonella. Washington, D.C.: American Council for an Energy-Efficient Economy, 1991.

10. *State of the World 1990, 1991, and 1993.* A Worldwatch Institute Report on Progress Toward a Sustainable Society. New York: W. W. Norton.

11. Richard Heede and the staff of the Rocky Mountain Institute. *Homemade Money—How to Save Energy and Dollars in Your Home.* Amherst, N.H.: Brick House Publishing Company, 1995.

12. Lovins, Amory B. "More Profit with Less Carbon." *Scientific American 293* No. 3 (September 2005), pp. 74–83.

Questions and Problems

1. Find the number of Btu conducted through a wall in 8 hours. The wall is 8 feet high by 24 feet long and has a total R-value of 16.2 hr · °F · ft^2/Btu, including the effects of the inner and outer air layers. The inside temperature is 68°F and the outside temperature is 16°F.

2. Find the number of degree days in the heating season at a location where the average outside temperatures are distributed as follows: 60 days at 55°F, 50 days at 45°F, 30 days at 35°F, and all the other days above 65°F.

3. A 1500 ft^2 wall has an R-value of 11 hr · °F · ft^2/Btu, including the effects of inside and outside air layers. How many Btu are lost through this wall in a 5600 degree day heating season?

4. Calculate the R-value of a brick-faced wall that has the following layers: 4 inches of brick, 0.75 inch of insulating sheathing, 3.5 inch of urea foam, 0.25 inch of plywood, inside air layer R-value of 0.68, and outside 0.17.

5. A house in Reno, Nevada (6040 degree days), consists of the following:

   (a) 544 ft^2 of single-pane windows.

   (b) 544 ft^2 of walls with total R-value, including air layers, of 16.31.

   (c) 480 ft^2 of stone walls, 13 inches thick. Add air layers.

   (d) 1632 ft^2 of roof with R-value of 8.0, including air layers.

   Neglecting the heat lost through the concrete slab floor, air infiltration, and solar energy input, calculate the total number of Btu needed to heat the home for one heating season.

6. Calculate the cost of heating the home in problem 5 for one season with a natural gas furnace. Assume that the cost of natural gas is \$13.00/10^6 Btu, and the furnace is 65% efficient.

7. Calculate the savings in fuel cost for one season if the house in problems 5 and 6 has the roof insulation increased to an R-value of 19, including air layers. Would the purchase of such insulation appear to be a wise choice if it costs \$0.45/ft^2?

8. What would be the payback period for installation of double-paned windows (originally single paned) in Fairbanks, Alaska, if fuel costs \$15.50/10^6 Btu, the furnace is 75% efficient, and the cost of double-paning is \$4.00/ft^2?

9. Estimate the percentage savings in the fuel bill if the thermostat is set back from 68°F to 50°F for 12 hours of the day, and left at 68°F for the other 12 hours, assuming an outside temperature of 15°F all day.

10. How many Btu would be contributed to the heating of a house by the metabolism of four occupants over the six-month heating period? Assume each person generates 100 W and they are in the house 50% of the time.

11. List the ways in which your home could be made more energy efficient.

12. Enumerate some changes you could make in your personal life that would lead to appreciable savings in energy.

13. Estimate the life-cycle energy costs for a few common household appliances and compare them to the initial appliance cost. See Table 7.5.

14. Estimate the annual energy cost in dollars for an electric blanket. Assume that it has a 135-watt heating element, which is on 50% of the time for each 8-hour night, 160 nights each year. Does it make sense to cover an electric blanket with ordinary blankets?

15. How could you improve the design of the typical gas-fired residential water heater? Consider electric ignition, a flue damper, insulation, tempering tank, insulated pipe connections, separate "instant" heater, and so forth.

16. Consider the relative ease and cost-effectiveness of measures you might take to conserve energy in your present means of transportation as compared to your home.

17. You can purchase a 15-watt compact fluorescent lamp for $4 that is rated to last 10,000 hours. This lamp provides the same light output as a 75-watt incandescent lightbulb that costs $0.50 and lasts 1000 hours. Based on an electric energy cost of $0.10/kWh, calculate the money saved over the life of the fluorescent lamp by using it instead of the conventional lightbulb. Include the cost of the lightbulbs.

18. How much would be saved over an eight-year period by each household that signed up for the Niagara Mohawk Conservation Program kit?

Multiple Choice Questions

1. For a 20-day period the outside temperature is 20°F and the inside temperature is 65°F. How many degree days is this?
   a. 1800
   b. 900
   c. 450
   d. 45
   e. 9000
   f. 4500
   g. 200
   h. 2000

2. For a typical house and ordinary weather conditions, the inside air is completely replaced by outside air about once every ___ due to infiltration around doors, windows, and through other cracks.
   a. 1 minute
   b. 2 minutes
   c. 100 hours
   d. 2 days
   e. 1 day
   f. 30 seconds
   g. 1 hour
   h. 20 hours

242 Chapter 7 **Energy Conservation**

3. In the United States space heating represents about how much of our residential energy consumption?
   a. 1%
   b. 5%
   c. 24%
   d. 48%
   e. 80%
   f. 98%
   g. 99%
   h. 100%

4. What is the exact meaning of the designation *R*-20 as used in describing quality of insulation?
   a. 20 (Btu/hr · ft^2 · °F)
   b. 20 (hr/Btu · ft^2 · °F)
   c. 20 (hr · ft^2 · °F/Btu)
   d. 1/20 (hr · ft^2 · °F/Btu)
   e. 20 (watt · hr/day)
   f. 20 (cal/sec)
   g. 20 (hr · ft^2 · K/Btu)

5. How many Btu/hr are lost through a 100-square foot wall that is made up of an 8-inch thick brick wall with an *R* value of 1.6 and 2 inches of foam insulation with an *R* value of 10.5? The inside temperature is 65°F, and the outside temperature is 35°F.
   a. 290 Btu/hr
   b. 29 Btu/hr
   c. 537 Btu/hr
   d. 248 Btu/hr
   e. 35 Btu/hr
   f. 112 Btu/hr
   g. 224 Btu/hr
   h. 1750 Btu/hr

6. For usual winter conditions, the dominant thermal resistance of a single-pane window is in _____
   a. the glass
   b. the internal air layer
   c. the external air layer
   d. infiltration
   e. none of the above

7. The efficiency of the furnace averages about ___% in existing homes, but the best new furnaces have efficiencies of about ___%.
   a. 25, 75
   b. 55, 70
   c. 90, 95
   d. 85, 98
   e. 15, 50
   f. 40, 60
   g. 30, 70
   h. 60, 95

8. The pilot light in a gas-fired furnace or water heater uses 1 ft^3/hr of natural gas. In terms of heat energy, this is equivalent to a ___ watt lightbulb burning continuously.
   a. 10
   b. 30
   c. 100
   d. 150
   e. 300
   f. 500
   g. 3000
   h. 10,000

9. On a cloudy winter day with an outside temperature of 0°F, what percent savings in fuel is realized if the thermostat setting is lowered from 75 to 65°F?
   a. 5%
   b. 15%
   c. 45%
   d. 85%

10. The energy equivalent of how much oil is saved, according to Table 7.7, by recycling a case (24 cans) of aluminum beverage cans rather than using them once and throwing them into the trash is _____

    a. 42 barrels

    b. 0.8 gallons

    c. 0.0194 gallons

    d. 13.3×10^{-6} barrels

CHAPTER 8

Transportation

(*Source*: Ed Pritchard/Stone/Getty Images)

8.1 Introduction

When determining our priorities, personal transportation usually comes right after the obvious essentials of food, drink, shelter, and clothing. This can be seen in the ways we use our dollars, and our time. Many obviously put their personal transportation even higher on the priority list. The classified ad sections of the Sunday newspapers have page after page of transportation items, often more than is advertised for anything else. Our reliance on transportation goes far beyond just moving ourselves around. Almost every material item we have or use came from someplace far away, by truck, train, plane, pipeline, or ship. Ours would be a very narrow existence without reliable transport of both people and goods.

This dependence on transportation is not new. Even the ancient civilizations had among their highest goals the achievement of fast and reliable transportation. Well before recorded history, camels and horses and other animals provided swifter and stronger land travel than did the human foot, and rafts and boats of all descriptions made waterways favorite avenues of commerce. We now remember many ancient cultures by the lingering artifacts of their transportation—canals, roadways, sea lanes, and even heavily worn footpaths.

The search for better means of transportation has continued into modern times. Not long after the steam engine was invented, it was adapted to power vehicles. The same is true of the internal combustion engine. There is a temptation to attribute the successes and failures of modern transportation to our learning to exploit the fossil fuels, but that is certainly not the whole story. Centuries ago sailing ships driven by the wind traveled freely over most of the globe, and the early steam-powered trains and boats were fueled by firewood—hardly a modern fuel.

We now have more than 220 million cars, trucks, and buses in the United States, and each one of them has a parking space at both ends of every trip it takes. The hundreds of millions of parking spots need to be connected by roads, nearly all with a paved surface. The resources of land, labor, material, and energy consumed by building and maintaining this infrastructure are enormous. The land alone is more than 1% of the country, more than the land area of many states and even many nations. In addition to the land area given over to roads and parking spots alone, there is a larger effect in the way our burgeoning system of roadways continues to infiltrate the landscape. This leads to urban sprawl, loss of productive farmland, loss of wildlife habitat, and a pattern of low population density inconsistent with efficient delivery of many community services. There is also the argument that more roads mean more traffic, more traffic means more fuel consumption, more fuel consumption means more fuel tax money going into building more roads, more roads mean more traffic, ... The cycle continues and feeds itself.

More of us by far drive to work alone than get there by any other means, and it has become common for workers to spend an hour or more of their time commuting to and from their jobs each day. Every large city of the world has become afflicted with traffic congestion, noise, and pollution. Precious land is being given over to highways, parking lots, airports, garages, and railroad facilities. Although there are individual exceptions, even the world's best mass transit systems have generally not been able to compete with the convenience, flexibility, speed, cost, privacy, or prestige of the personal automobile.

We cannot speak of a national transportation system without considering its dependence on energy. The fuel alone to power transportation accounts for 27% of the total national energy budget, and the construction and maintenance of the roadways and manufacture of the vehicles accounts for a significant additional fraction. Table 8.1 shows how fuel energy is used in various modes of transport in the United States, and Table 8.2 shows the fuel energy efficiency for several

Table 8.1 Energy Used in U.S. Transportation (2002)

Mode	Btu (10^{12})	Percent
Automobiles	9,326	35.4
Motorcycles	24	0.1
Buses	191	0.7
Light trucks	6,842	26.0
Other trucks	5,027	19.1
Air	2,213	8.4
Water	1,185	4.5
Pipeline	935	3.5
Rail	621	2.4
Total	**26,364**	
(4.7×10^9 bbl petroleum)		

Source: Transportation Energy Data Book, Edition 24, ORNL-6973, December 2004.

Table 8.2 Energy Efficiency Estimates for Various Forms of Transportation

Passenger Transportation	Passenger-Mile/10^6 Btu
Bicycle (8 mph)	3200
Walking (3 mph)	1900
Bus, intercity	1100
Bus, transit	240
Automobile	280
Train	210–370
Airplane, commercial passenger	270

Freight Transportation	Ton-Mile/10^6 Btu
Ocean oil tanker	12,500
Pipelines	3300
Railroad	2900
Waterway	2100
Truck	385
Aircraft	32

Source: Richard C. Dorf, *The Energy Factbook*, McGraw-Hill, 1981; *Transportation Energy Data Book*, Edition 24, ORNL-6973, December 2004.

modes, for both passenger and freight transport. The automobile clearly is the leading form of passenger transportation. Because of their large number, trucks in the United States burn much more fuel than do trains, even though trucks are several times less energy efficient than trains in terms of ton-miles per Btu.

8.2 Power and Energy Requirements

Can there be a way to move a car down the road without burning so much fuel? Countless numbers of us have pondered this question, wondering if we would make it to the next filling station, and if we could then afford to fill the tank. Should we go fast or slow? Maybe try to coast down hill or between periods of acceleration?

Unfortunately, perhaps, it's a scientific fact that there is no practical way to accelerate a car, or even keep it moving along at constant speed, without expending energy. While it is true from Newton's laws that an idealized description of motion would permit an object to move along at constant speed without any motive force being applied, this is not the case for vehicles in the real world. Energy must be provided to make up for the energy lost between the tires and the road, in the flexing of the tires, between the vehicle and the air it shoves aside, and in the friction of the wheel bearings.

We can arrive at an understanding of how much energy must be provided to a vehicle if we start by looking at the force required to move the car forward, and then derive the necessary power and energy from this force.

The total force that must be provided to a practical vehicle can be seen as the sum of four terms. The first of these is the force needed to accelerate the vehicle, meaning to increase its velocity. This force, the acceleration force, is

$$F_a = ma = m \frac{\Delta v}{\Delta t},$$

where m is the vehicle's mass, and the acceleration a is written as the change in velocity, Δv, divided by the time interval, Δt, during which it took place. Of course F_a will be equal to zero if the vehicle is moving at constant velocity. If we are working in the British system of units, so that the force comes out in pounds, we must use ft/sec for velocity, sec for Δt, and $m = w/g$, where w is the vehicle weight in pounds and $g = 32$ ft/sec^2.

If the car is to be able to climb up hills, this will take another force which we will call the *hill-climbing force*, F_h. This force depends on the mass m of the vehicle, the slope of the hill, s, and the acceleration of gravity, $g = 32$ ft/sec^2, all in the British system if we want force in pounds:

$$F_h = msg,$$

where the slope, s, is the vertical rise of a section of road divided by the horizontal distance in which it takes place. A road having a 5% grade, rising 5 feet in every 100 feet of horizontal distance, would have a slope of $s = 5/100 = 0.05$.

If a vehicle is moving along at constant velocity on a level road, both F_a and F_h would be zero.

The third type of force, F_r, is needed to overcome the rolling resistance due to energy losses in flexing the tires and turning the wheel bearings as well as to compensate for energy lost to the road surface. This force is complicated and differs from one vehicle to the next and from one set of driving conditions to another. It can even depend on choice of lubricants, and it is known to vary with temperature. There are various ways of expressing the rolling resistance mathematically. One choice that works reasonably well for typical vehicles, speeds, and driving conditions gives the force as a product of vehicle mass times its speed, multiplied by a number that takes into account units and dimensions as well as typical characteristics of the tires, road surface, bearings, etc. At highway speeds, F_r is generally smaller than and therefore not as important as the other force terms. In equation form, the rolling resistance force may be written as:

$$F_r = C_r\, m\, v,$$

where C_r is a constant which takes into account type of tires, bearing quality, and other factors. For a typical case, $C_r = 0.007$, to get F_r in pounds, with mass(w/g) and velocity(ft/sec) in the British unit system. If v is in miles/hour, mass in (lb/g), $g = 32$ ft/sec^2, then $C_r = 0.01$ to get force in pounds.

The fourth force, F_{ad}, is necessary to overcome aerodynamic drag on the vehicle. It is usually negligibly small at low velocities but increases rapidly as the velocity increases. It is the most important drag term at speeds greater than about 40 mph, and is responsible for most of the decrease in fuel economy experienced at higher speeds. The aerodynamic drag force depends on air density, wind direction and speed, vehicle speed, vehicle shape, vehicle size, and even the smoothness of the outer surface. It has been determined from engineering studies that, for situations where no wind is present, and with v given in miles per hour, a good approximate equation for the aerodynamic drag force, in pounds, is

$$F_{ad} = \frac{C_D A_f\, v^2}{370},$$

where C_D is the *aerodynamic drag coefficient*, which expresses how well the car is streamlined so that it will move through the air with a minimum of energy being used to push air out of the way or creating turbulence. It is related to the sleekness of the vehicle shape, not the vehicle size. Some representative values of C_D are given in Table 8.3. The symbol, A_f, is the *frontal area* of the vehicle given in square feet. The vehicle can be thought of as cutting its frontal area out of the air just as a cookie cutter cuts its area out of dough. A_f includes projecting ski racks, mirrors, door handles, tires, and antennae.

On a level road, for constant speeds above about 40 miles per hour, aerodynamic drag represents the most important force. The energy delivered to the drive wheels from the gasoline used in highway driving goes mostly into overcoming air resistance.

Table 8.3 Some Values for the Aerodynamic Drag Coefficient (C_D)

Object	C_D
Square flat plate (worst)	1.17
Rectangular block	1.00
Ordinary truck	0.70
1972 Dodge Polara Wagon	0.60
Streamlined truck	0.55
1981 Cadillac Eldorado	0.55
Porsche 928	0.45
Jaguar XKE	0.40
Ford Escort	0.39
Datsun 280 ZX	0.39
Porsche 924	0.34
1992 Ford Taurus	0.32
1997 Audi A8	0.29
2005 Toyota Prius	0.26
2005 Honda Insight	0.25
VW research vehicle	0.15
Teardrop (theoretical best)	0.03

If the total motive force is to be found, for a situation that may require including up to all four components, the total force, F_T, is given by adding together the individual forces:

$$F_T = F_a + F_h + F_r + F_{ad}.$$

Example 8.1

Calculate the aerodynamic drag force in pounds for a car having a drag coefficient of 0.39, a frontal area of 28 square feet, and a speed of 75 mph.

Solution

Use the equation given for the aerodynamic drag force:

$$F_{ad} = \frac{C_D A_f v^2}{370} = \frac{0.39 \times 28 \times (75)^2}{370} = \textbf{166 pounds}$$

Once the force needed to propel a vehicle is determined, it is a simple matter to calculate the required power. The power expended by any force is the product of the force times the velocity with which an object moves under the influence of the force. This can be written as:

$$\text{Power}\left(\frac{\text{ft} \cdot \text{lb}}{\text{sec}}\right) = \text{force (lb)} \times \text{velocity}\left(\frac{\text{ft}}{\text{sec}}\right),$$

or, if horsepower is preferred,

$$\text{Power (hp)} = \text{force (lb)} \times \text{velocity}\left(\frac{\text{ft}}{\text{sec}}\right) \times \left(\frac{1 \text{ hp}}{550 \text{ ft} \cdot \text{lb/sec}}\right).$$

These equations may be used to determine the power necessary to overcome the total drag forces on a vehicle or any one of the individual drag forces.

For most vehicles the power available at the drive wheels is just a small fraction of the power released by the combustion of the fuel. The various losses are shown in Figure 8.1. The net result of these losses is that only about 15% of the fuel energy is available for propelling the vehicle, with the range of actual efficiencies being about 10 to 20%.

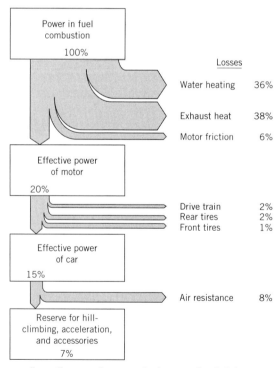

Figure 8.1 A power flow diagram for a typical rear-wheel drive automobile operating under standard urban operating conditions. The large losses to water heating and exhaust gases are thermodynamically unavoidable, but their exact proportions will depend on engine design.

Example 8.2

For the vehicle in Example 8.1, how many horsepower must be available at the drive wheels to overcome the air resistance?

Solution

In making this calculation we first convert the velocity in miles per hour to feet per second:

$$75 \frac{\text{miles}}{\text{hour}} \times \left(\frac{5280 \text{ ft}}{\text{mile}}\right) \times \left(\frac{1 \text{ hour}}{3600 \text{ seconds}}\right) = 110 \frac{\text{ft}}{\text{sec}}$$

Then, using the equation given above, and the force of 166 pounds from Example 8.1, we can calculate:

$$\text{Power (hp)} = \text{force (lb)} \times \text{velocity}\left(\frac{\text{ft}}{\text{sec}}\right) \times \left(\frac{1 \text{ hp}}{550 \text{ ft} \cdot \text{lb/sec}}\right)$$

$$= 166 \text{ lb} \times 110 \left(\frac{\text{ft}}{\text{sec}}\right) \times \left(\frac{1 \text{ hp}}{550 \text{ ft} \cdot \text{lb/sec}}\right) = \mathbf{33 \text{ hp}}$$

Now that we have worked through the force and power required for a vehicle, it's a fairly straightforward matter to calculate the energy requirement. Once the energy is known, by making an estimate of the efficiency of the engine and power train, we can compute the amount of fuel needed for a trip of any distance.

The energy required to move a vehicle a given distance is the same as the work done by the motive force. Work is defined as the product of force times distance, leading to:

$$\text{Energy (ft} \cdot \text{lb)} = \text{force (lb)} \times \text{distance (ft)}.$$

An equivalent approach would be to note that energy is given by power multiplied by the time during which it operates. Then:

$$\text{Energy (ft} \cdot \text{lb)} = \text{power}\left(\frac{\text{ft} \cdot \text{lb}}{\text{sec}}\right) \times \text{time (sec)}.$$

Or, if we know the force, the velocity, and the time taken for the trip:

$$\text{Energy (ft} \cdot \text{lb)} = \text{force (lb)} \times \text{velocity}\left(\frac{\text{ft}}{\text{sec}}\right) \times \text{time (sec)}.$$

This gives us three different ways to calculate the energy supplied to the vehicle, depending on what is known. The fuel energy required will be considerably larger than what is calculated using these equations because of inefficiencies in the engine and drive train.

Example 8.3

Calculate the miles per gallon expected for a car which weighs 3200 lb, has $C_D = 0.35$ and $A_f = 28$ ft², on a level road, with no wind, at a steady speed of 75 mph. Assume that the delivered energy is 18% of the fuel energy of gasoline. Use 9.7×10^7 ft·lb/gallon as the heat of combustion for gasoline.

Solution

Start by calculating the force.

Because the road is level the hill-climbing force, F_h, is zero. Because the speed is not changing the acceleration force, F_a, is also zero. This leaves only the forces expended against rolling resistance and air resistance.

$$F_{Total} = F_r + F_{ad} = 0.01mv + \frac{C_D A_f v^2}{370}$$

$$= \left(0.01 \frac{3200}{32} \times 75\right) + \left(\frac{0.35 \times 28 \times (75)^2}{370}\right)$$

$$= 75 \text{ lb} + 149 \text{ lb} = 224 \text{ lb}$$

Per mile, the energy expended is the product of force times distance:

$$224 \text{ lb} \times 1 \text{ mile} \times 5280 \frac{\text{ft}}{\text{mile}} = 1.18 \times 10^6 \text{ ft} \cdot \text{lb}$$

At 18% efficiency, the *input* energy must be:

$$\frac{1.18 \times 10^6 \text{ ft} \cdot \text{lb/mile}}{0.18} = 6.56 \times 10^6 \text{ ft} \cdot \text{lb/mile}$$

In terms of gallons:

$$\frac{6.56 \times 10^6 \text{ ft} \cdot \text{lb/mile}}{9.7 \times 10^7 \text{ ft} \cdot \text{lb/gal}} = 0.068 \text{ gallon/mile}$$

Taking the reciprocal of this:

$$\frac{1}{0.068 \text{ gal/mile}} = \textbf{14.7 miles/gallon}$$

From Example 8.3 it may be concluded that if the various resistances to vehicle motion are understood, and if the efficiency of the engine and drive train are known, then one can calculate quite reliably how much fuel is required to travel a given distance. These conclusions can help engineers focus on specific items to improve fuel efficiency. We have seen that the largest inefficiency is due to losses of heat energy from the automobile engine, which is a heat engine. Part of this loss of heat energy is unavoidable: the Carnot analysis of heat engines that we discussed earlier has shown that. The loss of energy to air resistance is also important and explains all the effort that has gone into streamlining: smoothing out door handles, the fitting of window glass, air dams under the front of the car to reduce the drag from the lumpy underside of the vehicle, devices on semitrucks to smooth the transition from the cab to the trailer. All of these things pay off in reduced fuel consumption, especially at highway speeds. The streamlining of modern aircraft probably represents the ultimate concentration on reducing aerodynamic drag for passenger conveyances.

The analysis in Example 8.3 also applies to an understanding of the very large demand for transportation energy in the United States. For every vehicle-mile driven there is going to be a certain amount of energy consumed, and this energy amount is calculable for a given set of conditions. When we consider the 220 million plus registered vehicles in the United States, and combine that with the number of miles that each is driven per day, we see a huge appetite for fuel. Reduction of this demand can focus on reducing vehicle-miles or it can concentrate on improving the energy efficiency of the vehicles.

Improvement of the vehicles' energy efficiency can be directed to several factors revealed in the four force terms we have just dealt with. For any trip the motive energy required is given by the product of the force times the distance. The first three of the force terms (F_a, F_h, F_r) are proportional to the vehicle's mass, and the fourth (F_{ad}) is proportional to the vehicle's frontal area. This means lighter and smaller cars are more efficient. The third and fourth terms (F_r, F_{ad}) increase as the velocity increases, indicating that slower speeds reduce gas consumption.

Another area of improvement in energy efficiency lies in the automobile engine and gear train. Over the past 25 years automobiles have shifted almost entirely from carburetors to fuel injection and from breaker-point ignition to electronic ignition. Engine operation is continuously monitored by microprocessor-based sensors that provide feedback corrections to both ignition and fuel systems to keep their operation optimum. Automatic transmissions now have overdrive gears and lock-up features that improve their efficiency. Better lubricants and bearings reduce friction losses. The newer and lighter materials used for the bumpers, body, and chassis means that engines can now be smaller for the same level of vehicle performance. One simply cannot buy a new passenger car with an engine of the large size common in the 1950s, 60s, and 70s. There is no longer either the need or demand for large-sized engines. Performance has not suffered significantly. The record of these improvements is shown by fuel efficiency graphs such as Figure 8.2.

In this section we have presented an analysis that suggests that a vehicle's fuel economy must worsen as speed increases. In order to make the analysis tractable, it relies on several simplifying assumptions that can only be approximations to what a more sophisticated approach would involve. The Federal Highway Administration has sponsored a project with Oak Ridge National Laboratory to measure how fuel economy actually depends on the speed of various vehicles. Some of the results are summarized in Figure 8.3 where it can be seen that the fuel economy improves with speed in the lower range. This effect is not

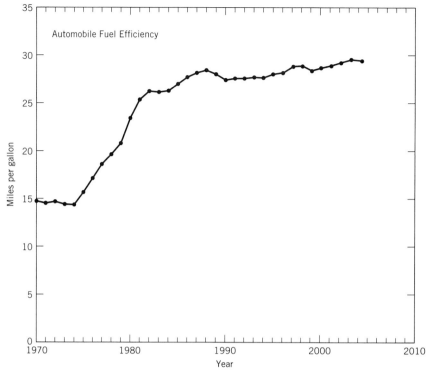

Figure 8.2 New car fuel efficiency in the United States since 1970. These are sales-weighted averages including domestic and imported passenger automobiles. Trucks, buses, light trucks, and sport-utility vehicles are not included. The increase since the minimum in 1974 is due to federally mandated standards for fuel economy as well as to technical advances in automobiles. (*Sources*: Oak Ridge National Laboratory, Light-Vehicle MPG and Market Shares System, Oak Ridge, TN, 1997; *Transportation Energy Data Book*, Edition 24, ORNL-6973, December 2004.)

predicted by our simple analysis. It is due to the engine and transmission operating at lower efficiency at the lowest speeds. At the higher speeds, the aerodynamic drag force becomes dominant, and fuel economy, as expected, decreases as speed increases. These measurements make it apparent that high vehicle speeds do indeed lead to higher fuel consumption, and therefore greater carbon dioxide emissions for any highway trip. It appears that speeds in the range of 30 to 60 miles per hour produce the best fuel economy.

8.3 *Electric Batteries, Flywheels, Hybrids, Hydrogen, Alcohol*

In the previous section of this chapter we've seen that the need for an energy source for transportation is very real. If mass transit is the issue, then overhead

Electric Batteries, Flywheels, Hybribs, Hydrogren, Alcohol 255

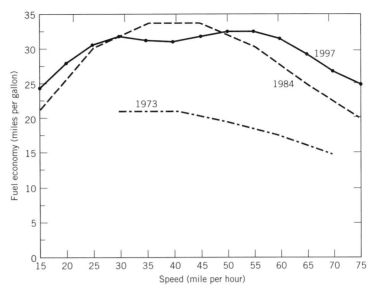

Figure 8.3 The dependence of fuel economy on speed. The results labeled 1973 are for an average of 13 vehicles from the model years 1970 and earlier. The 1984 test data are for 15 1981–1984 automobiles and light trucks; the 1997 data are for 9 1988–1997 automobiles and light trucks. Every one of the 37 tested vehicles showed a loss of fuel economy at 75 mph compared to 55 mph. (*Source*: *Transportation Energy Data Book*, Edition 24, December 2004. Stacy C. Davis and Susan W. Diegel. ORNL-6973, Oak Ridge National Laboratory.)

electric lines can power trains or trolley buses, but for individual free-roaming automobiles, trucks, and airplanes, the energy source must be carried aboard the vehicle. Gasoline, jet fuel, and diesel fuel have served this purpose, but they are all a part of the fossil fuel resource now in decline. Natural gas and propane are becoming more common for cars and trucks, but they are also fossil hydrocarbons. The more we use them, the more quickly they will be gone.

The search is now underway for alternative energy sources for transportation. Some that are under serious development or consideration will be discussed in this section. Some of the push for using alternative energy sources is stimulated by the new laws.

For many years the state of California has led the way in establishing policy to control vehicle emissions. Other states, and even other countries, have used the California regulations as guidelines in setting their own standards. In 1990 California adopted a regulation that would have required that 10% of new vehicles be emission-free by 2003. In spite of serious efforts to market electric vehicles, it turned out that vehicle manufacturers were unable to meet that goal, and the regulations were relaxed to accommodate low-emission rather than zero-emission vehicles. In 2003 the California Air Resources Board adopted new rules that would require more hybrid and fuel-cell vehicles effective with the 2005 model year. Most recently, in California, the Pavley bill, named for its sponsor, has been put forth. This bill addresses greenhouse gas emissions from vehicles and, in particular, sets stringent limits on carbon dioxide emissions. It is apparent that this approach is equivalent to setting fuel economy standards for fossil fuel–consuming vehicles. Objections have been raised by federal officials and others who argue that the setting of fuel economy standards is the responsibility of the federal government alone and is not permitted for states. The future of the Pavley bill remains uncertain.

8.3.1 Electric Vehicles

Battery-powered electric vehicles are not new. Early in the 20th century, electric cars powered by lead–acid batteries outnumbered combustion-powered vehicles, but the vastly superior energy density of gasoline soon led to its dominance in transportation, and for a time electric cars all but disappeared. There now are roughly 50,000 electric vehicles on the road in the United States, growing 13% per year over the past decade, and nearly 230 million gasoline-powered cars and trucks. The higher efficiency of the electric vehicles (more than 40%) compared to that of the combustion-powered vehicles (about 15 to 25%) somewhat compensates for the fact that the energy stored in a battery is less than 1% of that in an equivalent weight or an equivalent volume of gasoline. However, because of the huge difference in energy density, gasoline-powered vehicles have a much greater range, and battery-powered airplanes are unheard of. It seems unlikely that this situation will change even with the best of batteries under development. The advantages of the electric car lie not in range or power, but in quietness, freedom from maintenance, the possibility of regenerative braking, reduction of local emissions, and, most of all, the ability to derive energy for transportation from any energy source capable of driving an electric generator. Even in vehicles, batteries can form a part of a load-leveling storage capacity for large electric utilities if the vehicle batteries are charged during off-peak hours when electric power can be less expensive.

On a *weight* basis, gasoline stores energy at a density of 13,000 Wh/kg, and lead–acid batteries store only about 25 Wh/kg. Thus a charged lead–acid battery has an energy disadvantage of 520 times less energy density than gasoline. Only some of this disadvantage is recovered by the higher efficiency of getting the energy from the battery to driving the car. If we look at the relative energy densities on a *volume* basis, we find about 12,000,000 Wh/m^3 for gasoline, and approximately (25 Wh/kg × 15 kg/0.008 m^3) = 47,000 Wh/m^3 for a lead–acid battery. This is a factor of about 250 in favor of gasoline. Any way you look at it, batteries are at a huge disadvantage, even considering the newest types not yet on the market which are capable of perhaps twice the energy density of lead–acid batteries. The battery situation has not changed much from 100 years ago.

The issue of the *primary* energy source for electric vehicles is often overlooked in public discussions of their advantages. On analysis, the energy supply situation is sobering. From Table 8.1 we see that transportation fuels amount to about 26 × 10^{15} Btu each year in the United States. If only 10% of this were to be displaced by electric vehicles, the added load on our electric generating system would be enormous. Using the 10% number for estimating purposes, and ignoring energy conversion efficiencies for the moment, we find that electric energy in the amount of 2.6×10^{15} Btu/yr, which is equal to 0.76×10^9 MWh/yr, would be needed. Dividing this *energy* need by the 8760 hours in each year, we find an electric *power* demand of 8.7×10^4 MW = 87,000 MW. This is equivalent to the output of an additional (87) large power plants running continuously. The electric generating capacity of the nation now is about 950,000 MW, and the electrical energy produced annually is about 4.0×10^9 MWh (corresponding to an average duty factor of about 50% for the generating plants). Thus, replacing only 10% of our transportation fuel energy (amounting to 0.76×10^9 MWh per year) by electricity would increase the total load on our electric generating capacity by an estimated 19%. Replacing *all* of our motor fuel by electricity would lead to extreme demands for electricity—triple our present electricity generation. This discussion could be refined by including the efficiencies of fuel-burning engines and battery charging and discharging, but the conclusions would not change dramatically.

From the point of view of efficient use of fossil fuel energy, it may be a reasonable balance to use liquid petroleum fuels or natural gas in power plants to make electricity with 38% efficiency, then use the electricity in a battery-powered vehicle at an estimated 40% efficiency, considering both battery charging and discharging, for an overall efficiency of about $0.38 \times 0.40 = 0.15 = 15\%$. This compares reasonably well with the 15 to 25% efficiency of gasoline-powered internal combustion engines in automobiles. Such an approach would also reduce point-of-use vehicular emissions.

We have become accustomed to refueling our cars in a matter of only a few minutes. It is not hard to put a billion joules of fuel energy into a gas tank in this time. The situation is quite different for charging electric batteries. The rate at which batteries can receive energy is determined by the permissible charging current. There is a practical limit on how much charging current can be forced through a battery without shortening its lifetime. It is possible to charge an ordinary 12 volt battery at 100 amperes, but to do so will drastically shorten its useful lifetime. As a routine practice, a charging current of 10 amperes would be reasonable. This charging current would put energy into the battery at a rate of 10 amperes $\times$ 12 volts = 120 watts. (In practice, the charging voltage would be about 13 or 14 volts.) The efficiency of the charging process is variable depending on the charging rate and the condition of the battery. For typical conditions, perhaps 80% of the power taken from the charging source, or about 100 watts in this example, produces stored energy in the battery; the rest is dissipated as heat.

Suppose an electric car carries 20 lead–acid batteries weighing 25 kilograms each, for a total of 500 kilograms of batteries capable of storing energy at a density of 25 watt-hours per kilogram. This gives a total stored energy of 12,500 watt-hours, equal to 45 million joules. This would be about a dollar's worth of electric energy if paid for at the same rate charged to residential users. At a charging power of 100 watts into each individual battery, or 2000 watts total, this would mean a charging time, for completely discharged batteries, of a little more than six hours, significantly longer than our usual practice of filling a car's tank with gasoline in just a few minutes. This is a serious limitation for electric cars. For local use, where one can do a day's normal driving without a recharge, there may not be a problem because the batteries can be recharged at home overnight. However, for long-distance travel, there will be difficulties. One solution may be to replace the entire set of discharged batteries with another complete set of charged batteries at each "re-fueling" stop. This takes time and labor and it could be expensive depending on how the trade of one set of batteries for another is accounted for financially.

Another limitation for the present generation of electric vehicles has to do with the service life of the batteries before replacement is required. The only battery now available at a reasonable cost on a commercial basis for electric cars is the lead–acid battery. These batteries have not changed fundamentally in the last 100 years. They are heavy, and they have a limited lifetime. In electric vehicle use, it is expected that they may need to be replaced every two or three years, at a cost of $1,000 or more each time. This would require owners of electric cars to budget a substantial amount each year for new batteries.

Currently, the main stumbling block for electric cars is their limited range. The few electric cars now available to the consumer will probably have advertised ranges of less than 100 miles, and the stated range may be valid for new batteries, fully charged, at a fairly low steady speed, with limited use of accessories, and without headlights. Today's lead–acid batteries are simply inadequate to store the amount of energy most consumers demand in an automobile.

The essential advance that the electric vehicle industry is looking for will most certainly require new battery technology. Numerous new types of batteries have been under intense investigation for decades, but none has yet proved superior to the old lead–acid design. There have been problems of cost, weight, lifetime, operating temperature, availability of special materials, toxicity, and so forth. Some of the advanced types under investigation and development carry the designations nickel–cadmium, nickel–metal hydride, nickel–iron, sodium–sulfur, sodium nickel chloride, lithium polymer, lithium iron disulfide, lithium aluminum disulfide, lithium–ion, metal–air. Each has its problems. It is not certain that any will soon displace the lead–acid battery in electric cars for the mass market.

8.3.2 *Flywheel-Powered Vehicles*

Most of us are familiar with the flywheel in one or more of its common forms: a potter's wheel, a child's top, a gyroscope, or a toy automobile. In these examples a wheel-shaped mass is set into rotation through human effort and the rotational energy continues to sustain this motion until it is dissipated by frictional forces. The flywheel is obviously an energy storage system; it takes energy from some source and delivers it later when needed, unless the rotational motion has first ceased because of friction either in the bearings or against some surrounding medium such as air. In recent years, flywheel-based units have become commercially available for emergency standby power to provide a continuous supply of electricity to hospitals, computer installations, and other users in the event of a power interruption. Flywheels have also been demonstrated for powering urban transit buses. There is emerging interest in using flywheels for powering a new generation of automobiles. They are not yet in a practical form ready for the market, but there is continuing progress in this direction.

The energy stored in a rotating flywheel is the sum of the kinetic energies of the individual mass elements that make up the flywheel. When the wheel is in rotation, each of these mass elements has kinetic energy proportional to the square of its velocity, and the velocity is proportional to the rotational speed (as in revolutions per second) and to the distance of the mass element from the axis of rotation. Thus the energy stored is greater for higher rotational speeds and if most of the mass is far from the rotational axis. This latter consideration favors a configuration shaped like a bicycle wheel but not a shape like a solid spinning ball.

The limit on how much energy can be stored in a flywheel is determined by its strength. The faster it is made to rotate, the more the energy stored. However, if it is rotated too fast, it tears itself apart. The best materials in terms of maximum energy storage are those with the greatest tensile strength and therefore the ability to resist the forces that try to pull the flywheel apart. The most dense materials are not necessarily the best for high-energy density storage at high rotational speed. One material, fused silica fiber, has a ratio of strength to mass about 20 times greater than that of steel and, as a consequence, promises a specific energy (in watt-hours per kilogram) about 20 times that achievable with a steel flywheel and about 40 times greater than that of a lead–acid battery. With some of the newer materials, such as the silica fibers or graphite fibers, the useful energy that can be carried aboard a flywheel-powered vehicle would, with its drive train system, have an energy density comparable to that of the fuel plus engine and drive train of a petroleum-powered vehicle having similar acceleration, speed, and range. The flywheel-powered vehicle has the advantage that during downhill driving and braking, part of the kinetic energy of the vehicle is put back into the flywheel energy storage by regenerative braking. This advantage is shared by battery-powered vehicles.

For a flywheel to be a generally useful energy storage device aboard a vehicle, it not only must store the required energy, but also must have a sufficiently long run-down time between periods of use so that the energy will be available when needed. The required run-down times are achieved by using the best possible bearing systems and by operating the flywheel in a vacuum chamber to eliminate air drag on the rapidly rotating flywheel. Run-down times of 6 to 12 months are considered achievable, possibly with noncontact magnetic bearings. It is expected that flywheel-powered cars could be recharged much more quickly than is the case for storage batteries. Some estimates put the recharge time for flywheels about ten times faster than for batteries for the same stored energy.

Flywheel-powered cars would impose the same very large demand on our electricity-generating capacity as was estimated for battery-powered vehicles. This is a forbidding challenge. It is often tempting to think of these alternative technologies as new sources of energy and as emission-free. Neither of these assumptions is correct because the burden of providing energy for transportation and of controlling emissions is transferred to the electrical industry. In a sense the pollution problem is merely transferred from the point of energy use to the point of energy generation. This gets it out of the cities.

The general arrangement by which a flywheel may be coupled to the drive wheels of a car is shown in Figure 8.4. This setup would be suitable for energy input to the flywheel from an electric outlet, from a solar panel, from an on-board gasoline or diesel generator, or from regenerative braking.

Table 8.4 shows some of the characteristics of flywheel energy storage compared to lead–acid batteries and other transportation energy sources. Flywheels and batteries appear to have very limited energy density compared to that of gasoline, but this is an unfair comparison. The efficiencies of recovering energy from flywheels, batteries, and gasoline and converting it to the mechanical energy needed to power the car are about 80%, 60% and 20%, respectively. In addition, the engine, radiator, exhaust system, transmission, and drive train of a gasoline-powered vehicle add much more weight to the vehicle than do the electric motors used with flywheels and batteries.

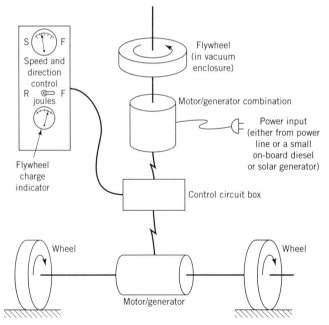

Figure 8.4 A schematic view of a flywheel-powered motor vehicle. The motor/generator units can be operated in two modes, so that the flywheel can be charged either from an external source or whenever the vehicle is being braked electromagnetically.

Electric Batteries, Flywheels, Hybribs, Hydrogren, Alcohol

Table 8.4 Energy Density

Medium	Wh/kg[a]
Hydrogen	38,000
Gasoline	13,000
Flywheel, fused silica	870
Flywheel, carbon fiber	215
Flywheel, steel	48
Lead–acid battery	25

[a]Not including weight of motors and generators or conversion efficiency.

The downside of flywheel energy storage aboard vehicles is the danger of catastrophic failure, which could throw off fragments of the flywheel at dangerously high velocities. The two approaches being taken to limit this threat are to make the flywheels small, about the size of a coffee can, and to make them of a fiber material that would not throw large fragments on disintegration. The small flywheels would store sufficient energy by rotating at very high speed, like 60,000 revolutions per minute. If the flywheels are made of fiber materials, it is possible to contain the fragments from a disintegrating flywheel entirely within the surrounding housing. A second potential problem with flywheel energy storage aboard vehicles has to do with the gyroscopic forces that are generated whenever the axis of rotation is tipped. This force on the vehicle can be eliminated by mounting small flywheels in pairs rotating in opposite directions. A vehicle could have several pairs of these small flywheels aboard, each storing a kilowatt-hour of energy.

Example 8.4

We have just learned of the possibility of having several small flywheels aboard a vehicle, each storing a kilowatt-hour of rotational energy. If all of the 1 kWh of this energy in one small flywheel unit could somehow be made to lift a 2000 pound vehicle into the air (or propel it up a hill), how high (in feet) would it rise?

Solution

From the table of Energy Unit Conversion Factors inside the front cover, there is 3.78×10^{-7} kWh per ft · lb, or 2.6×10^6 ft · lb per kWh. If the car is to be lifted into the air, it gains potential energy (ft · lb) in the amount of its weight in pounds times its elevation in feet.

In the form of an equation:

$$w \text{ (lb)} \times h \text{ (ft)} = 1 \text{ kWh} = 2.6 \times 10^6 \text{ ft} \cdot \text{lb}$$

Rearranging gives

$$h \text{ (ft)} = \frac{2.6 \times 10^6 \text{ ft} \cdot \text{lb}}{w \text{ (lb)}} = \frac{2.6 \times 10^6 \text{ ft} \cdot \text{lb}}{2000 \text{ lb}} = \mathbf{1300 \text{ feet}}$$

8.3.3 Hybrid Vehicles

Many of the problematic limitations of vehicles powered by batteries are being overcome by interesting variations on their design. Development work has been underway for years on hybrid vehicles, and they have now made their way into the marketplace.

The first hybrid vehicles were on the road more than 100 years ago. One model, produced by a French manufacturer, was the gasoline–electric hybrid 1903 Krieger passenger car, which had a gasoline engine, an electric motor, and a battery pack, similar in many ways to our modern hybrid vehicles. The Krieger is said to have had front-wheel drive and power steering. Another, a hybrid truck, was made and marketed in 1909 by the Standard Gas and Electric Power Company of Philadelphia, Pennsylvania. Their sales literature included a photograph of the truck and said, "The general principle involved in the combined power unit is the use of a 2-cylinder gasoline engine with a small electric dynamo mounted on the engine shaft between the flywheel and the engine cylinder. It is so designed that as the speed of the engine increases, the dynamo becomes a generator, storing electric power during this period in a storage battery, while when the speed of the engine goes down, due to heavy pulls, hills or bad roads, it is automatically converted into a motor and takes current from the storage battery, giving a great surplus of power for momentary heavy overloads, or a less amount of power for longer periods." This reads much like current descriptions of the most modern hybrid vehicles.

The early hybrids essentially disappeared from the market along with electric-only cars early in the 20th century. Gasoline-only cars became more reliable, less expensive, and soon dominated the marketplace. And it has stayed that way for most of a hundred years.

Within the past several years hybrid cars have again become available, first from Toyota and Honda, and now from Ford and several other manufacturers. These vehicles are basically gasoline powered. They have no primary source of power other than gasoline. But these designs have achieved improved fuel economy and reduced emissions by using a combination of batteries, an electric motor, and the gasoline engine working together under the management of a sophisticated control system. Their performance is the equal of a traditional gasoline-powered car; the price is somewhat higher, the range is at least as great, the emissions are lower, and the fuel economy is generally much better. One example of a hybrid vehicle is shown in Figure 8.5.

Figure 8.5 A compact hybrid vehicle, the E1, developed by BMW. It uses a small internal combustion engine for long-distance travel and to recharge the vehicle's batteries. It is also designed to be used as an all-electric battery-powered vehicle. (*Source*: Courtesy BMW of North America, Inc.)

The idea behind the modern hybrid is that the gasoline engine can be rather small, and it can operate only when needed and at the speed which gives the best efficiency and the lowest emissions. Excess power can go to charging the battery. If more power is needed, energy can be drawn both from the battery and from the gasoline engine. Regenerative braking is a positive feature. Some of the energy which would otherwise go into heating up the brakes can be used to drive the generator and charge the battery. The gasoline engine does not run when the car is at a stop. The greatest gains in fuel economy are seen in stop-and-go city driving, where speeds are slow, trips are shorter, and brakes are used frequently. On the open highway, the gains in fuel economy are often smaller because the vehicle is then very much like an ordinary gasoline-powered car.

In the currently available hybrids the battery (nickel–metal hydride, NMH) stores only about 1 kWh of energy, equivalent to no more than a cup of gasoline in terms of stored energy. The battery thus serves only for short distances and slow speeds or to give a boost to the gasoline engine at times of heavy power demand. The battery is charged only by the gasoline engine or by regenerative braking. It is never plugged in to external power.

Innovative developments are now underway in several places to upgrade the standard hybrids to what is becoming known as a plug-in hybrid. In this design, the battery is replaced with a larger one (probably lithium–ion or NMH) having perhaps 10 kWh capacity, permitting trips of up to 30 miles on battery power alone. The battery can be plugged in and recharged overnight at home (for about $1) or during the day at work. Regenerative braking still boosts the battery charge. If daily trips are 30 miles or less, the gasoline engine may never run at all, and there would be zero emissions. The engine may be needed if the heater, air conditioner, or other accessories are used. In this way, the average fuel economy (miles driven divided by gallons of gasoline used) can be very high. It is possible that a tank of gasoline could last for months in ordinary commuter service. And the engine is always there to be used for longer trips, just like in the standard hybrid. The plug-in hybrid offers the best of both worlds, electric and gasoline. Although plug-in hybrids are not yet commercially available, the advantages appear to be so great that it seems likely, perhaps inevitable, that we'll soon see them at our local dealers. Their price may be quite high because of the large and expensive battery. Of course, the plug-in hybrid approach raises the same questions about electricity supply that were addressed in the earlier discussion of electric-only vehicles. The need for a primary energy source and the attendant emissions issues should not be overlooked.

Beyond passenger automobiles, hybrid technology is also finding its way into urban transit buses and delivery trucks. In these applications the engine is diesel, rather than gasoline, and the batteries are lead–acid, rather than NMH. As one example, the City of New York has already put into service 125 diesel–electric hybrid buses (Orion Bus Industries), following a lengthy evaluation of their performance. The city fleet of hybrid buses will soon be brought up to 325. These buses have been shown to reduce particulate emissions by 90%, nitrogen oxides by 40%, and greenhouse gases by 30%. Their fuel economy is 30% better than that of the traditional diesel buses. They use regenerative braking. Passengers report that the vehicles are more comfortable and quieter than the usual buses, and they have the same passenger capacity and equal or better performance.

8.3.4 *Hydrogen, Fuel Cells*

With the general realization that the fossil fuels, particularly oil and natural gas, are being depleted, there has been much speculation about a possible hydrogen economy. Hydrogen is different in an important way from oil and natural gas. It is not freely available in nature. It must be obtained from compounds to which it is bound. Once hydrogen is freed from these compounds, it can then be a means of storing and transporting energy. Rather than being a primary energy source, it is a secondary fuel. An understanding of the methods and economics of producing, storing, and using hydrogen is essential if the hydrogen economy is to become a reality.

Hydrogen is extremely abundant, accounting for perhaps 90% of the number of atoms in the universe. But in the earth's atmosphere hydrogen is rare as a free element, being present at a level of less than one part per million. This is because gaseous hydrogen is the least massive of all molecules; it is so light that at normal atmospheric temperatures hydrogen has a thermal velocity sufficient to escape the earth's gravitational field and drift off into space. Other gases such as oxygen and nitrogen cannot do that. Whatever hydrogen is naturally present on earth is essentially all tied up in chemical compounds such as water (H_2O) or methane (CH_4), and any hydrogen made available for use must be separated from those molecules.

Most of the hydrogen now produced is obtained from natural gas by a process called steam re-forming, using steam at temperatures up to 1600°F in the presence of a nickel catalyst. In a common scenario for a future hydrogen economy, it is envisioned that hydrogen will be produced mainly by electrolysis of water, requiring massive amounts of electric energy.

Hydrogen produced by electrolysis is very pure. The yield of hydrogen is about 7 ft^3 per kWh of electrical energy put into the electrolysis cell. This corresponds to an energy efficiency of about 67% based on a heat value of 325 Btu/ft^3 for hydrogen gas. The efficiency of less than 100% makes it clear that hydrogen is *not* a primary source of energy. It will never return as much useful energy as was consumed to create it, and energy from hydrogen will always have a higher cost than the energy used in its manufacture. Hydrogen is *not an energy resource* in the same sense as the fossil fuels are.

The source of electric energy for electrolysis could be nuclear reactors or some mixture of renewable energies such as hydroelectric, wind, solar cells, or biomass. Renewable energy sources are rather well matched to hydrogen production as the intermittent nature of some of the renewables will be compatible with production and storage of hydrogen.

Hydrogen (H_2) is a colorless, odorless, diatomic gas of low density; it weighs only one-eighth as much as natural gas, and as a fuel has nearly three times as much energy content per unit mass, but only about one-third as much per unit volume at normal temperatures. Hydrogen is well suited to the role of a *secondary* fuel and is a likely energy storage medium for use in transportation, and also for producing heat or for generating electricity through the use of fuel cells.

When hydrogen is burned as a fuel, for example, in an automobile engine, it combines with oxygen from the air to form pure water and heat energy; there are no other combustion products except possibly small amounts of nitrogen oxides formed from the air in the high-temperature combustion zone. The process is

$$2\,H_2 + O_2 \rightarrow 2\,H_2O + \text{heat energy}.$$

Thus, the fuel cycle (with hydrogen obtained by electrolysis) is a recycling of hydrogen (and oxygen) from water to water with energy put in at one stage and taken out at another. The hydrogen is not destroyed in the process in the same sense that hydrocarbon molecules in fossil fuels are destroyed by burning. The recycling time for the complete water–hydrogen–water cycle is likely to be short, a few days or weeks, in contrast to the fossil fuel–carbon dioxide–fossil fuel cycle time that is in excess of a million years.

Hydrogen can be stored as a liquid at very low temperature, below $-253°C$, and as a liquid it has an energy density per unit volume about a thousand times greater than as a gas at standard temperature and pressure. The necessary technology for cryogenic storage of hydrogen has been proved in the U.S. space program, where hydrogen is used as a rocket fuel.

There have been numerous demonstrations of vehicles powered by hydrogen fuel burned in a conventional engine. One example is shown in Figure 8.6. If hydrogen as a combustion fuel is ever to displace as much as 10% of our national transportation fuel budget, the same problem needs to be overcome as was encountered for electric batteries and flywheels, but here it may be more

Figure 8.6 Hydrogen-fueled BMW 735i being filled with liquid hydrogen at a solar-powered hydrogen fueling station in Germany. The hydrogen is obtained by electrolysis from water, using electricity from photovoltaic solar cells. The liquid hydrogen fuel tank in the car trunk holds enough fuel for the converted car to run for 300 kilometers. (*Source*: Martin Bond/Science Photo Library/Photo Researchers, Inc.)

difficult because of the low thermodynamic efficiency of the automobile engine. Certainly there would need to be a massive increase in the national electricity-generating capacity to provide hydrogen fuel for transportation in appreciable quantity.

The technology of fuel cells is a possibility for relieving the problem of the low thermodynamic efficiency of the internal combustion engine. Hydrogen–oxygen fuel cells have been known for a long time and have been used in many specialized applications. They have never quite become practical energy sources for powering vehicles, but the potential for wider use is there. In a hydrogen–oxygen fuel cell, these two gases are combined in the presence of a catalyst to produce electricity directly. The process is basically the exact reverse of the electrolysis that produced the hydrogen from water using electricity as an input.

A diagram of a hydrogen–oxygen fuel cell is shown in Figure 8.7. This is a polymer electrolyte membrane (PEM) fuel cell, also sometimes known as a proton exchange membrane (PEM) fuel cell, one of several types under development. A single cell of the type shown produces an output voltage of about 1 volt; many of these units are stacked together to produce higher voltages. The output current depends on the surface area of the electrodes exposed to the electrolyte.

Fuel cells operating on pure hydrogen can achieve 40 to 70% efficiency for producing electrical energy relative to the fuel energy content of the input hydrogen. Using the higher number, 70%, along with an efficiency of 90% for an

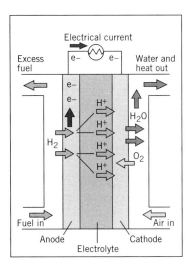

Figure 8.7 A polymer electrolyte membrane (PEM) hydrogen–oxygen fuel cell [sometimes known as a proton exchange membrane (PEM) fuel cell]. Hydrogen fuel enters from the left, and oxygen is supplied by air entering from the right. The hydrogen's electrons and protons are separated by a reaction involving a catalyst (typically platinum) at the anode. The electrolyte is a solid polymer which passes protons but not electrons. The electrons are diverted through an external load, thereby constituting the electric power output. At the cathode, the protons passing through the membrane are combined with oxygen from the air and with the returning electrons to produce water. Because the fuel cell is less than 100% efficient for producing electricity, some heat energy is generated directly. (*Source*: U.S. Department of Energy, Office of Energy Efficiency and Renewable Energy.)

electric motor, one could expect a combined efficiency above 60% for producing mechanical energy from hydrogen fuel. This is well above the 30% that can be achieved by the most efficient internal combustion engine burning hydrogen as a fuel, and the fuel cell operates with zero emissions.

Fuel cells have no moving parts; they are quiet and clean and produce only pure water and electrical energy. Their higher efficiency would result in much less demand for electricity-generating capacity, compared to what would be needed for hydrogen to be burned as a transportation fuel. An automobile powered by hydrogen–oxygen fuel cells is shown in Figure 8.8.

The energy density, by volume, of gaseous hydrogen, or any fuel stored as a gas, is very low compared to gasoline, and, for automobiles and other vehicles, liquid hydrogen has problems of safety and a limited holding time in cryogenic containers. For airliners, liquid hydrogen could be an attractive fuel, especially in consideration of the very high energy content on a weight basis. Liquid hydrogen has about three times the energy of jet fuel by weight, but for a given amount of stored energy, the liquid hydrogen containers would have to be about three times as large as those for jet fuel. Development work has been underway for many years on storage of hydrogen as a metallic hydride at an energy density comparable to that of liquid hydrogen, but at room temperature. This could open the way to the use of hydrogen as a practical automobile fuel. However, in spite of a large effort, not all the technical problems have yet been overcome.

There is widespread and serious concern about the safety of hydrogen as a fuel, with the burning of the dirigible *Hindenberg* in 1937 cited as an example of the danger. In some respects hydrogen is a more dangerous fuel than natural gas, because it forms an explosive mixture with air over a very wide range of concentrations, from 4 to 75% hydrogen; natural gas is flammable only in the range of 5 to 15% concentration in air. In addition, the ignition energy for hydrogen–air mixtures is very small (about 2×10^{-5} J), only about one-fifteenth

Figure 8.8 A hydrogen-powered fuel cell car, the Honda FCX, certified as a zero-emission vehicle by the California Air Resources Board. Several of these cars are in regular operation in California and Japan. (*Source*: Courtesy American Honda Motor Co., Inc.)

as much as for natural gas or for a gasoline–air mixture. This small ignition energy can easily be provided by a spark of static electricity. On the side of greater safety, any hydrogen from a leak will dissipate more rapidly into the atmosphere than will methane. Because of hydrogen's low density it will quickly float upwards. Early detection of leaks can be aided by adding an odorant to hydrogen, as is now done routinely with natural gas. Open hydrogen flames can be especially dangerous because they are nearly invisible; this may dictate adding an illuminant to the hydrogen gas to make the flames more easily visible.

While the idea of a hydrogen economy is attractive in many ways, the magnitude and complexity of the problems in switching from our present fossil fuel economy should not be underestimated. The time involved for a significant change is certainly decades. In the meantime, energy conservation measures, improved fuel efficiencies, and various renewable energy projects will help to bridge the gap.

8.3.5 *Alcohol as a Transportation Fuel*

The search for an alternative to the gasoline-powered automobile has ranged over many possibilities. Some of these possibilities are radical departures from our conventional liquid-fueled cars. A more conventional approach is the mere substitute of another liquid fuel for gasoline. The liquid fuel could be either ethanol ("grain alcohol," CH_3CH_2OH) derived mainly from biomass as discussed in Section 5.7, or methanol ("wood alcohol," CH_3OH) which can be derived from coal or natural gas or from renewable resources. In this section, we shall consider methanol as a motor fuel. Ethanol production and use have been discussed previously in Chapter 5.

There would be no sacrifice of vehicle performance in switching from gasoline to methanol. In fact, for the past 30 years all of the Indianapolis 500 race cars have been powered by methanol. This fact alone should remove all questions about acceleration, range, and top speed. The alcohol fuels offer a high energy density compared to most other alternatives to gasoline. The heat energy content of gasoline is about 125,000 Btu/gallon, that of ethanol is 84,600 Btu/gallon, and methanol has 64,600 Btu/gallon. Vehicle ranges of hundreds of miles on a single tank of alcohol fuel are possible.

When methanol burns in air, the flames are not easily visible. This presents a hazard because a person can unknowingly walk into a flame and become seriously burned. Addition of a colorant to the methanol can make the flames visible, increasing the safety of the fuel. It has been reported that "If an engine fire develops in a methanol-fueled Indy race car, the pit crew simply pours water on the fire to put it out." The greater safety of methanol in fires as compared to gasoline is one reason it is the only fuel used at the Indianapolis 500.

It is well known that methanol is acutely toxic to humans; this is a serious objection to its widespread use as a motor fuel. Blindness can occur from severe exposures. Whereas no cumulative toxic effect occurs for ethanol, that is not true for methanol. Daily exposure to methanol fumes is considered to be hazardous. It should be noted that frequent exposure to gasoline fumes is also a health hazard.

Methanol burns by combining with oxygen from air according to the equation:

$$2CH_3OH + 3\ O_2 \rightarrow 2CO_2 + 4H_2O + \text{heat energy},$$

producing only carbon dioxide and water. Aside from the carbon dioxide, there are no harmful emissions except for the possibility of nitrogen oxides formed from air in the combustion chamber of an internal combustion engine. This depends on combustion temperature and pressure.

Methanol can be made from any renewable energy resource containing carbon, such as municipal solid waste and biomass crops—even seaweed. It can also be made from nonrenewable fossil fuels such as natural gas and coal. The manufacture of methanol from coal would be one way to provide a liquid transportation fuel from our vast coal resource, but it would not eliminate the emission of carbon dioxide as a greenhouse gas. Methanol from biomass would be attractive in this regard, because then the carbon dioxide becomes a part of the plant-growing cycle.

In 2005, methanol cost close to $1 a gallon, and it takes nearly two gallons of methanol to equal the energy content of a gallon of gasoline, which now sells for about $3 per gallon. So methanol may present a cost advantage to the consumer.

Alcohol is not just another motor fuel for the distant future. Its use has been demonstrated for years in Brazil and elsewhere. Even the American automobile manufacturers are taking this energy source seriously, as shown in Figure 8.9.

270 Chapter 8 **Transportation**

Figure 8.9 Although this appears to be a perfectly normal car, it has some special features. It is the Ford Flexible Fuel Vehicle (FFV), an adaptation from a regular production Taurus, and it will operate on methanol, ethanol, gasoline, or any combination of the fuels. (*Source*: Courtesy Ford Motor Company)

The very strict California laws on automotive emissions are certainly stimulating some of the effort to bring these vehicles onto the market. Alcohol motor fuels are taxed by both state and federal governments, at a rate generally comparable to that of gasoline on an energy basis.

In addition to use as a combustion fuel powering ordinary automobile engines, methanol is likely to be an excellent energy source for vehicles powered by liquid-feed fuel cells. These devices are a remarkable new variation on the conventional hydrogen–oxygen gas-feed fuel cells described earlier under the discussion of hydrogen as a secondary energy source. The liquid-feed fuel cell is still under development. It is yielding efficiencies as high as 34% for converting the energy content of liquid methanol directly into electrical energy. If this electricity then powers an electric drive motor having 90% efficiency, the combination will offer about twice the energy efficiency of an internal combustion engine powered by methanol. This favorable efficiency is related to the fact that whereas the internal combustion engine is a heat engine subject to the limitations of the Carnot efficiency, the fuel cell is not. With further development, an efficiency of 40% is considered attainable for the direct liquid-feed methanol fuel cell. (Similar reports are now emerging from development laboratories for fuel cells fed by liquid gasoline, with suggestions that this technology might be on the market sometime after the first decade of the present century.) In the liquid-feed methanol fuel cell, electricity is produced directly by an electrochemical reaction of oxygen from air with hydrogen from the liquid methanol. The liquid-feed fuel cell is not yet on the market, but it is said to be simple, quiet, and inexpensive to manufacture, with no moving parts, it runs cool, and is highly efficient.

Today most of our methanol is manufactured from natural gas with 60% energy efficiency, yielding a convenient liquid fuel from a gaseous energy resource.

As stated earlier, the liquid methanol can then be used either as a combustion fuel or perhaps as a feedstock to a fuel cell. In both cases, the natural gas that went into methanol manufacture is the primary energy resource. The methanol may be seen as a secondary source or a carrier of energy.

As a liquid fuel for internal combustion engines, methanol appears to be facing a declining future. There now are only about 5000 (down from 18,000 in 1995) methanol-powered vehicles in use in the United States, as compared to about 150,000 (up from 1500 in 1995) U.S. vehicles powered by ethanol (E85). Also, it is reported that the Indianapolis 500 race cars will switch from methanol to ethanol over the next few years. One recent tabulation lists nearly 200 refueling sites in the United States for ethanol (E85) and none for methanol. Ten years ago, methanol was more widely used as a fuel than ethanol; now the trends are reversed. The reasons for this reversal involve the fact that methanol is primarily derived from natural gas, and ethanol from biomass. As the domestic natural gas reserves decline and the price increases, there is less support for promoting methanol as a liquid fuel. Also, compared to ethanol, methanol is far more toxic, and there are reports that it can be more corrosive to engine components.

In summary, methanol has been demonstrated as a liquid fuel for conventional internal combustion engines. It can also serve as a highly energy-dense storage medium capable of being re-formed into hydrogen gas for use in a standard fuel cell aboard a vehicle. It is a means by which natural gas, coal, and biomass can be the source of a combustible liquid fuel for transportation. Newly developed liquid-feed fuel cells offer the possibility of generating electricity directly from methanol, avoiding the Carnot limitation on efficiency. There is hope that this new development may offer a means of doubling the efficiency with which we use our fossil fuels for transportation.

8.4 Traffic Safety

8.4.1 The Record

Transportation, in general, certainly ranks among our most dangerous activities. In many countries, such as the United States, even wars have not come close to traffic in terms of total deaths. In the past few decades there have been an average of more than 20 million accidents, about 6 million injuries, and about 50,000 deaths each year in the United States. These numbers have come down somewhat in the last 15 years. Since the invention of the automobile, there have been more than 2 million traffic deaths in the United States. If the 50,000 annual death toll were to occur in a single yearly event, it would exceed any disaster ever recorded in this country, and each year we would experience a single disaster exceeded by only a few events in human history. One might seriously ask whether this is the equivalent of a single every-ten-years disaster that takes 500,000 lives, or whether the risk of traffic death surpasses the risk of energy-producing technologies, or whether the risk of traffic injury is not greater than that of the injurious effects of industrial pollutants in the environment.

In terms of the average life-shortening effect, traffic possibly surpasses the total of environmental pollutants because of the relative youth of its victims. This can be estimated by assuming that the average victim loses 30 years of life, and by considering the relative probability of traffic death compared with all other causes of death.

$$\text{Average life shortening} = (\text{years lost per traffic death})$$
$$\times (\text{probability of traffic death})$$
$$= 30 \text{ yr} \times \left(\frac{42,000}{2,400,000}\right)$$
$$= 0.53 \text{ yr} = 6.3 \text{ months}$$

In this estimate we have used the approximate values of 42,000 traffic deaths per year and 2.4 million total deaths per year in the United States.

Many studies show that the traffic accident death rate is correlated with the age of the driver. The results of one such study are shown in Figure 8.10, which compares the rate of fatal crash involvement with driver age. At driver ages below 25 it is as much as five times higher than for middle-aged drivers, presum-

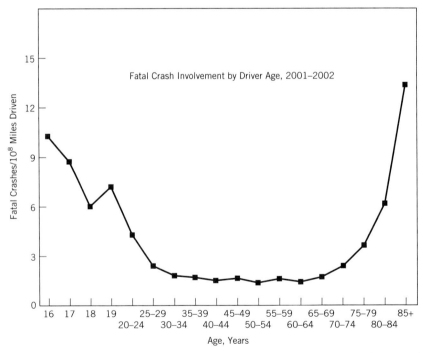

Figure 8.10 Fatal crash involvement by driver age. The high rates at the lower and higher ends of the age range are discussed in the text. (*Source*: Insurance Institute for Highway Safety, 2002.)

ably because of driver inexperience along with youthful lifestyle and less appreciation of driving risks. At the higher ages, above 75, the rate again goes sharply upward. This increase is reported to be due more to age-related fragility than to a tendency to get into serious accidents, although older drivers are markedly overinvolved in crashes. Apparently drivers at least risk are in the 25 to 70 age range.

When motor vehicle accidents, with 42,000 deaths in 2002, are compared to other causes of accidental death in the United States, it is apparent that nothing else comes close. Table 8.5 lists six of the leading causes of accidental death. Drownings and fires are each seen to be less than one-tenth as costly of life; only falls, which accounted for 16,000 deaths, begin to compare. On a per vehicle-mile basis, as shown in Table 8.6, motorcycles are by far the most hazardous of contemporary means of transportation.

The data of Table 8.6 can also be looked at in terms of deaths per passenger-mile instead of deaths per vehicle-mile. To see this, divide the given per-vehicle number by the number of passengers in the vehicle. Using the train as an example, let's estimate 100 passengers. That gives us a rate of 0.02 deaths per 100 million passenger-miles, to be compared to the 1.5 deaths per 100 million passenger-miles for an automobile with a single occupant. By this method, it becomes apparent that the risk of automobile travel ranks next after motorcycles, and all means of mass transportation, including airlines, are far down the list. According to the National Safety Council, the passenger death rate in automobiles in 2000 was 0.8 per 100 million passenger-miles. For buses, trains, and airlines, respectively, the corresponding rates were 0.05, 0.03, and 0.02.

There can be little question that if we are to make a serious effort to reduce accidental death rates, the dangers of automobile travel must be addressed with high priority. Airline fatalities in the United States totalled only 22 in 2003 and there were none in 2002, yet there seems to be public concern about airline safety far beyond the attention given to the risks of automobile travel.

Table 8.5 Fatalities in the United States by Type of Accident (2002)

Motor vehicle	42,000
Falls	16,000
Drowning	3,400
Fires	3,200
Electrocution	430
Lightning	75

Source: National Safety Council, 2005.

Table 8.6 Traffic Fatalities per Vehicle-Mile in the United States

Vehicle	Deaths per 10^8 Vehicle-Miles
Automobiles	1.5 (2002)
Automobiles on rural interstate highways	1.2 (2002)
Automobiles on urban interstate highways	0.6 (2002)
Bus	0.5 (2001)
Train	2.0 (2003)
Airline	0.000 (2002)
	0.313 (2003)
Motorcycle	34 (2002)
Horses	30 (1909)
(26×10^6 horses, 1.3×10^{10} miles, 3850 deaths)	

Sources: Statistical Abstracts of the United States, 1996; U.S. Bureau of Transportation Statistics, 2003.

8.4.2 Collisions

The immediate causes of injury and death in traffic accidents are numerous, ranging from broken glass, to fire, to crushing, to the hurling of victims against components of the car's interior, to being thrown from the car. All of these factors must be considered in designing a safer vehicle, but there is one basic factor in determining the survivability of a crash. That factor is the *acceleration* experienced by the passenger. It has been known since the days of Newton that acceleration of any object is related to the force acting on it by the fundamental equation, $F = ma$, arguably the most important single equation in all of physics, which states that the force on any object is given by the product of its mass times its acceleration. As an example, if a 150 pound person experiences a horizontal acceleration of 5 g (or five times the acceleration of the earth's gravity), that person will have an effective weight of $5 \times 150 = 750$ pounds in the horizontal direction in addition to 150 pounds vertically downward.

Since we understand acceleration to mean a change in velocity divided by the time in which the change took place, collisions of the type in which velocity is changed very abruptly produce the largest acceleration, and hence, the greatest forces. They are the most dangerous. An especially severe situation is presented when a vehicle strikes an immovable object, causing the vehicle's speed to be reduced from its initial value down to zero in a very short time.

There are several ways to express the maximum tolerable acceleration in a collision. Experiments with human volunteers have shown that an acceleration of 20 g is tolerable without injury. Published safety standards are in the range of 10 to 80 g, where the higher values can be experienced without injury only if the exposure is for a shorter time.

It is important for designers of vehicles and roadways to have some idea of what values of acceleration are survivable so they can do their design work accordingly. It should be noted that it is not sufficient to ensure that the passenger compartment of a vehicle has a tolerable acceleration. This consideration must apply specifically to the bodies of the passengers, and even more specifically to certain parts of the body. In some circumstances (see Example 8.5), the passenger compartment of a car may experience acceleration of a magnitude that is clearly survivable if the passengers are properly restrained by seat belts or air bags, but those passengers not so restrained will have no hope of survival. A second case in which the acceleration experienced by persons must be properly considered is that of an auto–pedestrian collision. This is actually two collisions; first the car hits the pedestrian, then the pedestrian hits the pavement. Even if the auto is designed to be free of protrusions and unyielding materials on the front end so that the first collision in this sequence is not fatal, the second may be.

Attempts to reduce the death and injury toll in automobile accidents must take into account that front-end collisions are responsible for most of the injury-producing accidents, followed by rollovers. This suggests that a serious effort must be given to designing the front ends of cars so that they will crumple in a way that results in a uniform acceleration over a relatively long distance, in order to keep the forces on the passengers low. It is equally important for the passengers to be physically restrained within the passenger compartment so that their acceleration will be controlled and they will not be thrown from, or about inside the vehicle during a rollover. It is difficult to overestimate the benefits of good passenger restraint systems; a published Swedish study found *no* deaths of seat-belted passengers in collisions up to 60 mph but reported deaths in collisions at speeds as low as 12 mph for passengers not restrained by seat belts. Similar statistics were reported by a study in the state of Utah.

The major shortcoming of seat belts is that they are not used on a regular basis by many drivers and passengers, at least in the United States. As demonstrated in several countries of the world, it is possible to enforce the wearing of seat belts through laws, and thus reduce the injury and death rate, but that is still not a universal practice in this country. Air bags provide much the same benefits as seat belts for front-end collisions even though the drivers and passengers take no deliberate actions to protect themselves. For various reasons, however, air bags were not standard equipment in all cars built in the 1990s in the United States. Objections to the early air bags included the arguments that they were generally ineffective in collisions other than the front-end type, their sudden inflation could produce hearing damage, their accidental inflation could cause an accident, there was no obvious way to test their readiness, and they did not produce a uniform deceleration during the course of a collision. There have also been instances of airbags causing injury and even death, particularly to children and smaller persons sitting close to the air bag. This very serious problem has been addressed by reducing the inflation rate to be slower than that which is known to cause injury or death, by advising drivers not to sit too close to the steering wheel, and by requiring that children sit only in the back seats. In spite of their admitted problems, there is growing evidence that air bags are an effective means of saving lives and reducing injury.

Air bags are now required equipment in cars and light trucks, and it is generally recommended that, even in vehicles equipped with air bags, seat belts should also be used. Rules regarding air bags continue to evolve, and these rules are now addressing the need for side-impact protection, head protection, and protection of back-seat as well as front-seat passengers. Several testing programs now evaluate and report on the effectiveness of the air bags installed in vehicles available to the public.

Example 8.5

An automobile traveling 55 mph crashes head-on into an immovable barrier. Assume a uniform rate of deceleration during the collision.

(a) How much crumple distance must be designed into the front end of the car so that the passenger compartment will not experience an acceleration of greater than 30 g? Use the equation $v_f^2 = v_i^2 + 2ax$, which relates the final velocity, the initial velocity, the acceleration, and the distance.

(b) What is the duration of the crash in seconds? Use the equation $v_f = v_i + at$, which relates the final velocity, the initial velocity, the acceleration, and the time.

(c) A careless passenger not wearing a seat belt stops in one-fifteenth the distance it took to stop the passenger compartment (because the passenger flies through the air before striking the already stopped dashboard and windshield). How many g would the passenger experience? Would the passenger be likely to survive?

Solution

(a) Start with a relationship between final velocity, initial velocity, acceleration, and distance:

$$v_f^2 = v_i^2 + 2ax$$

and let v_f be equal to zero because the car comes to a stop. Then

$$x = \frac{v_i^2}{2a}$$

Given $a = 30\ g$, $g = 32$ ft/sec^2, $v_i = 55$ mph $= 81$ ft/sec. Then

$$x = \frac{(81 \text{ ft/sec})^2}{2 \times 30 \times 32 \text{ ft/sec}^2} = \mathbf{3.4\ ft}$$

(b) From $v_f = v_i + at$, with $v_f = 0$, and a negative (because the car is decelerating), we get

$$t = \frac{v_i}{a} = \frac{v_i}{30 \times g} = \frac{81 \text{ ft/sec}}{30 \times 32 \text{ ft/sec}^2} = \mathbf{0.084\ seconds}$$

(c) In this case the stopping distance is

$$\frac{1}{15} \times 3.4 \text{ ft} = 0.23 \text{ ft} = 2.8 \text{ in.}$$

and using the same equation as was used in part (a), we solve for the acceleration to get

$$a = \frac{v_i^2}{2x} = \frac{(81 \text{ ft/sec})^2}{2 \times 0.23 \text{ ft}} = 14{,}263 \text{ ft/sec}^2$$

If we now divide by $g = 32$ ft/sec^2, we find that the acceleration experienced is

$$\frac{14{,}263 \text{ ft/sec}^2}{32 \text{ ft/sec}^2} = \mathbf{446 \ g}$$

or 446 times that of gravity. An unrestrained passenger would not survive the collision.

In addition to the measures indicated above with regard to controlling the accelerations that the passengers experience, it is equally important that automobiles be designed so that the passenger compartment maintains its integrity during a collision. The engine must not move back into the passenger space. The steering column must collapse rather than force the steering wheel back against the driver's chest.

In recent years it has become common practice for highway departments to place yielding structures, or even just arrays of yielding materials, such as empty oil drums, in front of unyielding concrete bridge abutments and other solid roadside objects to reduce the acceleration experienced by vehicles that may collide with these objects. Such measures, in combination with seat belts and air bags, make it possible that collisions even at 60 mph will be survived by all the passengers in an automobile. In the absence of these simple, inexpensive, and yet very effective measures, survival would be highly unlikely.

8.4.3 *Nontechnical Traffic Safety Measures*

An effective program of traffic safety certainly depends on more than the mere technical knowledge of possible means of improving the vehicles and roadways. Driver education and testing are obviously important, as are strict and effective measures to keep drinking drivers off the road. Traffic safety can be as much a legislative matter as it is a matter of engineering and education. Some of the impact of legislation is indicated in Figure 8.11, which suggests that a rising traffic death rate over a several-year period was reversed by passage of the Traffic Safety Act in 1966. This act included numerous specifications for the crashworthiness of automobiles. The legal requirement for seat belts, head restraints, padded dashboards, collapsible steering columns, hazard flashers, dual braking systems, and many other safety provisions made all these features a standard

278 Chapter 8 **Transportation**

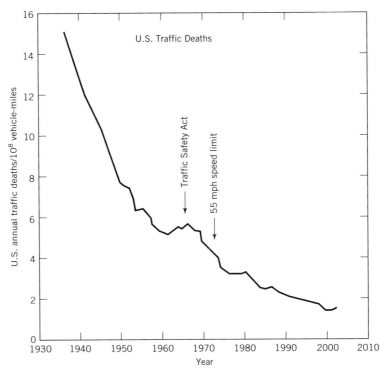

Figure 8.11 A 70-year record of U.S. traffic deaths per 100 million vehicle miles. From 1936 to 1960 there was a rapid decline due to improvement of vehicles, roadways, and driver licensing. After 1960 the rate started to turn up again until passage of the Traffic Safety Act. A further drop after 1973 is often attributed to the 55 mph speed limit. In 1988 the speed limit on some interstate highways was increased to 65 mph. In 1995 it became 75 mph in some locations. Over the time period shown on this graph the number of vehicle-miles in the United States each year increased more than ten times. (*Sources*: *Statistical Abstracts of the United States*; U.S. Bureau of Transportation Statistics, 2003.)

part of every new automobile on the market for the first time, even though they had been technically feasible for many years. It took the force of legislation to reduce traffic death and injury through these technical means. Of course, the full effect of such measures, which apply only to new vehicles, does not show up in accident statistics until several years after the regulations are put in force. A further reduction in the highway death rate may have been brought about starting in 1973 with the lowering of the speed limit to 55 mph, another purely regulatory action. It is certain that many thousands of lives have been saved and countless injuries prevented in these ways with relatively little sacrifice on the part of those using the roadways.

There is convincing evidence that these safety measures are effective. This is seen in Figure 8.11, and is reinforced by the statistic of only 42,600 U.S. traffic fatalities in 2003, about 20% lower than in 1980. The total 42,600 traffic fatalities in 2003 include about 5,400 pedestrians, bicyclists, and other vehicle non-occupants. In 1980, the non-occupant fatalities were more than 9,000. The total number of miles driven each year in the United States has doubled during that time.

8.5 Mass Transportation

Our present transportation system is emphatically centered on the private automobile. More than three-quarters of our more than 220 million motor vehicles are passenger-carrying vehicles. The attraction of the private automobile is obvious: the reliability, privacy, luxury, and flexibility it offers in arranging peoples' travel are unmatched by any other known system of passenger transport. And in many situations, the cost of automobile travel is competitive with buses, trains, planes, or any other alternative. Table 8.7 shows that more than three-quarters of American workers drive to work alone. It seems unlikely that Americans will be enthusiastic about any changes that will separate them from their private conveyances.

Why should anyone question the superiority of our system of private vehicles when it is so clearly what the public has chosen? The reasons are several. Many cities of the world are being increasingly plagued with problems of air pollution, primarily because of the automobile. This is more than just a problem of simple annoyance; human health is involved as is the deterioration of property. There is also a strong argument on behalf of the national economy. We are now exporting many tens of billions of dollars annually to purchase petroleum so that we may enjoy our cars and trucks. As our national reserves of petroleum diminish, this problem will likely become more severe. Aside from the matter of

Table 8.7 How Americans Get to Work

Transportation Mode	Percent
Drive alone	76.3
Carpool	11.2
Public transportation	5.2
Walk	2.7
Bicycle	0.4
Motorcycle	0.1
Other	0.9
Work at home	3.2

Source: Transportation Energy Data Book, Edition 24, ORNL-6973, 2004.

petroleum depletion, the proliferation of automobiles is enormously consumptive of other resources such as steel, aluminum, and other metals. This problem exists because our recycling system is far from complete in its recovery of these resources and because of the relatively short lifetime of automobiles. The private automobile is also a spectacularly dangerous mode of travel. All means of public conveyance are safer.

Most of the problems caused by a transportation system based on more than one private vehicle for every two citizens can be at least alleviated by the adoption of mass transit systems. Certainly something is lost in the way of convenience and privacy, but, on the other hand, we can be freed from problems of maintenance and parking, and the time in transit can be used for pursuits other than guiding a vehicle through a maze of traffic. Many of the advantages of mass transit can be quantified through a measure commonly known as *passenger efficiency*. This can be calculated for any vehicle as

$$\text{Passenger efficiency} = \frac{\text{number of passengers} \times \text{miles traveled}}{\text{fuel consumed}}.$$

Passenger efficiency is commonly reported in terms of passenger-miles per gallon of fuel, although in Table 8.2 passenger efficiency is given in passenger-miles per million Btu of fuel energy. For buses and trains with most of their seats occupied, the passenger efficiency is generally better than for an automobile, at least when the automobile carries only one or two passengers. However, a highly efficient small car carrying four passengers on the open highway can achieve a passenger efficiency comparable to that of the best mass transit systems. Fully loaded passenger aircraft can now achieve a passenger efficiency comparable to that of typical wheeled vehicles.

By shifting our transportation system to vehicles operating at high passenger efficiency, the air pollution problems related to fuel combustion will be relieved, as will the national balance-of-payments problem. Beyond this, trains and local buses can be powered by electricity produced at central power stations equipped with emission controls more effective than those operating on the individual internal combustion engines. These central power stations could even be based on the renewable energy sources: hydropower, geothermal energy, windpower, or solar energy. Today they are still more likely to be coal-burning or nuclear, but they need not be dependent on imported petroleum. Electric powered railways are common in Europe and are still found in our larger cities in the form of trains, subways, and cable cars, but they are less popular in the United States now than they were in the first half of the 20th century, before the automobile achieved its overwhelming dominance.

For a bus system or any other means of public transportation to be economically viable, it must be supported by a general public subsidy, by fares, or by some combination of the two. Very few local mass transit systems in the world are supported solely by fares. However, intercity bus systems that travel at higher speeds and with higher capacity factors have consistently earned profits based on passenger fares alone.

The difficulty of making a mass transit system pay for itself through passenger fares can be illustrated by looking at an example for buses. Suppose a bus costs $200,000 and remains in service for 300,000 miles. This reduces to $0.67 per mile as a capital cost. The interest cost on the financing of this capital investment may add another $0.10 per mile. With diesel fuel at $2.00 per gallon and a mileage rate of 4 miles per gallon, another $0.50 per mile is added. Maintenance costs, including tires, labor, lubrication, parts, and cleaning can be estimated at another $0.15 per mile. The driver's salary is estimated at $20.00 per hour, and with an average bus speed of 20 mph, this adds another $1.00 per mile. For the moment let's be optimistic and ignore the considerable costs of street construction and maintenance, publishing of schedules, office and garage space, supervisors, and many other overhead items. If we add the costs together we find that the operating cost is at least the following:

Item	Dollars per mile
Bus	$0.67
Financing	0.10
Fuel	0.50
Maintenance	0.15
Driver salary	1.00
Total	**$2.42**

The driver's salary accounts for about 40 percent of the total cost. This suggests that only limited savings can be effected by reducing the size of the vehicle at times and in places where a full-sized bus is not needed. At a passenger fare of $0.50, the bus must pick up at least 5 passengers per mile just to break even. If the fare is $1.00, the bus must board more than two passengers per mile. This may be easy to do in an urban setting during the morning and afternoon rush hours, but it will be difficult during most of the rest of the day, leaving the system to operate at a financial deficit during off-peak hours. It seems clear that local mass transit systems can be self-supporting only in areas with sufficiently high population density so that there will generally be several passengers boarding each bus per mile of its travel. Our prevalent pattern of low-density suburban sprawl certainly works against the financial success of any local mass transit system.

Local mass transit rail systems have the added handicap of having to finance their own roadways rather than using public streets. Economic feasibility studies typically report costs of 20 to 30 million dollars *per mile* to construct a rail system. These capital construction costs alone are a tremendous burden. If each mile of track were traversed 100 times a day, 365 days a year, for 20 years, the capital cost of the system per mile traveled by a light rail train can be estimated as

$$\text{Capital cost per mile of train travel} = \frac{25 \times 10^6 \text{ dollars/mile}}{100 \text{ trips/day} \times 365 \text{ days/yr} \times 20 \text{ yr}}$$

$$= \$34/\text{mile}.$$

By comparing this to the previous figure of less than $1 capital investment per mile for the bus system, it is apparent that one train must be the equivalent of a very large number of buses for the train system to be financially competitive. Of course the train system does carry more passengers per driver, and thus may save on a major operating cost. In general, rail systems have higher initial costs and lower operating costs than do bus systems. And, obviously, it is more difficult to rearrange routes and schedules with rail systems.

Key Terms

Passenger-miles
Passenger efficiency
Ton-miles
Acceleration
Rolling resistance
Aerodynamic drag
Aerodynamic drag coefficient
Frontal area

Terminal velocity
Flywheels
Hybrid vehicles
Plug-in hybrids
Fuel economy
Energy density
Power density
Fuel cells

Suggested Reading and References

1. Davis, Stacy C., and Diegel, Susan W. *Transportation Energy Data Book*, Edition 24, ORNL-6973. Oak Ridge National Laboratory, December 2004.
2. Bartlett, A. A. "The Highway Explosion." *Civil Engineering*, December 1969.
3. Ross, Marc. "Energy and Transportation in The United States." *Annual Review of Energy*, 1989, p. 131.
4. Post, Richard F. "A New Look at an Old Idea, The Electromechanical Battery." *Science and Technology Review* (April 1996). Lawrence Livermore National Laboratory.
5. Post, R. F., and Post, S. F. "Flywheels." *Scientific American*, **229**, No. 6 (December 1973), pp. 17–23.
6. Nadis, Steve, and MacKenzie, James J. *Car Trouble*. World Resources Institute. Boston: Beacon Press, 1993
7. Gray, Charles L., Jr., and Alson, Jeffrey A. "The Case for Methanol." *Scientific American* (November 1989), pp. 108–114.
8. Sperling, Daniel. "The Case for Electric Vehicles." *Scientific American* (November 1996), pp. 54–59.
9. Burns, Lawrence D., McCormick, J. Byron, and Borroni-Bird, Christopher E. "Vehicle of Change." *Scientific American*, **287**, No. 4 (October 2002), pp. 65–73.
10. Ashley, Steven. "On the Road to Fuel-Cell Cars." *Scientific American*, **292**, No. 3 (March 2005), pp. 62–69.

11. Bezdek, Roger H., and Wendling, Robert M. "Fuel Efficiency and the Economy." *American Scientist*, **93** (March–April 2005), pp. 132–139.
12. Crabtree, George W., Dresselhaus, Mildred S., and Buchanan, Michelle V. "The Hydrogen Economy." *Physics Today*, **57,** No. 12 (December 2004), pp. 39–44.

Questions and Problems

1. We know from the Principle of Energy Conservation that energy cannot just disappear. In this chapter we are told that most of the energy used to propel a car at highway speeds goes into overcoming air resistance. Then what happens to this energy? Follow it as far as you can.
2. A parachute jumper weighs 200 pounds (including gear), has a parachute with an area of 200 square feet, and a drag coefficient of 1.0. What is the highest downward speed (the terminal velocity), in miles per hour, that this jumper can attain? At the terminal velocity the downward force (the weight) is equal to the upward force (the aerodynamic drag force).
3. Describe several measures that can be taken to improve the safety of streets and highways. Do not include changes or improvements in vehicles and drivers, but only changes to the fixed elements of streets and highways.
4. The large automobile manufacturers are now marketing or are announcing serious intentions to market a hybrid automobile of one type or another. Summarize the characteristics of at least one of these designs, using recent information from newspapers, magazines, television, the Internet, and so forth.
5. Estimate the cost to the consumer for transportation by various means on a dollars per passenger-mile basis. Include airliners, trains, buses, privately owned autos, car pools, taxis, bicycles, walking, horses, and whatever else you'd like to include.
6. Discuss the probable environmental impact of various modes of transportation on a per passenger-mile basis.
7. Estimate your chances of not surviving the drive to a new job in Seattle from your university graduation in Miami.
8. If you drive alone to the airport 40 miles away, how do the chances of not surviving the drive compare to the odds of not surviving the airline travel on the 2000 mile trip you are about to take?

Multiple Choice Questions

1. In the United States, about _____ of our national energy consumption is accounted for by fuel used in transportation.
   - **a.** 1%
   - **b.** 5%
   - **c.** 8%
   - **d.** 27%
   - **e.** 50%
   - **f.** 75%
   - **g.** 95%
   - **h.** 99%

2. How much force is needed for a 2700-pound car to accelerate at 3 ft/sec^2? Neglect all other forces such as aerodynamic drag.
   a. 253 lb
   b. 84 lb
   c. 8100 lb
   d. 810 lb
   e. 402 lb
   f. 2700 lb
   g. 900 lb
   h. none of the above

3. A box with a frontal area of 2.5 ft^2 and with a drag coefficient of $C_D = 0.9$ is fastened on the roof of your car. What is the aerodynamic drag force in pounds due to the box alone at a velocity of 88 ft/sec^2?
   a. 0.4 lb
   b. 0.5 lb
   c. 6.3 lb
   d. 9.2 lb
   e. 11 lb
   f. 22 lb
   g. 47 lb
   h. 94 lb

4. A total force of 88 pounds is needed to have a vehicle move at a constant 95 ft/sec. What horsepower must be delivered to the drive wheels?
   a. 18
   b. 83
   c. 76,000
   d. 138
   e. 800
   f. cannot be calculated because the mass is not given

5. On I-70 just west of the Eisenhower tunnel, there is a long 7% downgrade. What is the terminal (maximum) velocity of a large truck coasting down this hill? Assume that aerodynamic drag is the dominant retarding force, $w = 60,000$ lb, $A_f = 100$ ft^2, and $C_D = 0.7$.
   a. 109 mph
   b. 119 mph
   c. 129 mph
   d. 139 mph
   e. 149 mph
   f. 159 mph
   g. 169 mph
   h. 179 mph

6. The aerodynamic drag coefficient, C_D, is determined mostly by the _____ of a vehicle.
   a. speed
   b. velocity
   c. size
   d. shape
   e. area
   f. mass
   g. weight
   h. energy

7. At highway speeds, vehicles exert most of their driving force against _____ .
   a. aerodynamic drag
   b. rolling resistance
   c. acceleration
   d. deceleration
   e. engine bearings
   f. powering accessories
   g. tire flexing
   h. brake drag

Multiple Choice Questions

8. On the average, each American experiences a life-shortening of about _____ due to traffic accidents.
   a. 6 seconds
   b. 6 minutes
   c. 6 hours
   d. 6 days
   e. 6 weeks
   f. 6 months
   g. 6 years
   h. 6 decades

9. A 150-pound person experiencing an acceleration of 30 g has an effective weight of _____ .
   a. 5 lb
   b. 30 lb
   c. 150 lb
   d. 960 lb
   e. 1500 lb
   f. 4500 lb
   g. 6000 lb
   h. 7500 lb

10. In terms of increasing Btu per ton-mile for freight transportation, the correct ordering would be _____ .
    a. railroad, truck, airplane
    b. railroad, airplane, truck
    c. truck, railroad, airplane
    d. truck, airplane, railroad
    e. airplane, truck, railroad
    f. airplane, railroad, truck

11. A diesel-powered commuter train uses 5 gallons of fuel for each mile of travel and carries 800 passengers. The passenger efficiency is _____ passenger-miles per gallon.
    a. 0.00625
    b. 1600
    c. 5
    d. 800
    e. 160
    f. 320
    g. 400
    h. 4000

12. If a bus averages 250 miles travel for each 8-hour shift on its route, and if the driver's pay, including all benefits, is $16/hr, how many passengers, at 50 cents each, must be picked up each mile just to pay the driver?
    a. 0.3
    b. 0.5
    c. 0.7
    d. 1.0
    e. 2.0
    f. 3.0
    g. 4.0
    h. 5.0

13. The total operating cost of a certain urban passenger train is $19.65 per mile. The average speed is 15 miles per hour. How many passengers at 75 cents each must board per hour in order to have the system break even financially?
    a. 28
    b. 35
    c. 41
    d. 393
    e. 4140
    f. 280
    g. 145
    h. 64

14. In terms of traffic fatalities per passenger-mile, the order from the safest to the least safe is _____ .
    a. auto, airline, bus, motorcycle, horse
    b. airline, bus, auto, motorcycle
    c. bus, train, motorcycle, auto
    d. train, auto, bus, horse
    e. horse, airline, bus, auto, motorcycle

15. In terms of traffic fatalities per vehicle-mile, the order from the safest to the least safe is _____ .
    a. auto, airline, bus, motorcycle, horse
    b. airline, bus, auto, motorcycle
    c. bus, train, motorcycle, auto
    d. train, auto, bus, horse
    e. horse, airline, bus, auto, motorcycle

CHAPTER 9

Air Pollution

(*Source*: Courtesy NASA)

9.1 Spaceship Earth

It is only in recent decades that there has been widespread awareness that our atmosphere and oceans can no longer be considered as infinite. For most of man's time on this planet it had been tacitly assumed that the wastes of our society could be dumped into the environment with effects too small to be important. The rate of dumping wastes has increased with increasing population and the expanding technical base for our way of life. Our ability to sense the effects on the environment has also improved. There is now clear evidence that we are seriously polluting the one atmosphere that we have. Mankind has only one home, earth, a small planet hurtling through space, and it is clear that we must take care of it as it takes care of us.

The sources of atmospheric pollution are many and have far-reaching results. For the first time, truly global effects have been uncovered, and the difficulties of achieving international agreements to mitigate the problems are also being revealed.

9.2 The Earth's Atmosphere

Many of our most severe problems of environmental pollution involve the earth's atmosphere. The atmosphere weighs about 5.7×10^{15} tons, about one-millionth the weight of the earth. The earth's surface covered by the atmosphere has an area of about 200 million square miles, and the farthest reaches of the atmosphere extend up to hundreds of miles over the entire surface, although most of the air is in a very thin layer over the earth's surface. Half of the air is below 18,000 feet altitude above sea level. It is no more than a layer of paint on a basketball. Near the earth's surface air has a density of about 1.3 kg/m^3. Atmospheric pressure at sea level is 14.7 lb/in.^{2} (1.01×10^5 N/m^2), where this is just the weight of the overlying atmosphere.

This density, and the corresponding pressure, gradually decrease with altitude. By 50,000 feet the pressure has been reduced to 1.6 lb/in.^{2} from the sea level value of 14.7 lb/in.^{2}, and by 600 miles altitude the atmospheric pressure is essentially zero. There is a structure to the atmosphere, in that some properties such as temperature gradient vary from one altitude to another. The names of the various regions of the atmosphere along with the temperature variations are shown in Figure 9.1. It is the troposphere which affects us most directly and with which we are mainly concerned. We shall see, however, that all regions of the atmosphere are important, as we can, with various pollutants, affect processes that take place right up to the mesosphere. These pollutants at the very high altitudes can affect conditions down into the troposphere where we live.

The major permanent constituents of the atmosphere are listed in Table 9.1. In addition to these permanent gases, there are a number of others, such as water vapor, carbon dioxide, methane, carbon monoxide, ozone, and ammonia, that fluctuate with time, altitude, and location. The concentration of water vapor in the air can be as high as 3%, but it is usually less than 1%, depending strongly on local conditions. Likewise, carbon dioxide is present, now at an average level of about 0.038%, but its concentration varies with time of year and location. Some of the gases that are present in relatively small amounts in the atmosphere play vital roles in absorption of certain parts of the spectrum of solar radiation, in the complex chemical reactions that take place in the upper atmosphere, and in the heat balance of the earth. Man is changing the concentrations of some of these gases, and the results of these changes are not always predictable. Thus,

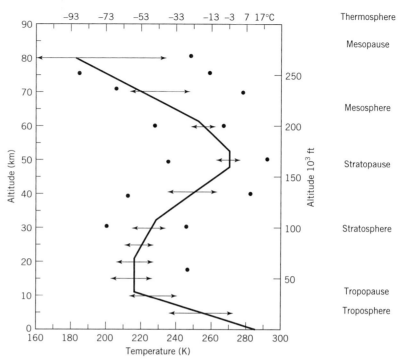

Figure 9.1 The temperature of the atmosphere as a function of altitude. The arrows indicate the normal range of temperature variation and the dots the extreme values. The names given to the various regions of the atmosphere are shown on the right.

Table 9.1 Major Permanent Constituents of the Atmosphere

Gas	Percent by Volume	Parts per Million
Nitrogen (N_2)	78.08	
Oxygen (O_2)	20.95	
Argon (Ar)	0.93	
Neon (Ne)		18.2
Helium (He)		5.2
Krypton (Kr)		1.1
Hydrogen (H_2)		0.5

we are in some instances doing giant experiments in our very thin and delicate atmosphere with our pollutants; the results will not be known for many years, and the processes may not be reversible.

9.3 *Thermal Inversions*

The old saying "the solution to pollution is dilution" has some basic truth in it. Although the real long-term solution to pollution problems is clearly to reduce the amount of emitted pollutants, there can still be something to be gained by dispersing the pollutants far from the point of emission. This is important to those who may live near the power plant or industry that produces the pollutants. The purpose of the tall smokestacks we see at coal-burning power plants and some other industries is to emit the pollutants into the atmosphere at an elevation where their chance of dispersal by being carried aloft, rather than hovering near the ground, is greatly enhanced.

Up to an altitude of about 10,000 meters, the temperature of the atmosphere normally decreases with increasing altitude. As shown in Figure 9.1, the temperature decreases from roughly 20°C at ground level to about −60°C at 10,000 meters. Above the troposphere, the temperature gradually increases through the stratosphere until about 50,000 meters where it starts to decrease again. The temperature profile varies considerably with time of year and location, but normally a negative temperature gradient exists near the earth, and this has important consequences for the dispersal of pollutants. It is commonly recognized that warm air will rise in the presence of surrounding cooler air. This observation derives from the fact that as air temperature increases its density decreases, requiring a parcel of warm air to float upward in a surrounding atmosphere of cooler, more dense, air. If a parcel of warm polluted air, for instance, from a smokestack, is released into the lower levels of the atmosphere under normal meterological conditions, it will rise in the atmosphere to as much as 10,000 meters, at which height it normally presents no immediate problems to the people on the ground below.

Not all meterological conditions, however, are conducive to this upward motion of the warmed polluted air. There is a relationship between temperature and altitude that is determined by the basic thermodynamic principles governing the behavior of a gas. The generally prevalent temperature–altitude relationship in the lower atmosphere is known as the *adiabatic lapse rate* (ALR). In thermodynamics, the word *adiabatic* describes any process in which no heat energy is either gained or lost by some defined volume of gas. The word *lapse* indicates that temperature decreases with increasing altitude. If a given parcel of air, warmer than its surroundings, starts to rise in the atmosphere, and if it can be considered to do so without exchanging heat energy with the neighboring air, it will expand and cool at the adiabatic lapse rate. Expansion occurs because as the air rises, the air parcel is under less pressure, owing to the thinner burden and lesser weight of the overlying air. The adiabatic lapse rate for dry air is

−1°C/100 m. If the air is moist, the ALR is less; it can be as low as −0.35°C/100 m. An approximate average ALR is −0.65°C/100 m. (It may be easier to remember a rate of approximately 3°F per 1000 feet. This is what we commonly see in going up or down mountains, or as an airplane is climbing or descending.) The ALR, in simple terms, is the rate at which the temperature of a volume of air will naturally tend to decrease as altitude increases, or increase as altitude decreases.

If the prevailing temperature profile, because of unusual meterological circumstances, is such that the atmospheric temperature decreases more rapidly with altitude than the ALR, then any parcel of air released near ground level and warmer than its surroundings will rise indefinitely into the upper atmosphere. This is because as it cools at the ALR, it will always be warmer, and thereby less dense, than the surrounding air. This unstable condition is obviously desirable because it leads to good vertical mixing and a relatively pollution-free lower atmosphere.

If the existing temperature profile and the ALR happen to be the same, there will be a neutral condition that neither forces the warm air upward nor traps it near the earth. On the other hand, it is possible that the existing air temperature profile may be such that the temperature decreases more slowly with altitude than indicated by the ALR. Under this condition, a volume of warm air released near ground level will rise in the ambient cooler air, cooling at the ALR as it rises, until at some level it is no longer warmer than its surroundings, at which point it will then cease to rise. An extreme condition exists when the air temperature actually increases with altitude. This corresponds to a very stable condition where any polluted warm air released near the ground will be trapped and not vertically dispersed. This condition is known as a *thermal inversion*. There are several causes for thermal inversions, and their duration varies from a few hours to many days. Their occurrence in cities such as Los Angeles, Denver, and Mexico City is a major contributor to the air pollution problems of those cities. Figure 9.2 illustrates these conditions of stability.

One of the most notable incidents of a thermal inversion occurred in London, England, on December 5, 1952. A thermal inversion developed, enveloping the city in fog with essentially no vertical movement of air higher than 150 feet. The sulfur dioxides and particulates from burning coal, in addition to the other pollutants of a large city, accumulated for four days. After 12 hours of the inversion, people began coughing and complaining of respiratory ailments. During the next four days, an estimated 4000 deaths occurred beyond those normally experienced for a four-day period. The majority, but not all, of the people who succumbed were over 55 years old. Two similar, but not as severe, episodes also occurred in London in 1956 and 1962. Many cities have been plagued for years with this type of smog, primarily sulfur oxides and particulates. In addition to causing various respiratory problems, this type of smog, now known as *classic smog*, puts a coating of soot over buildings and other objects.

The smog problems of cities such as Los Angeles and Denver differ from those of London in that their smog does not include sulfur oxides from coal burn-

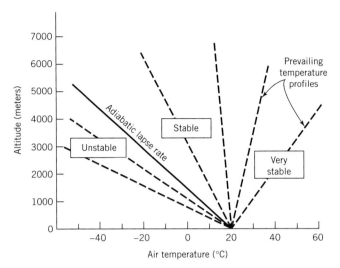

Figure 9.2 Some possible temperature profiles in the lower atmosphere. The region to the left of the adiabatic lapse rate is unstable and will lead to appreciable vertical mixing of polluted air released near the ground. The region to the right represents stable conditions and air stagnation.

ing as a major factor. In these cities, the smog is formed by the interaction of solar energy with the types of primary air pollutants emitted by automobiles and trucks. This type of smog is known as *photochemical smog*, and its effects are also most severe during thermal inversions.

In general, all major air pollution incidents that have lead to documented elevated levels of human mortality have occurred during periods of thermal inversions.

Many of the causes of thermal inversions are now well understood, and their occurrence can usually be predicted by meterologists. One such cause is a *high pressure subsidence*. When a high pressure region of the atmosphere subsides, or moves downward toward the earth where the pressure is greater, the air mass will be compressed and its temperature will rise. This relatively dense warm air then will continue to move toward the earth until it meets the higher density air near the surface. A temperature profile will result such as shown in Figure 9.3b. This temperature profile shows that the air tends to get cooler with increasing elevation above the ground, up to the point where the high pressure air mass is sitting, and then the temperature gradient reverses and the air becomes warmer with altitude. Some mixing of the air close to the earth's surface will occur, but none of the warm polluted air can break through the lid provided by the thermal inversion. It was this type of inversion that caused the problems in London, and other notable incidents in Donora, Pennsylvania; New York; and the Meuse Valley in Belgium.

Thermal Inversions

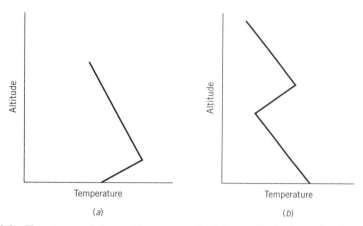

Figure 9.3 Two types of thermal inversions. In (a), a radiative inversion, because of radiative cooling during the night, the air near the ground surface is cooler than the air above it. In (b), a high pressure subsidence, a high pressure mass of air subsides toward the earth and is compressed and heated in the process, causing a thermal inversion layer some distance above the ground.

Example 9.1

Find the height to which polluted air 10°C warmer than its surroundings released at ground level will rise if the prevailing air temperature profile is +5°C/100 m (a thermal inversion).

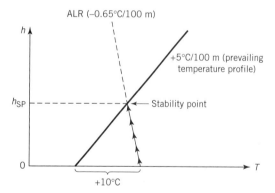

Solution

For each one meter of height, the lines converge by $(0.05 + 0.0065)$°C. To reach the stability point, they must converge by 10°C. This will occur at a height of

$$\frac{10°C}{0.0565°C/m} = \mathbf{177\ m}$$

The polluted air will rise to this height and accumulate there unless dispersed by winds.

Another type of inversion called a *radiative thermal inversion* is much more frequent but less troublesome than those caused by high pressure air masses. On a clear night, the earth's surface radiates thermal energy into space, thus cooling both the surface and the air near the surface. The energy being radiated came from the sun the previous day. After a night of cooling, the air near the surface will be cooler the next morning than the air above it, and a thermal inversion such as shown in Figure 9.3a will result. As the morning progresses, the sun will warm the surface of the ground and the lower atmosphere, and the thermal inversion will disappear by mid-afternoon. Such radiative inversions are almost a daily occurrence in Denver and similarly situated cities.

The city of Los Angeles has a reputation for air pollution that is not to be envied. Later in this chapter we will discuss in some detail the rather complicated reactions and pollutants that are involved, but the problem originates just as much in the geography and weather patterns experienced by that city. A large part of the problem stems from global air circulation patterns, two of which meet at about the latitude of Los Angeles, sending air down toward the earth. This air is warmed by compression and forms a lasting thermal inversion. The San Gabriel Mountains to the east of Los Angeles act to deter winds that would help to move the air from this area. In addition to these effects, there is the usual land–sea air movement. During the daytime the heated air over the land, which has been warmed by the sun, tends to rise, and cooler air from over the sea comes in to replace it. This results in an onshore breeze. At night, however, the surface of the sea does not cool as rapidly as the land, the wind direction is reversed, and the polluted air returns.

There are also other meterological causes of thermal inversions, but those already mentioned are the most significant. It would be tempting to blame air pollution on the weather, but that would be self-defeating. The real culprit is us. In the next sections of this chapter we will learn what the main pollutants are, how they come about, and what can be done to reduce their concentration in the atmosphere.

9.4 Carbon Monoxide

Most of our serious air pollution is produced either directly or indirectly by the combustion of fuels. In addition to the ideal combustion process that combines carbon (C) and oxygen (O_2) to form carbon dioxide (CO_2), there can also be incomplete combustion of the carbon that leads to the formation of carbon monoxide:

$$2 \text{ C} + O_2 \rightarrow 2 \text{ CO}$$

The formation of carbon monoxide takes place when the oxygen present during combustion is insufficient to form carbon dioxide. A prime source of CO is the gasoline-fueled, spark-ignited internal combustion engine, where the burning of gasoline takes place at high pressure and temperature but not in an overabundance of oxygen. Table 9.2 lists the amounts of CO and other major pollutants that are estimated to be emitted in the United States from various sources. We see that motor vehicles account for more than 50% of the carbon monoxide.

Carbon monoxide is a colorless, odorless gas that is toxic at high concentrations. Its toxicity stems mainly from its ability to form a stable compound with hemoglobin called *carboxyhemoglobin*. Hemoglobin is the substance in red blood cells that carries oxygen to the tissues. Carbon monoxide has an affinity for hemoglobin 200 times more than that of oxygen. Because of this, CO tends to block the normal distribution of oxygen in the body and leads eventually to suffocation. The effect of carbon monoxide on people is a function of the concentration and duration of exposure. A concentration of 100 parts per million (ppm) in air for 10 hours will lead to headaches and a reduced ability to think clearly. Concentrations of 300 ppm for 10 hours lead to nausea and possibly loss of con-

Table 9.2 Major U.S. Air Pollutants (10^6 tons/yr) in 2003[a]

	CO	SO_2	NO_x	VOC[b]	PM2.5[c]	PM10[c]
Fuel combustion, electric utilities	0.53	10.93	4.46	0.05	0.14	0.68
Industrial fuel combustion	1.38	2.22	2.78	0.17	0.15	0.31
Other fuel combustion	3.00	0.60	0.73	0.88	0.43	0.46
Chemical and allied product manufacturing	0.33	0.33	0.10	0.22	0.02	0.05
Metals processing	1.42	0.29	0.09	0.07	0.03	0.14
Petroleum and related industries	0.77	0.75	0.64	0.80	0.20	0.45
Waste disposal and recycling	1.85	0.03	0.14	0.43	0.36	0.39
Highway vehicles	58.81	0.26	7.38	4.43	0.13	0.19
Off-highway	24.45	0.44	4.10	2.57	0.28	0.31
Miscellaneous	14.35	0.12	0.31	6.44	4.45	20.06
Total	**106.89**	**15.97**	**20.73**	**16.06**	**6.19**	**23.04**

[a]Adapted from data from U.S. Environmental Protection Agency.
[b]VOC stands for volatile organic compounds. Their main source is solvent utilization, included under miscellaneous.
[c]PM2.5 designates particulates smaller than 2.5 microns; PM10 designates particulates smaller than 10 microns.

sciousness. At 600 ppm, after a similar length of time, death can result. At 1000 ppm, unconsciousness occurs in 1 hour and death in 4 hours. It has not been determined if there are serious effects from chronic low-level exposure. There are, however, a number of people with respiratory diseases or anemia who are susceptible to even relatively low concentrations.

Carbon monoxide is generally considered to be the most serious air pollutant in cities. In Los Angeles County, in a recent year, there were over 3 million cars, and more than 8000 tons of CO emitted every day from internal combustion engines. This amounts to about 5 pounds per vehicle per day. The carbon monoxide levels in cities have ranged from a few parts per million to more than 100 ppm on occasion, depending on the traffic volume, car speeds, and various aspects of the weather such as the temperature profile discussed previously.

The concentration of carbon monoxide has been of continuing concern in cities because of the increasing number of cars. The CO level in downtown areas, parking garages, and on freeways during rush hours has at times reached a point where the general health of those exposed is endangered. The present National Ambient Air Quality Standard (NAAQS) maximum permitted concentrations are 9 ppm for an 8-hour period and 35 ppm for 1 hour. These limits should not be exceeded more than once per year, but they have been exceeded in the past more frequently in cities such as Los Angeles, Denver, Cincinnati, and Detroit.

The basic air quality standards for the United States were set by the Clean Air Act of 1970 with important amendments in 1990. The Environmental Protection Agency (EPA) has responsibility for setting the National Ambient Air Quality Standards (NAAQS). The existing standards are shown in Table 9.3. The primary standards are limits set to protect public health, including the health of sensitive populations such as asthmatics, children, and the elderly. The secondary standards are limits set to protect public welfare, including protection against decreased visibility and against damage to animals, crops, vegetation, and buildings. In 2003 the Bush administration introduced to Congress a new plan called the Clear Skies Act. This new act proposed to regulate the emissions of sulfur dioxide, nitrogen oxides, and mercury from power plants by putting emissions caps on those substances, with the first phase effective in 2010 and the second phase in 2018. There has been considerable concern expressed that Clear Skies would be in fact a weakening of the Clean Air Act, and after two years, Clear Skies has not yet been enacted.

It should be noted that the standards in Table 9.3 are stated in terms of the number of micrograms (μg) or milligrams (mg) per cubic meter (m^3); this is the mass of the pollutant in every cubic meter of air. The standards are also stated in parts per million (ppm) by volume or, equivalently, by number of molecules. To understand the equivalence of mass and volume measures of concentration, it is useful to recall that a volume of 22.4 liters at standard temperature and pressure (STP, 0°C and 760 mmHg) will contain 1 mole, that is, Avogadro's number of molecules of any gas.

Table 9.3 National Ambient Air Quality Standards

Pollutant	Averaging Time	Concentration
Carbon monoxide (CO)		
Primary	1 hour[a]	35 ppm (40 mg/m^3)
Primary	8 hour[a]	9 ppm (10 mg/m^3)
Nitrogen dioxide (NO$_2$), Primary and Secondary	Annual arith. mean	0.053 ppm (100 μg/m^3)
Sulfur Oxides		
Primary	Annual arith. mean	0.03 ppm (80 μg/m^3)
Primary	24-hour[a]	0.14 ppm (365 μg/m^3)
Secondary	3-hour[a]	0.5 ppm (1300 μg/m^3)
Particulates (PM10)		
Primary and Secondary	Annual arith. mean[b]	50 μg/m^3
Primary	24-hour[a]	150 μg/m^3
Particulates (PM2.5)		
Primary and Secondary	Annual arith. mean[c]	15 μg/m^3
Primary	24-hour[d]	65 μg/m^3
Ozone (O$_3$), Primary and Secondary	1-hour[e]	0.12 ppm (245 μg/m^3)
	8-hour[f]	0.08 ppm (163 μg/m^3)
Lead, Primary and Secondary	Calendar quarter	1.5 μg/m^3

[a]Not to be exceeded more than once a year.
[b]For each monitor within an area.
[c]Three-year average from single or multiple monitors.
[d]Three-year average of 98th percentile for each monitor.
[e]Not applicable for most areas after June 2004.
[f]Three-year average of fourth-highest daily maximum 8-hour average at each monitor.
Source: U.S. Environmental Protection Agency, October 2004.

Example 9.2

If carbon monoxide is present in air at a concentration of 1 ppm, how many CO molecules are there in 1 m^3, and what is the mass of CO in 1 m^3?

Solution

Air at standard temperature and pressure (STP) has Avogadro's number (6.02 × 10^{23}) of molecules per mole, which occupies 22.4 liters or 0.0224 m^3. Therefore the molecular density of air is

$$\frac{6.02 \times 10^{23} \text{ molecules}}{0.0224 \text{ m}^3} = 2.69 \times 10^{25} \text{ molecules/m}^3$$

At 1 ppm of CO, there are in 1 m³ of air

$$(1 \times 10^{-6}) \times (2.69 \times 10^{25}) = \mathbf{2.69 \times 10^{19}} \textbf{ molecules CO}$$

Since CO has a molecular weight of 28, the corresponding mass density is

$$2.69 \times 10^{19} \frac{\text{molecules}}{\text{m}^3} \times \frac{28 \text{ g/mole}}{6.02 \times 10^{23} \text{ molecules/mole}} \times 10^3 \frac{\text{mg}}{\text{g}} = \mathbf{1.25 \text{ mg/m}^3}$$

Figure 9.4 shows the daily variation of carbon monoxide concentration at a downtown location in Denver as it occurred several years ago. We see that the levels of CO encountered greatly exceed the 8-hour standard of 9 ppm. The two main peaks occur somewhat after the main commuter traffic at the beginning and end of the working day. Because of its high altitude, Denver had a severe problem in meeting the national standards; the automotive emission controls were primarily set for sea level, and they were not as effective at 5000 ft altitude.

Figure 9.5 demonstrates the success of air quality programs in Denver. Since 1975, the State of Colorado has mandated the use of oxygenated gasoline during the winter months. A program of regular vehicle emissions inspections was put into place. There have been technical improvements in automobile engines, public transportation has been improved, and controls on fixed-source emissions have been strengthened. The graph shows that the second 8-hour maximum carbon monoxide concentration for each year is now only one-fifth of what it was 30 years ago. The second 8-hour maximum is the second highest concentration averaged over an 8-hour period observed within a year. If the 9 ppm 8-hour

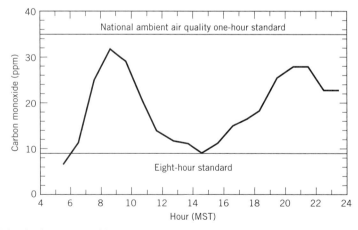

Figure 9.4 Carbon monoxide concentrations averaged over a one-hour period during a weekday at a downtown location in Denver, Colorado. The standard shown of 35 ppm is also for a one-hour averaging time. (*Source:* State of Colorado, Department of Health.)

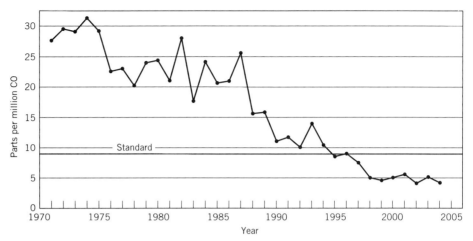

Figure 9.5 The second 8-hour maximum concentration of carbon monoxide observed each year in Denver from 1970 to 2004. There have been no violations of the NAAQS carbon monoxide standard for the past decade, owing to the success of air quality control programs. (*Source:* Colorado Air Quality Control Commission, Report to the Public, 2003–2004.)

NAAQS standard is exceeded only once per year, there is no violation. It is the second exceedance that counts. Denver has not been in violation of the carbon monoxide standard since 1995. Prior to that time, exceedances were commonplace, and by large amounts.

Carbon monoxide in the atmosphere has an estimated half-life of 0.2 years and is probably converted to CO_2 by interaction with OH molecules in the tropopause. About 290 million tons of CO are injected into the world's atmosphere every year, the majority in the Northern Hemisphere. The average atmospheric concentration seems to be stable at about 0.08 ppm.

The problems of carbon monoxide atmospheric pollution are basically local problems in cities where, because of short-term high traffic volume and weather patterns, excessively high concentrations are reached only for relatively short periods. There are two apparent solutions to the problem. The replacement of the personal car with mass transportation systems, particularly those that use electric power, is an option that many cities are pursuing. Because of the high capital cost of such systems and because of the great convenience many people find in the use of their own cars, this approach is not being widely accepted. The other main approach is to use technical measures for reducing carbon monoxide emissions from automobiles. Since automotive emission control involves more air pollutants than just CO, we discuss that topic after learning about some of the other ways we are fouling the air.

With the modern emphasis on tightening up houses to reduce air leakage in the interest of energy conservation, as described in Chapter 7, there has been a growing problem of carbon monoxide in homes increasing to dangerous levels.

The sources of CO in the home are most often the gas appliances (cooking stove, oven, clothes dryer, furnace, etc.) being used without adequate ventilation. Inexpensive CO detectors are now widely available for the home. They measure the CO level in ppm and sound an alarm if the level becomes dangerous. According to the Occupational Safety and Health Administration (OSHA), the maximum permissible indoor concentration for continuous exposure for healthy adults in any 8-hour period is 50 ppm.

9.5 The Oxides of Nitrogen

If a nitrogen–oxygen mixture such as air is heated to over 1100°C, the nitrogen and oxygen will combine to form nitrogen oxide (NO). If the cooling process is slow, the reaction will reverse, and the NO will decompose back into N_2 and O_2. However, if the cooling takes place rapidly, as is the case for internal combustion engines, the nitrogen oxide will not decompose; it will remain as NO in the exhaust gas. The reaction that forms NO does not depend directly on the fuel used. The vital ingredients are only the nitrogen, the oxygen, and the high temperature. The exhaust of an internal combustion engine running at high speed without emissions controls will contain about 4000 ppm of NO. The exhaust of a coal-fired steam generator will have 200 to 1200 ppm of NO. Nitrogen oxide is a colorless gas, toxic in high concentrations, but its toxicity is generally considered to be minor compared to another oxide of nitrogen, nitrogen dioxide, NO_2.

Nitrogen dioxide is produced in the same manner as nitrogen oxide, although for most combustion processes, NO is much more prevalent. Nitrogen dioxide is also produced, however, when NO is released into the atmosphere. In one process, NO reacts with ozone, O_3, and becomes NO_2. After about 10 hours, 50% of the NO will have been converted to NO_2. Because of its greater toxicity compared with NO, NO_2 is of much greater importance environmentally, although frequently the nitrogen oxides are treated together as NO_x. Nitrogen dioxide is a reddish-brown gas that accounts for the brownish color of the familiar smog in cities. The major sources of NO_x in the United States are listed in Table 9.2. Although motor vehicles are important, other combustion processes are also seen to contribute major amounts of nitrogen oxides.

Nitrogen dioxide can be smelled at about 0.5 ppm in air, and at 5 ppm it begins to affect the respiratory system. At concentrations of 20 to 50 ppm, there is a strong odor, one's eyes become irritated, and damage to the lungs, liver, and heart has been observed. At 150 ppm, serious lung problems occur with 3- to 8-hour exposures. Many feel that chronic lung damage will occur at concentrations as low as 5 ppm for day-long exposures. The National Air Quality Standard is 0.053 ppm annual arithmetic mean (see Table 9.3). There are alerts in cities when the concentration reaches 4 ppm. In a city such as Los Angeles, about 750 tons of NO_x are put into the atmosphere every day, about 500 tons from internal combustion engines and 250 tons from electric power plants. The major effects of the nitrogen oxides, however, are indirect. In the presence of water vapor in the atmosphere, they are partially converted to nitric acid, HNO_3. The consequences of this formation of acid are considered later in the discussion of acid rain. Another indirect consequence of NO_x in the atmosphere is the important role it plays in photochemical reactions and the formation of smog.

Under solar radiation the following reaction occurs:

$$NO_2 + \text{sunlight} \to NO + O.$$

The sunlight absorbed in this reaction is in the ultraviolet and blue portions of the spectrum, hence the reddish-brown color of nitrogen dioxide. The consequences of this reaction, however, go far beyond reduced visibility, as the resulting atomic oxygen can react with molecular oxygen, O_2, to form ozone, O_3, a very strong oxidant. The reaction can be represented as

$$O + O_2 \to O_3.$$

Among other consequences, the ozone then can react with NO in the following way:

$$O_3 + NO \to NO_2 + O_2,$$

and the original nitrogen dioxide is converted to another nitrogen dioxide. The reaction cycle can go on again as long as there is sunlight. More ingredients are needed in this discussion, however, to explain fully the photochemical smog that engulfs our cities.

9.6 *Hydrocarbon Emissions and Photochemical Smog*

A little more than 60 years ago in about 1943, people in Los Angeles began to experience a new kind of air pollution. It damaged plants, caused eye irritation, cracked stressed rubber, and generally decreased visibility. For several years, the origin and nature of this pollution was a mystery. After an intensive period of research, in the early 1950s, A. J. Haagen-Smit and his colleagues first arrived at an understanding of the basic processes that were responsible. Although the main contributing chemical reactions have been identified by Haagen-Smit and other investigators, the entire process is complicated, involving many substances and chemical reactions, and research continues at the present time.

We have seen that a combination of nitrogen dioxide and sunlight leads to the formation of ozone. A number of other reactions involving various hydrocarbons also result in strong oxidants such as ozone. The basic ingredients of what is now called *photochemical smog* are sunlight, NO_2, and hydrocarbons. Another necessity is a meterological condition in which the ingredients have time to interact before being dispersed. Most of the nitrogen dioxide and hydrocarbons are related to automobile emissions. A city such as Los Angeles, with its sunlight, traffic, and thermal inversions, has all of the conditions to form photochemical smog in great abundance. The smog first became an important issue during the 1940s when Los Angeles was growing rapidly and cars were becoming the main mode of transport. Although local automotive emission controls began in 1966 and the federal Clean Air Act became law in 1970, the effects of the controls have been largely offset by the continued increase in vehicle miles traveled.

As the name implies, hydrocarbon molecules are made up of atoms of hydrogen and carbon, but atoms of oxygen and chlorine may also be involved. In a study of the Los Angeles air, 56 different species of gaseous hydrocarbons were identified, but apparently the number observed is limited only by the sensitivity of the analytical techniques used.

The various hydrocarbons enter the atmosphere from a number of different sources (see Table 9.2). Some of the most important are the following:

(a) Auto exhaust and partially unburned gasoline.
(b) Gasoline evaporated in various steps in production, refining, and handling.
(c) Organic solvents used in manufacturing, dry cleaning fluids, inks, and paints.
(d) Chemical manufacturing.
(e) Incineration of various materials, industrial dryers, and ovens.

Once the hydrocarbons are liberated into the atmosphere, they can combine with atomic oxygen or with ozone formed by the interaction of sunlight with nitrogen dioxide, as described previously. These reactions result in the formation of complex molecules of hydrogen, carbon, oxygen, and nitrogen, known as *peroxyacyl nitrates*, abbreviated PAN. The PANs are strongly oxidizing and account for many of the harmful properties of smog.

The measure of the intensity of photochemical smog is the total oxidant concentration. The standards (see Table 9.3) relate to ozone. The ozone standard now calls for 0.12 ppm (245 μg/m^3), based on a 1-hour average, not to be exceeded more than once a year. Cities in the United States, Los Angeles in particular, have a very difficult time meeting this standard. Figure 9.6 shows the concentrations of hydrocarbons, NO, NO$_2$, and O$_3$ as a function of time during a weekday. At the beginning of the day, NO and HC show a large increase due to commuter traffic. After an hour or so, the NO has been converted to NO$_2$, and the photochemical reactions can now begin with the strong morning sun. The resulting oxidants peak about an hour later and then diminish during the day. It can be seen by comparing Figure 9.6 and the National Ambient Air Quality Standards (Table 9.3) that Los Angeles has had a difficult time meeting these standards, but there have been improvements in recent years. A view of the city under smoggy conditions is shown in Figure 9.7.

What are the harmful effects of photochemical smog? If one asks a typical resident of one of the afflicted cities, he or she would certainly complain about the eye irritations that affect about three-fourths of the population. The eyes smart, and tears form to wash away the irritants. It is the oxidants, particularly the PANs, and others called *aromatic olefins*, *aldehydes*, *formaldehyde*, and *acroleins* that cause the problems, but interestingly, ozone does not seem to be

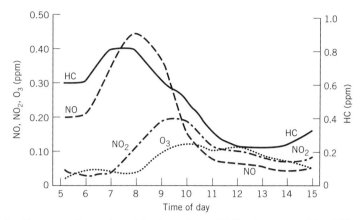

Figure 9.6 Concentrations of total hydrocarbons (HC), nitrogen oxides (NO and NO$_2$), and ozone (O$_3$) measured at a downtown location in Los Angeles for different hours of the day.

directly involved in the eye irritation. Another feature of photochemical smog widely experienced is the odor, which is largely caused by ozone. The threshold for detection of the ozone odor is 0.02 ppm, and it is readily noticed at 0.2 ppm or more. Single exposures of a few hours to ozone in the range of 80 to 400 parts per billion (ppb, where 1000 ppb = 1 ppm) have noticeable effects even on young

Figure 9.7 A smoggy morning in downtown Los Angeles. (*Source*: Ulf E. Wallin/The Image Bank/Getty Images)

people. Increased coughs, reduced athletic performance, and respiratory problems have been documented at this level. With continued exposure, many residents apparently become less sensitive to the ozone odor.

Beyond these irritations, researchers feel that there are undesirable consequences for the respiratory system. So far, there have been no definitive epidemiological studies that directly associate photochemical smog with acute respiratory disease and death, as was the case in London and other locations with classic smog. It is certainly known that conditions such as chronic sinus trouble, hayfever, bronchitis, asthma, and other respiratory problems grow worse in an atmosphere with severe photochemical smog. It is also suspected by some researchers that photochemical smog may contribute to lung cancer and chronic pulmonary disease. In animal studies, it has been shown that 1 ppm of O_3 can cause lung diseases.

There appear to be two types of plant disease, called *smog injury* and *grape stipple* (or *weather fleck*), related to photochemical smog. Smog injury, mostly from the PANs, refers to a collapse of the cells on the upper part of the leaf that causes the leaf to appear water-soaked. Grape stipple refers to the blotched or stippled appearance of affected leaves such as tobacco or grape. Even low levels of ozone appear to be capable of producing such an effect. The loss of agricultural production because of smog damage to plants is a documented problem with serious economic consequences.

The deterioration of materials exposed to photochemical smog is another serious effect of smog. The cracking and disintegration of stretched rubber has been mentioned as one of the initial manifestations of smog. Effects of this type are generally serious for materials such as paints and fabrics which eventually lose their strength and color from the oxidants in the air.

The visibility problem that plagues many cities is mainly due to aerosols and particulates, and we shall postpone discussion of these for the moment. The common brown cloud, however, is NO_2 undergoing photochemical reactions, and it is a necessary component of photochemical smog. Overall, it is difficult to assess the damage in terms of human misery and the economic consequences of photochemical smog and carbon monoxide. It is obvious, from any point of view, that our atmosphere is too limited to absorb the pollutants that are being dumped into it and still let us pursue a reasonably healthful existence. In the next section we discuss the role the automobile plays in generating the pollutants and summarize some steps that can be taken to reduce the harmful emissions.

9.7 *Reduction of Vehicle Emissions*

Over most of the years of our petroleum-powered transportation system, the internal combustion engine, and hence, the automobile, has been the main source of carbon monoxide, nitrogen oxides, and hydrocarbons in many cities. The automobile has also been a heavy contributor to the widespread pollution of the environment with lead. Prior to the 1970s the automobile engine was a truly enormous source of pollution, but fortunately the emissions from this source have been reduced by a large amount. One measure of the effectiveness of automobile air pollution controls for carbon monoxide is shown in Figure 9.5. Another example of similarly effective controls will be shown shortly for lead emissions into the environment. There are many impressive examples of success in the fight against air pollution from vehicles.

The importance of the internal combustion engine as a generator of emissions is indicated in Table 9.2. Here we see that most of the carbon monoxide emissions come from highway vehicles, as do substantial portions of the particulates, volatile organic compounds, and nitrogen oxides. Table 9.4, for a different year and from a different source with different statistics, tells a similar story—broken down by percent from transportation relative to the total U.S. emissions. Transportation is seen to be a major source of carbon monoxide (82%), nitrogen oxides (55%), and VOC (42%), but it contributes a mere 4% in sulfur dioxide emissions.

It is clear that even with the large improvements that have been made, emissions from vehicles remain important. Many of the improvements that are now serving us so well can be traced back to advances in electronics. The older cars were quite inefficient because there was no accurate control of fuel–air mixtures or ignition timing. These functions are now all controlled by inexpensive microprocessors that sense the engine's performance constantly and make adjustments as needed to ensure high efficiency and low emissions. Some of the gain in efficiency is reflected in the significantly improved fuel economy of today's passenger cars compared to those of 30 years ago (see Figure 8.2).

Environmental emissions from the older cars were not all from the exhaust stream. The gasoline tanks were vented directly to the atmosphere so that fumes were continuously released by evaporation. The same thing occurred from the

Table 9.4 U.S. Emissions, 2001

Pollutant	All Sources (10^6 tons/yr)	Transportation (10^6 tons/yr)	Percent
CO	120.76	99.50	82
NO_x	22.35	12.41	55
VOC[a]	17.96	7.50	42
PM2.5[b]	7.38	0.45	6
PM10[b]	24.10	0.53	2
SO_2	15.97	0.70	4

[a]Volatile organic compounds.
[b]PM2.5 designates particulate matter smaller than 2.5 microns; PM10 designates particulate matter smaller than 10 microns.
Source: Transportation Energy Data Book, Edition 24, ORNL-6973, December 2004.

float bowls of carburetors, and this was often worse because the carburetors were typically located atop the hot engine, enhancing the evaporation. Additional fumes from the engine crankcase, which is usually hot during engine operation, were deliberately vented to the atmosphere. These latter factors resulted in especially large hydrocarbon emissions. Cars made in recent decades have various controls for hydrocarbon emissions. Carburetors have been mostly replaced by fuel-injection systems, gasoline tanks are sealed against release of fumes, and crankcase fumes are recirculated back into the engine. Only a few decades ago, there were no programs of regular emissions inspections as now are required in various states.

At the same time as engine improvements have been made, there have been advances in processing the exhaust gases in various ways to ensure that emission standards are met. The most widely used device is the catalytic converter installed on the exhaust pipe under the car. This device causes the NO in exhaust gas to be converted into harmless N_2 and O_2. The decomposition of NO proceeds slowly at room temperature, but at higher temperatures in the presence of a catalyst such as platinum, the process takes place much more readily. The converters that use platinum are widely employed in this country, but it was quickly recognized that the platinum catalyst would be poisoned by the lead universally used as an anti-knock additive in gasoline. The use of lead was an inexpensive way to increase the octane rating of the fuel and thereby decrease the tendency for preignition in high-compression engines. For a 50-year period, from about 1923 to 1974, lead compounds were present in gasoline at a level of more than two grams of lead per gallon, leading to massive emissions of lead into the environment. By the 1970s, anti-knock compounds accounted for about 20% of the lead used in the United States, and automobile exhaust was responsible for about 90% of the airborne lead. Since that time, both because of the poisoning of catalytic converters and to reduce environmental emissions, the sale of lead-containing fuels has been greatly restricted by regulations put into effect by the Environmental Protection Agency, and leaded fuels were banned entirely starting in January 1996. Figure 9.8 shows the decline of lead emissions resulting from the EPA regulations. Lead emissions from highway vehicles have now been essentially eliminated. This is one of the great success stories in environmental science and policy. Other types of converters, some containing mixtures of chromium, iron, copper, and so forth, are also used to promote the oxidation of hydrocarbons and carbon monoxide in the exhaust stream.

The 1990 Clean Air Act amendments called for three alternative fuel programs. The first was a summer reformulated gasoline program, which has been required after 1995 in the nine worst ozone areas. The second is a provision for a winter oxygenated fuel program for areas not in compliance with the CO standards. The third is a clean-fuel fleet program started in 1998 in the 20 current ozone noncompliance areas. This means that some fraction of the new car fleet must operate on low-emission alternative fuels.

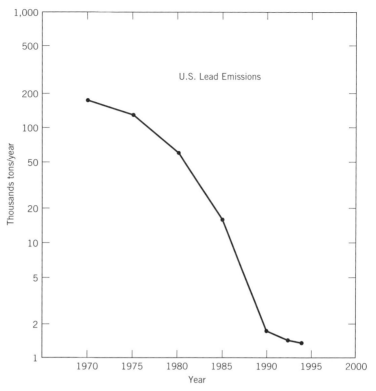

Figure 9.8 Lead emissions into the environment from highway vehicles in the United States. From 1970 to 1994 these emissions were reduced from 200,000 tons/year to less than 2,000 tons/year. By 1997 they were essentially down to zero. (*Source: Transportation Energy Data Book*, Edition 16, 1996. Stacy C. Davis and David N. McFarlin, ORNL-6898. Oak Ridge National Laboratory.)

9.8 Sulfur Dioxide in the Atmosphere

Since sulfur is present to some extent in almost all fossil fuels and since the United States and the rest of the world are heavily dependent on fossil fuels for energy, it should not be surprising to learn that sulfur dioxide, SO_2, is an important atmospheric pollutant. When fossil fuel is burned, the various sulfur compounds in the fuel are converted to SO_2, a colorless and nonflammable gas. Table 9.2 lists the main sources of SO_2, and we see that the automobile is not a major source of this pollutant. The largest SO_2 sources are stationary, particularly coal- and oil-burning electric power plants. The next most important sources are industrial processes, with the smelting of copper, zinc, and lead being the leading contributors. The sulfur content of coal and petroleum varies widely, but it is generally in the range of one-half percent to a few percent by weight.

About one-third of the sulfur compounds in the atmosphere come from man-made sources, about 93% in the Northern Hemisphere. The natural sources of atmospheric sulfur are mainly the decay of terrestrial and marine organic matter, where the sulfur is in the form of hydrogen sulfide, H_2S. Of comparable importance are the sulfates released in the form of aerosols in ocean spray. The H_2S is converted into SO_2 in a day or two, presumably by interaction with ozone in the troposphere. Of course there are various avenues by which SO_2 is removed from the atmosphere. Dry deposition, precipitation, and plant uptake are the principal removal processes. It is estimated that the total SO_2 in the atmosphere is about 11 million tons. As can be seen from Table 9.2, the U.S. contribution of man-made SO_2 to the atmosphere is about 16 million tons every year. The problems of SO_2 are related more to its regional rather than its global effects. As mentioned earlier, sulfur dioxide and particulate matter are the essential harmful ingredients of the classic smog that has long plagued cities such as London and New York.

It is difficult to separate the effects of SO_2 from those of H_2SO_4 (sulfuric acid) that results from SO_2. When sulfur dioxide enters the atmosphere, it is oxidized to sulfur trioxide, SO_3, in a relatively short time (a few days), and the SO_3 can then combine with moisture to form H_2SO_4 or sulfate salts. The problem presented by acid rain is discussed separately. The local effects of SO_2 are somewhat intermixed with those of H_2SO_4, but both substances are known to be irritants of the respiratory system. In combination with particulates, measurable increases in the rate of illness and death have been observed in cities where the SO_2 concentration is around 0.1 or 0.2 ppm. Concentrations of 0.01 ppm for a year, ranging to concentrations of a few ppm for 30 seconds, are thought to affect the health of people, particularly through the respiratory and cardiovascular systems. The experience in the Netherlands has been that there is a definite increase in the mortality rate from lung cancer and bronchitis at SO_2 concentrations of 0.04 ppm. It is hard to isolate the effects directly due to sulfur dioxide, however, because particulates and moisture are usually present in the same environment and they play a synergistic role.

Building materials such as marble, limestone, and mortar are severely affected by sulfur dioxide because the carbonates present are, to some extent, exchanged for sulfates originating from SO_2. The sulfates are soluble in water and are washed away with time by rain. Buildings and statues of great historic and aesthetic value throughout the industrial world are suffering from such deterioration. The oxide coatings on metals (rust, for example) partially lose their ability to protect the metal in an environment of SO_2 and moisture. Organic metals and paints are also weakened and discolored by the effects of sulfur dioxide. Plants, too, are affected by SO_2. Various crops and trees suffer appreciable damage to their leaves and internal cells at exposures of 0.01 ppm for 1 year and 1 ppm for 1 hour.

Example 9.3

Consider a 1000 MW$_e$ coal-burning power plant operating 24 hours a day at an efficiency of 33%.

(a) How many tons of coal are burned daily?
(b) If the coal has 1% sulfur content, how many micrograms of SO$_2$ are released daily by this plant?
(c) Now consider a thermal inversion with no winds so that the emitted SO$_2$ is confined to a volume 1 km high × 10 km × 10 km. What SO$_2$ concentration, in μg/m^3, will be achieved in one 24-hour day?

Solution

(a) Using the table of Energy Equivalents inside the front cover,

$$24 \frac{\text{hr}}{\text{day}} \times 1000 \text{ MW} \times 1000 \frac{\text{kW}}{\text{MW}} \times \frac{1}{0.33} \times \frac{1 \text{ ton coal}}{7800 \text{ kWh}} = 9324 \frac{\text{tons coal}}{\text{day}}$$

(b) For coal with 1% sulfur, 0.01 × 9324 = 93 tons of sulfur are released a day, and this results in 186 tons SO$_2$ because SO$_2$ is half sulfur (atomic weight of S = 32) and half oxygen (2 × 16 = 32). In units of micrograms, this is

$$186 \text{ tons} \times \frac{1 \text{ tonne}}{1.104 \text{ ton}} \times 10^3 \frac{\text{kg}}{\text{tonne}} \times 10^3 \frac{\text{g}}{\text{kg}} \times 10^6 \frac{\mu\text{g}}{\text{g}}$$

$$= \mathbf{1.68 \times 10^{14}} \; \boldsymbol{\mu}\textbf{g SO}_2 \textbf{ released daily}$$

(c) To find the SO$_2$ concentration, we need the volume of the air mass:

$$\text{Air volume} = 1 \text{ km} \times 10 \text{ km} \times 10 \text{ km}$$
$$= 10^3 \text{ m} \times 10^4 \text{ m} \times 10^4 \text{ m} = 10^{11} \text{ m}^3$$

Therefore,

$$\text{SO}_2 \text{ concentration} = \frac{1.68 \times 10^{14} \; \mu\text{g}}{10^{11} \text{ m}^3} = 1680 \frac{\mu\text{g}}{\text{m}^3}$$

is achieved in one day. This is 21 times the U.S. Primary Air Standard of 80 μg/m^3, about 5 times the annual maximum standard of 365 μg/m^3, and comparable to the concentration experienced during the famous 1952 London smog episode that caused 4000 excess deaths in a few days.

To accomplish a reduction in SO$_2$ emissions, one must look principally to the coal-burning power plants. There are at least three useful directions to pursue: burning coal with less sulfur content, removing the sulfur before burning the coal, and removing the SO$_2$ from the stack gases. There is, in fact, a vast amount of low-sulfur coal (0 to 1% sulfur content) in the United States, 90% of it in the West. The coal is largely located in the Rocky Mountain region, in areas such as the Powder River Basin in Wyoming and Montana. Although this coal has less sulfur than eastern coal, which has as much as 3 to 5% sulfur, the western coal is largely of the subbituminous and lignite types, which have a lower energy content per ton than the eastern hard coals. The eastern coals are mostly bituminous with some anthracite. For a given energy output of a power plant, it is not always certain that western coal will result in lower total SO_2 emissions. There is also the problem that the western coal is located at relatively long distances from the large eastern cities, and hence a larger transportation cost, in both dollars and energy, is involved. Another related problem is that the efficiency of fly ash precipitators is reduced when low-sulfur coal is burned.

There have been attempts to remove the sulfur from the coal before it is burned. With high sulfur coals, iron sulfide, FeS_2, is the most prevalent sulfur compound. It is possible to remove most of the FeS_2 by crushing the coal to a fine powder, washing it with water, and taking advantage of the density of FeS_2, which is about four times more dense than pure coal. Sulfur can also be removed from coal by hydrogenation, particularly when the coal is to undergo liquefaction. These processes, however, are expensive, and in many cases not fully effective because of the variety of sulfur compounds present.

Sulfur dioxide can also be controlled after the coal is burned. It can be removed from the stack gases by a number of processes. One of the most common processes causes the SO_2 to combine chemically in a flue gas scrubber with an alkaline substance such as limestone, lime, or dolomite to form $CaSO_4$. In this particular process, the resulting material can be just discarded as waste. In other processes the by-products are processed to remove the sulfur for resale. There is a vigorous research effort underway to refine the techniques of SO_2 removal and to provide new and better processes because technical and economic problems discourage routine use of existing removal methods.

Title IV of the 1990 Clean Air Act amendments provides a two-phase program for further control of SO_2 emissions. In Phase I the act imposes an annual emission limit of 2.5 lb/10^6 Btu on 110 utility plants. In Phase II a lower emission standard of 1.2 lb/10^6 Btu would be applied to utility plants that were in operation with more than 25 MW_e before 1990. These new requirements are calculated to reduce the SO_2 emissions by 10 million tons from the 1980 emission levels.

9.9 *Particulates as Pollutants*

Particulates are very different in nature from the gaseous air pollutants we have discussed. The particulates can be solids or liquids, and, as such, they can have a certain size as well as a chemical composition. The term *aerosol* is often used to describe either solid or liquid matter suspended in the atmosphere; however, we shall use the term *particulates*.

There are a number of natural sources of particulates, such as salt from ocean spray, dust from fields, volcanic ash, and forest fires. Worldwide, the natural sources produce about 14 times as much particulates as are produced from man-made sources. Although we tend to focus on man-made pollutants as sources of our atmospheric problems, the fairly recent eruptions of Mount St. Helens in Washington State, El Chichón in Mexico, and Mount Pinatubo in the Phillipines have been dramatic reminders that nature, on occasion, can outdo us by many orders of magnitude.

Man-made particulates are often emitted in areas where the population density is high, and thus they tend to have more far-ranging consequences for human health than do particulates from a forest fire in a remote corner of Montana, for example. The largest contribution to man-made particulates is made by the fly ash from coal combustion. Petroleum combustion contributes only about one-twentieth of the particulate emissions from coal burning. Iron and steel mills, cement manufacturing, and the burning of wood and other materials are all important sources, as are agricultural processes. Table 9.2 indicates the important sources of man-made particulate matter in the United States and shows that the total emissions are 6.2 million tons of particulates smaller than 2.5 microns and 23.0 million tons of particulates smaller than 10 microns per year.

Figure 9.9 displays some of the different types of particulates and their sizes. One of the main threats to health presented by particulates results from their deposition in the lungs. Various mechanisms resist this, such as filtering out of large particles in the nose, the trapping of particles in the mucous membranes of the airways, and sneezing and coughing to remove particles. However, the

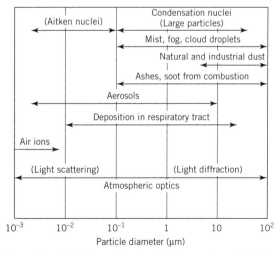

Figure 9.9 The sizes of particulate matter in the atmosphere. Particles larger than about 1 μm are effectively trapped in the nasal passages and in the trachea. The smaller particles (10^{-1} to 10^{-2} μm) can find their way to the lungs and cause harm to the respiratory system.

smaller particles can get through these traps to the sensitive air sacs (the alveoli) that are important in the proper functioning of the lungs. The particles reaching the inner lung can directly interfere with the respiratory system, or the particles may be toxic or carry a toxic substance with them. It is believed that some particulates transport to the lungs chemical free radicals, which can have effects similar to those of ionizing radiation. Lung cancer is a known occupational hazard for chimney sweeps and coal miners. Epidemiological studies have associated the presence of particulates with bronchitis and emphysema. General deterioration of health appears for long-term exposure to particulate concentrations of about 80 $\mu g/m^3$.

Example 9.4

A published epidemiological study has indicated that in the United States in the 1960s, the excess death rate due to particulate air pollution was 0.04 × the particulate concentration in $\mu g/m^3$, and that due to sulfur oxides was 0.07 × the sulfur oxide concentration in $\mu g/m^3$, both expressed as annual death rates per 10,000 population. Apply these findings to the estimated 2005 U.S. population to predict the corresponding annual excess death rates in the United States, *if* the average particulate concentration is 75 $\mu g/m^3$ and the average SO_2 concentration is 80 $\mu g/m^3$. Compare these concentrations to the Air Quality Standards given in Table 9.3.

Solution

Particulates

$$(296 \times 10^6 \text{ persons}) \times \frac{0.04 \text{ annual deaths}}{10^4 \text{ persons} \times \mu g/m^3} \times 75 \ \mu g/m^3$$

$$= \textbf{88,800 deaths per year}$$

Sulfur dioxide

$$(296 \times 10^6 \text{ persons}) \times \frac{0.07 \text{ annual deaths}}{10^4 \text{ persons} \times \mu g/m^3} \times 80 \ \mu g/m^3$$

$$= \textbf{165,800 deaths per year}$$

The concentrations of the air pollutants used in this problem are consistent with the Air Quality Standards of Table 9.3.

In addition to their effects on human health, particulates in the atmosphere can scatter and absorb an appreciable amount of sunlight. It is known that the volcanic ash put into the atmosphere by the eruption of Tambora in the Dutch East Indies in 1815 resulted in a general lowering of the global temperature for several years afterward. In one year during this period, there was frost every month in New England; there was no real summer. On a more local level, cities now receive about 20% less sunlight than do areas with less industry and fewer power plants. Coupled with this is a general reduction of visibility and enhancement of fog formation. In addition to the soiling of clothing and buildings by the deposition of particles of soot, corrosion of metals and degradation of other materials are caused by particulates that have had sulfuric acid and other corrosive liquids condensed on them. Paint, masonry, electrical contacts, and textiles are all affected.

The usual measure of the amount of particulates in the atmosphere is in terms of micrograms per cubic meter (μg/m^3) for particles less than 10 microns in diameter. The standards (see Table 9.3) are that the annual mean should not exceed 50 μg/m^3, and the 24-hour maximum that should not be exceeded more than once a year is 150 μg/m^3. For particulates smaller than 2.5 microns, the annual mean should not exceed 15 μg/m^3, and the 24-hour mean should not exceed 65 μg/m^3. The standards for particulates do not consider chemical composition. Cities in the United States have an average of about 100 μg/m^3, considerably above the standard, and heavily polluted areas are at times as high as 2000 μg/m^3. During the deadly episode of classic smog in London in 1952, the particulate concentration reached 1700 μg/m^3, and the SO$_2$ level reached about 0.7 ppm. Urban areas in the United States have an average of about 45 μg/m^3 particulates. Very remote areas (wilderness, deserts, national forests, etc.) average 2 to 20 μg/m^3 particulates. The large cities such as New York, Chicago, Pittsburgh, and Philadelphia are among the most highly polluted areas as far as particulates are concerned, having concentrations in the range of 155 to 180 μg/m^3.

There are natural processes that tend to remove the particulates put into the air by either man-made or natural sources. Gravity acts on all particles and tends to move them toward the earth. This sedimentation, or gravitational settling process, is really only effective for large particles that have a radius greater than 20 microns. Particles smaller than this are removed to some extent when they are near the ground by impact on various objects such as trees and buildings. In the higher atmosphere, small particles can coalesce and form nucleation centers about which raindrops form. This brings the coalesced particles to earth. There is also a washout process where precipitation intercepts some of the larger particles in the air beneath clouds and thus brings them to earth. The natural residence time for particles in the atmosphere ranges from days to years. Unfortunately, we cannot rely on nature to clean up the air as fast as we are polluting it.

As shown in Table 9.2, particulate air pollution arises from a number of sources, and the controls or devices for reducing emissions from these various sources differ considerably. Coal-burning power plants are a major source of particulates and we shall discuss the control devices used by this industry. After coal is burned, the gaseous effluents go off; along with them, small particulates, or fly ash, go up the stack and are released to the atmosphere. The heavier residue, or bottom ash, left from combustion of the coal must be physically removed from the collection areas at the bottom of the boilers and disposed of as waste.

A number of different devices are used by power plants to remove the fly ash from the stack emissions. In order to be fully effective, the devices must remove more than 99% of the fly ash. This requires an extensive system of controls, usually involving two separate types of devices, to bring the effluent up to the standards. The 1979 Federal New Source Performance Standards initially required that 90% of the sulfur dioxide and all but 0.03 pounds of particulates per million Btu of thermal energy must be removed. The standards were strengthened to 95% and 0.02 pounds by the mid 1980s. Even 0.03 or 0.02 pounds of particulates per million Btu can represent the emission of enormous numbers of smaller particles, and it is these smaller particles that are the most damaging to human health.

The exact devices used to remove fly ash depend on the type of coal being burned and other factors, but filters through which the stack gases pass are often used. A porous woven fabric material is used in a structure called a *bag house*, where the filters are automatically shaken and air is blown through them in the reverse direction to clean them. The fly ash collects at the bottom of the bag house and is carted off for disposal. One version of a bag house is shown in Figure 9.10. Devices of this type have removal efficiencies of (in terms of $\mu g/m^3$) greater than 99% and are quite effective for particles smaller than one micron. The high-temperature gases must be cooled before they are passed through the filters. Sometimes the filters are preceded by a settling chamber, which lets the heavier particles settle out by gravity prior to entering the bag house.

Electrostatic precipitators are also widely used because they have high removal efficiencies and no moving parts. The stack gases are passed through a chamber where, in a region of high electric fields, some of the gas molecules are ionized. Some of these ions become attached to particles of fly ash, causing them to have an electric charge so that they can be collected on metal plates maintained at a high electric potential.

A third type of control device, the cyclone separator, is arranged so that the stack gases spiral upward with a circular motion. The heavier particles hit the

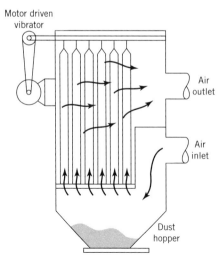

Figure 9.10 A bag house with a motor-driven vibrator. Vibration of the filter bags frees the fabric of the particulates which then collect at the bottom.

walls; they settle out and are collected at the bottom. Cyclone separators are not highly efficient for submicron particles, but they can be used for high-temperature moisture-laden gases. There is also a family of devices called scrubbers, or wet collectors, which remove particles by having them come into contact with water. In the simplest version, the stack gases are passed directly through water. Spraying water in fine droplets into the stack gases is a more effective arrangement. The wastewater, of course, must be treated to remove the particulates, and this is rather costly. One such device is shown in Figure 9.11.

9.10 Acid Rain

The emission into the atmosphere of sulfur oxides and nitrogen oxides from various man-made sources has been discussed in some detail. The formation of acids, primarily sulfuric acid and nitric acid, from these pollutants and the resulting damage caused by the acidic rain formed is a story of growing importance and interest. In terms of global atmospheric problems, many regard the overall ramifications of the acid rain problem second only to carbon dioxide and the greenhouse effect. Boundaries between states and countries are in no way barriers to the flow of pollutants that travel many hundreds and even thousands of miles before returning to the earth as acid rain.

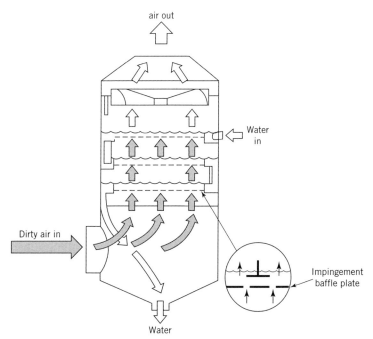

Figure 9.11 An impingement scrubber. The particulates are trapped out of the air by the water which passes downward through the several stages as shown.

Before relating some of the interesting and important details of the acid rain story, it will prove useful to review the definition of pH as a measure of acidity. The pH scale ranges from 0 to 14, with the midpoint, 7.0, taken as the neutral (neither acid nor alkaline) point. Values less than 7.0 represent an excess of hydrogen ions, and hence, acidity. Values of pH above 7 represent alkaline (or basic) soils or liquids. The pH scale is logarithmic. Hence a change of 1 point on the pH scale corresponds to a change by a factor of 10 in the excess hydrogen ion (H^+) concentration. Vinegar, for example, has a pH of about 2.5 and obviously it is quite acidic. Milk of magnesia, with a pH of 10.5, is strongly alkaline.

Pure rainwater has a pH of around 5.6, somewhat lower than the pH of 7 for a neutral solution. This slight acidity comes about mainly from the formation of carbonic acid (H_2CO_3) from the carbon dioxide normally present in the atmosphere. The problem of acid rain arises because of the further reduction in the pH by acids formed from the sulfur oxides and nitrogen oxides that originate primarily from fossil fuel burning, smelters, and other industrial processes. The name *acid rain* is somewhat a misnomer because 10 to 20% and at times more than 50% of the acidity comes from the dry deposition of particles of sulfur oxides and nitrogen oxides. Some of these particles are so small that they do not fall quickly out of the atmosphere but remain suspended for a very long time. The air in which they are suspended interacts with the surface of the earth and the particles are caught by surface objects, where they contribute to the acidity by mixing with water. The dry deposition of acid is just as harmful as acids brought to the surface by precipitation.

Prior to the 1950s, no effective effort was made to monitor the acidity of rain and the surface waters in the United States; however what evidence there is from that time indicates that the pH of the rain or snow was around 5.6 or higher. At this level of acidity, plant and animal life in the lakes does not seem to be adversely affected.

Since that time, a general increase in the acidity of rain and snow has been documented by a number of investigators, and, particularly in Western Europe and the northeastern section of North America, alarming increases have been noted. These increases appear to be directly related to increased emission of sulfur and nitrogen oxides. Sulfur dioxide is associated with about 70% of the hydrogen ion concentration. In North America, the SO_2 is primarily from coal-burning power plants in Ohio, Indiana, Missouri, Pennsylvania, and Illinois, among other states in the region. The United States is the largest emitter of SO_2 in the world, accounting for about 16 million tons a year. In Europe the primary sources of SO_2 are the Ruhr Valley of Germany, England, France, and the Low Countries. The important factors involved are the sulfur concentration in the coal, the prevailing wind directions, and the precipitation patterns. In the case of Europe, lakes in Norway and Sweden are where the effects have been felt the most. In North America, the lakes in the Adirondack Mountains of New York, upper New England, and the provinces of Ontario and Quebec in Canada have been the most adversely affected. Figure 9.12 shows the pH of surface waters in North America. These values were measured in the spring when the acidity is the most extreme.

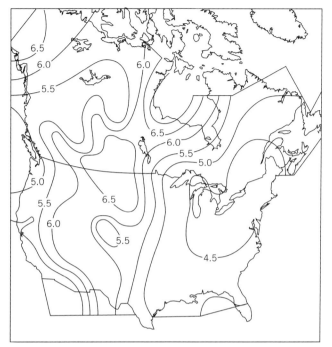

Figure 9.12 Approximate pH values measured in the spring of the year in the surface waters of various regions of the United States and Canada.

After acid rain is received, its effects on the ecosystem depend on the type of soil and rock. A limestone rock base in the soil or water can largely neutralize the acidity. On the other hand, if the bedrock is granite, gneiss, quartzite, or quartz sandstone, there is little buffering action since the waters contain few dissolved minerals. Large sections of the northeastern United States and eastern Canada have bedrock of these latter types and hence are quite vulnerable to the harmful effects of acid rain.

If one looks at a lake that has been affected by acid rain, it appears to be clear, fresh, unpolluted water. Unfortunately, when the pH goes below 5 or so, the effects on the plants and fish are very real even if the water appears clear. Effects on the reproduction process of fish are among the first consequences of a lowering pH level. The newly hatched fish, fish fry, will generally not survive a pH level in the region of 4.5 to 5.0, or if they do survive they are often deformed. The acid also tends to upset the body chemistry of the fish, causing a calcium deficiency that leads to deformed, humpbacked, or dwarfed fish. Another effect of lower than normal pH is the release of aluminum from the soil surrounding a lake, leading to clogging of fish gills and gradual suffocation.

Similar effects are seen for frogs, salamanders, clams, crayfish, and many of the aquatic insects. Aquatic plants are also adversely affected by acidity. Eventually, after the fish and plant life of a lake have disappeared because of the increase of acidity, the lake can support only a thick mat of algae, moss, or fungus at the bottom of the lake. It has been noted that the bottom of a dead, acid-killed lake looks as if it has been covered with Astroturf.

In the early 1970s 50% of the lakes above 2000 feet of elevation in the Adirondack Mountains of New York were found to have a pH lower than 5. In the period of 1920 to 1937 only 4% of the lakes had a pH that low. By 1970 about 200 of these lakes were considered dead; that number must now be considerably higher. A similar series of dead lakes was found in the La Cloche Mountains near Sudbury, Ontario, where the International Nickel Company has large smelters. Some 1400 streams in the Mid-Atlantic highlands and 90% of the streams in the New Jersey Pine Barrens have become acidified by acid rain. The forests of the Appalachians and the Shenandoah and Great Smoky Mountains National Parks have also suffered due to acid rain taking vital plant nutrients from the soil through chemical reactions.

During the winter months, snow with its acid content accumulates in the mountains. When the spring runoff occurs, this acid is released to the rivers and lakes, and the pH can suddenly be decreased in bodies of water by as much as 2 units in the span of a few weeks. These two units of pH decrease correspond to a hundred-fold increase in the excess hydrogen ion concentration. This chemical shock is even more devastating to the flora and fauna than would be a steady concentration of the same low pH.

In 1980 the U.S. government established the National Acid Precipitation Assessment Program (NAPAP). The goal of this program was to evaluate the causes and effects of acid rain and to propose ways to control it. It was a massive 10-year undertaking involving 2000 scientists and costing $570 million. Many measurements of water acidity were made and mathematical models created, for example, for acid deposition on a regional basis. Unfortunately, in spite of the program being scientifically sound, there were many delays in issuing the final report, and much political controversy with the administration of the program. This delay has reduced the usefulness of the effort in providing guidance for legislation on control. The 1990 Clean Air Act amendments, which reduce SO_2 emissions, were drawn up without appreciable benefit from NAPAP.

The 6000-page final NAPAP report presents a picture that is perhaps not quite as disastrous as that presented by some of the earlier reports, but there is no question that a significant problem exists. In brief, the report states that acid rain has adversely affected aquatic life in about 10% of the eastern lakes and streams. The decline of red spruce trees at high elevations is associated with acid rain, and acid rain has contributed to corrosion of buildings and materials. Acid rain has also reduced visibility in the Northeast and parts of the West.

There is evidence that progress is being made, as the concentration of sulfur compounds in the air in the Northeast has decreased and the rate at which lakes in the area are acidifying has slowed. The progress is mainly due to control measures of the Clean Air Act of 1970 and some additional controls imposed by states.

There is far greater agreement that there is a problem than there is on some definite solution. Because of the long-range transport of the air pollutants, the people affected are not the people who are generating the pollutants. Reduction of acid rain in the Adirondacks and eastern Canada can come about by requiring reduced SO_2 emissions from coal-burning plants in the U.S. Midwest. The cost in increased electric power rates to the residents of the Midwest to install flue-gas desulfurization units will be considerable, and such programs have not been enthusiastically received. The Environmental Protection Agency (EPA) has ruled that 70 to 90% of the gaseous sulfur must be removed from all new coal-based power plants. Unfortunately, most of the sulfur emissions come from older power plants that will not be affected by the EPA requirement.

The 1990 Amendments to the Clean Air Act provided some innovative measures for the reduction of emissions, particularly SO_2, to lessen the acid rain problems. Title IV of the Act was designed to reduce the SO_2 emissions by 10 million tons a year. Three incentives were put forth to accomplish this: avoided emissions, conservation and renewable energy, and reduced utilization. The details of the program are a bit complicated, but essentially it rewards utilities for not emitting SO_2 with *allowances*; one allowance equals one ton of SO_2 or a reduction of 500 MWh of electric energy. These allowances can then be traded or sold. If a plant is emitting 5000 tons of SO_2, it must have 5000 allowances or be fined $2,000 per ton for non-compliance. Phase I of the program went into effect in 1995 for 110 coal-burning plants in 21 eastern and midwestern states. Phase II went into effect in the year 2000. It includes all coal-burning power plants over 25 MWe of capacity.

9.11 *Summary of Air Pollution in 2005*

In 1963 Congress passed the original Clean Air Act (CAA). However, it was the amendments in 1970 that provided the laws for major reductions in air pollution. Figure 9.13 shows the history of the emissions for the three pollutants NO_x, SO_2, and Volatile Organic Compounds (VOC). Since 1970, there have been significant decreases in SO_2 and VOC, but NO_x has not declined appreciably. The decreases in sulfur dioxide and VOC, and the holding steady of nitrogen oxides, have taken place as population has increased, so on a per-capita basis this can be seen as an improvement. Carbon monoxide emissions have also come down from 128 million tons in 1970 to about 90 million tons in 2003, according to one compilation. Another data set from the Environmental Protection Agency gives the drop in carbon monoxide emissions during this time as from 204 to 107 million tons. Emission of particulate matter under 10 microns (PM10) has dropped by about a factor of four since 1940.

Airborne lead emissions from transportation have been discussed separately in this chapter, and it is apparent that this source of air pollution is being greatly reduced through cooperative efforts of government regulators and the automotive industry. There is still some release of airborne lead but it is declining.

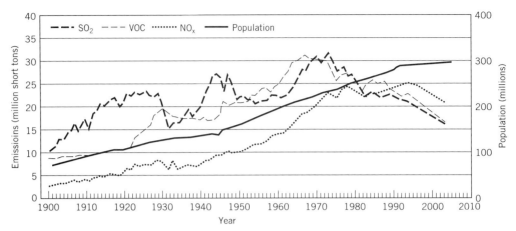

Figure 9.13 A century-long record of annual air pollution emissions compared to the population of the United States. (*Sources:* Trends in National Emissions, United States Environmental Protection Agency, Office of Air Quality Planning and Standards, October 1995, September 2004; U.S. Census Bureau, *Statistical Abstracts of the United States*, 2003. For NO_x, these three sources do not show exact agreement in the years following 1970, and some averaging is shown here.)

Electric power plants continue to be a main source of air pollutants. Of these, the coal-fired plants are the dominant producers of electricity, accounting for 51% of production (see Table 9.5). As shown in Table 9.5, they also are the dirtiest, emitting the largest amounts of CO_2, NO_x, and SO_2 per megawatt-hour of electricity produced. Carbon dioxide is not a pollutant in the usual sense of the word, but its growing concentration in the atmosphere is of great concern. This will be discussed in the next chapter.

Table 9.5 National Average Emission Factors

Fuel Type	Percent of Generation[a]	CO_2[a] (lb/MWh$_e$)	NO_x[b] (lb/MWh$_e$)	SO_2[b] (lb/MWh$_e$)
Coal	51.0%	2100	8.8	17[c]
Fuel oil	3.2%	2000	4.2	12
Natural gas	15.2%	1300	4.6	0
Other[d]	30.6%	0	0	0

[a]U.S. Energy Information Administration, July 2000.
[b]Energy Efficiency and Renewable Energy—Opportunities, from Title IV of the Clean Air Act, EPA 430-R-94-001, February 1994.
[c]Emission rate for coal is an average of the 1990 value of 22 lb/MWh$_e$ and a projected value for 2000 of 12 lb/MWh$_e$.
[d]Other includes nuclear and hydroelectric and other renewable energy sources.

The Clean Air Act Amendments of 1990 have not yet made their final impact on the level of emissions, as some of the new regulations did not go into effect until after 1996. Further emission reductions resulting from these regulations are expected to be evident in the coming years.

Current regulations regarding particulates refer only to sizes less than 10 microns in diameter with concentrations limited to no more than 150 $\mu g/m^3$ averaged over any 24-hour period and with an annual average of less than 50 $\mu g/m^3$. For particles smaller than 2.5 microns (included in the standard for particles smaller than 10 microns) the allowable concentrations are reduced to 65 $\mu g/m^3$ for 24 hours and 15 $\mu g/m^3$ for an annual average. It is believed that the smaller sized particles (less than 2.5 microns) are the ones that offer the most serious threat to health since they lodge more deeply in the lungs.

Key Terms

Troposphere	Aerosols
Thermal inversion	Particulates
Adiabatic lapse rate	VOC
Classic smog	Sulfur oxides
Photochemical smog	Nitrogen Oxides (NO_x)
High pressure subsidence	pH
Radiative thermal inversion	Carbonic acid
Carboxyhemoglobin	Acid rain
Avogadro's number	Catalytic converter
Peroxyacyl nitrates (PAN)	Oxygenated fuels

Suggested Reading and References

1. Reisner, Marc. *Cadillac Desert*. New York: Penguin Books, 1986.
2. Harte, John, and Socolow, Robert H. *Patient Earth*. New York: Holt, Rinehart and Winston, 1969.
3. Hodges, Laurent. *Environmental Pollution*. New York: Holt, Rinehart and Winston, 1973.
4. Likens, G. E., Wright, R. F., Galoway, J. N., and Butler, T. J. "Acid Rain." *Scientific American*, **241**, No. 4 (October 1979), pp. 43–54.
5. Mohnen, Volker A. "The Challenge of Acid Rain." *Scientific American*, **259**, No. 2 (August 1988), pp. 30–48.
6. Balzhiser, Richard E., and Yeager, Kurt E. "Coal Fired Power Plants for the Future." *Scientific American*, **257**, No. 3 (September 1987), pp. 100–107.
7. Turiel, I. *Physics, the Environment, and Man*. Englewood Cliffs, NJ: Prentice–Hall 1995.

8. "National Air Quality and Emissions Trends Report, 2003," United States Environmental Protection Agency, Office of Air Quality Planning and Standards, Research Triangle Park, NC, 2004.

9. "Transportation Energy Data Book, Chapter 12—Criteria Air Pollutants," Stacy C. Davis and Susan W. Diegel, Oak Ridge National Laboratory, ORNL-6973, December 2004.

Questions and Problems

1. What is the total force in pounds on the top surface of a box that has been evacuated (perfect vacuum inside), if the top surface is 10 inches wide by 20 inches long?
   (a) At sea level?
   (b) At 50,000 ft altitude?

2. Under normal conditions what is the temperature at the top of a 20,000 ft mountain if it is 70°F at sea level?

3. Describe the basic differences between the London smog problems in 1952 and the Los Angeles type smog that occurs almost daily.

4. What are the ways in which sulfur dioxide emissions from coal-burning power plants can be reduced?

5. As the result of a thermal inversion the prevailing air temperature profile increases 1°C/100 m above the ground level. To what maximum height will polluted air rise if it is released at ground level 15°C warmer than the atmosphere? The adiabatic lapse rate is −0.65°C/100 m.

6. How many molecules are there in 1 m^3 of air at standard temperature and pressure?

7. A concentration of 1 ppm of NO corresponds to how many μg/m^3? The atomic weight of nitrogen (N) is 14 and that of oxygen (O) is 16. How many molecules of NO at 1 ppm are there in 1 m^3 of air?

8. The primary standard for SO_2 in the atmosphere is 365 μg/m^3 averaged over a 24-hour period. How many parts per million by number of molecules is this?

9. In the year 2020, it is conceivable that 20 gigawatts of electricity could be continuously generated in a single western state by coal-burning power plants.
   (a) At 38% efficiency for generating the electric power, how many tons of coal would be burned per year?
   (b) If the sulfur content of the coal averages 1%, how many tons of SO_2 would be released per year?
   (c) If a standard of 0.1 pounds of particulates per million Btu is adhered to, how many tons of particulates would be released per year?
   (d) How many tons of CO_2 would be released per year, assuming that the coal is 100% carbon?

10. In problem 9b, assume that the correct answer is 1.18×10^6 tons of SO_2 released per year. If a thermal inversion persists for a 24-hour period over the whole state and the released SO_2 can rise no higher than 1000 m, what concentration of SO_2 (μg/m^3) will be present in the air? Assume that the state is 500 km × 500 km in area. Will the concentration of SO_2 be above the 24-hour standard?

11. How many pounds of SO_2 could be released if 1 ton of coal is burned under Phase I of the 1990 Clean Air Act Amendments? Phase II? Assume 2.7×10^7 Btu/ton for the coal.

12. If the concentration of hydrogen ions (microequivalents per liter) is 300 for a lake in which the pH is 3.5, what is the hydrogen ion concentration when the pH is 4.5?

Multiple Choice Questions

1. The adiabatic lapse rate is $-1°C/100$ meters and the existing temperature profile is $-1.5°C/100$ meters. Under such conditions, which is true?

   (a) It is a condition of great stability and a thermal inversion will surely result.

   (b) It is a condition of great stability and thermal inversions will not result.

   (c) It is an unstable condition conducive to clean air and no thermal inversions.

   (d) It is a condition of stability and one can expect some thermal inversions.

2. Due to a high pressure subsidence a thermal inversion layer is formed above the ground as shown in the following graph. The ALR is $-0.65°C/100$ m. To what height above the ground will polluted air rise if it is released at ground level at a temperature $1.6°C$ warmer than its surroundings?

   a. 104 m
   b. 40 m
   c. 200 m
   d. 250 m
   e. 260 m
   f. 280 m
   g. 300 m
   h. ≥ 300 m

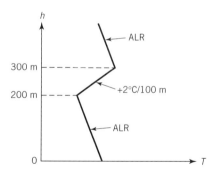

3. Use the same information in problem 2, except polluted air is released at ground level at a temperature of $5°C$ warmer than surroundings.

   a. 104 m
   b. 40 m
   c. 200 m
   d. 250 m
   e. 260 m
   f. 280 m
   g. 300 m
   h. ≥ 300 m

324 Chapter 9 **Air Pollution**

4. Carbon monoxide is dangerous to human health because _____ .
   a. suffocation occurs because of its great affinity for hemoglobin in the blood
   b. the greenhouse effect will warm up the earth to the point where life will be affected.
   c. it will reside in the lungs and induce cancer
   d. it is an important ingredient of photochemical smog

5. If you are exposed to an atmosphere containing 100 ppm of CO for 10 hours, what will result?
   a. death
   b. no apparent symptoms
   c. coma or violent sickness
   d. headache and reduced mental acuity

6. The National Ambient Air Quality Standard for NO_2 is 0.05 ppm. This corresponds to about _____ $\mu g/m^3$. Nitrogen has an atomic mass of 14; oxygen has atomic mass of 16.
   a. 8.2
   b. 100
   c. 204
   d. 160
   e. 260
   f. 365

7. The two emissions from coal-burning power plants most important to human health and air pollution are _____ .
   a. CO and SO_2
   b. NO_x and particulates
   c. SO_2 and particulates
   d. CO and NO_x
   e. CO_2 and particulates
   f. HC and NO_x
   g. SO_2 and NO_x
   h. radioactivity and SO_2

8. The most environmentally harmful emissions from automobiles are _____ .
   a. CO, SO_2, and particulates
   b. hydrocarbons, CO_2, and SO_2
   c. hydrocarbons, CO, and NO_x
   d. CO, particulates, and NO_x
   e. CO_2, SO_2, and NO_x
   f. CO, SO_2, and NO_x
   g. CO, O_3, and NO_x
   h. hydrocarbons, particulates, and CO

9. Pick the set of answers which best characterizes the major fuel, time of occurrence, principal constituents, and principal effect of photochemical (Los Angeles-type) smog:
   a. petroleum, early morning, (CO, SO_2), bronchial irritation
   b. petroleum, midday, (O_3, NO, NO_2, CO), eye irritation
   c. coal, midday, (O_3, NO, NO_2, CO), eye irritation
   d. petroleum, early morning, (O_3, NO, NO_2, CO), eye irritation

10. Pick the set of answers which best characterize the major fuel, time of occurrence, (principal constituents), and principal effect of London-type smog:
    a. coal and petroleum, late evening, (particulates, CO, sulfur compounds), eye irritation
    b. coal and petroleum, early morning, (particulates, CO, sulfur compounds), bronchial irritation

    c. natural gas, midday, (O$_3$, NO, CO), bronchial irritation

    d. natural gas, early morning, (SO$_2$, CO, NO$_2$), eye irritation

11. The purpose of an automobile's catalytic converter is to _____ .

    a. decompose and oxidize pollutants

    b. filter out pollutants   **d.** recycle pollutants

    c. condense pollutants   **e.** trap pollutants

12. The primary source of man-made SO$_2$ is _____, and its basic harm to people, animals, vegetation, and material is through the formation of _____.

    a. gasoline burning, HNO$_3$

    b. hot springs and geothermal power, H$_2$SO$_4$

    c. coal burning, H$_2$SO$_4$

    d. coal burning, hydrocarbons

13. If there are 0.28 ppm of SO$_2$ in the air at STP, how many SO$_2$ molecules are there in 1 m^3? The atomic weight of sulfur is 32; that of oxygen is 16.

    a. 0.75×10^{18} **e.** 1.68×10^{20}

    b. 1.50×10^{19} **f.** 7.52×10^{18}

    c. 7.52×10^{15} **g.** 6.02×10^{23}

    d. 3.01×10^{20} **h.** 3.01×10^{19}

14. The pH of normal rain is _____; it contains _____.

    a. 4.5, H$_2$SO$_4$ **d.** 7.0, carbonic acid

    b. 7.0, no acid   **e.** 6.5, HNO$_3$

    c. 5.6, carbonic acid   **f.** 7.5, H$_2$SO$_4$

CHAPTER 10

Global Effects

(*Source*: Jim Stamates/Stone/Getty Images)

10.1 Introduction

This chapter will deal with two examples of our influence on the global atmosphere and climate. The first example, the depletion of ozone in the stratosphere or "a hole in the sky" as it is sometimes called, is well documented and corrective action is underway. What is surprising about the ozone problem is that the release of relatively minor gases can have such unexpected and important consequences. What other serious effects on our thin and fragile atmosphere will surface with time are, of course, unknown. There is a lesson in the story of ozone depletion.

The second example of our large-scale impact is the very serious possibility of global climate change caused by greenhouse gases. It is generally accepted that global temperature increases and changes of climate are being caused by man's actions, and most scientists involved in research in this area feel that corrective actions should be undertaken *now* while the problem may still be tractable. A beneficial measure would be to reduce the rate at which we emit carbon dioxide into the atmosphere. As we have seen in earlier chapters, the production of CO_2 is an inherent consequence of burning fossil fuels. Is the world's population ready to adjust to a cutback in the rate of producing energy from fossil fuel? Although there may be other corrective measures that can be initiated, the reduction of our rate of burning fossil fuels must be at the heart of any serious attempt to reduce CO_2 emissions. With our present practices, we are conducting a huge experiment on our atmosphere with unpredictable consequences.

10.2 Ozone Depletion in the Stratosphere

The observed depletion of ozone in the stratosphere, particularly over the Antarctic in the spring months, is one of the most significant examples of human interference with the earth's atmosphere. This hole in the ozone is a well-measured and reasonably well-understood problem, whose causes are directly linked to a class of widely used industrial gases.

In the troposphere ozone is a harmful pollutant, and as we have seen, it plays an important role in the formation of photochemical smog. In addition, it is one of the greenhouse gases. However, in the stratosphere at altitudes of 10 to 30 km, it is vital to our well-being, because it is a very strong absorber of ultraviolet radiation from the sun. If ozone were not present in the stratosphere, the ultraviolet radiation on the earth's surface would be sufficient to cause a large rise in the incidence of skin cancer and to damage the ocean phytoplankton, which are basic to the food chain.

In 1985, British scientists working in Halley Bay, Antarctica, announced a startling drop in the ozone concentration that they had been measuring in a vertical column extending upward from their ground-based instrument. From 1956 to 1968 the mean springtime ozone concentration in such a vertical column had been relatively constant at about 300 Dobson units (DU, a milliatmosphere-centimeter of ozone), but by 1984 it had dropped to 200 DU. Later measurements showed that by 1991 it was less than 150 DU. These initial observations have since been verified by measurements made from satellites and high-altitude aircraft by groups from various countries. Ozone depletion has also been noted in the Arctic and even in the mid-latitude regions. Overall the depletion represents a reduction of about 3% in the global stratospheric ozone but a 50% reduction in the Antarctic ozone, and the reduction continues on its course. This trend is evident in Figure 10.1.

Ozone is formed in the stratosphere when ultraviolet radiation dissociates O_2 into two free oxygen atoms. One of these oxygens then combines with O_2 to form ozone, O_3. Ozone itself is also dissociated by the ultraviolet radiation, but a net balance remains in equilibrium with the rates of formation and destruction. In a series of studies, it has been found that the main cause of man-made depletion of the ozone below its natural concentration is the injection into the atmosphere of man-made chlorinated fluorocarbons (CFCs). Two of the more widely used CFCs are Freon-11 ($CFCl_3$) and Freon-12 (CF_2Cl_2). These chemicals are used in air conditioners, refrigerators, some building insulation, some plastic foam products, solvents used in the electronics industry, and spray cans. Measurements in Antarctica and elsewhere showed that the amount of CFCs in the atmosphere had tripled from 1970 to 1980. The CFCs can be transported high up into the stratosphere where ultraviolet radiation dissociates the molecule into free chlorine atoms and other molecular fragments. The free chlorine can then act as a catalyst in dissociating O_3 in the following way.

First the chlorine interacts with the ozone:

$$Cl + O_3 \rightarrow ClO + O_2,$$

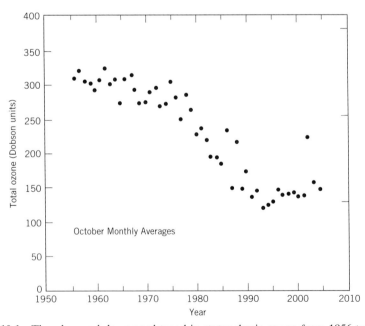

Figure 10.1 The observed downward trend in statospheric ozone from 1956 to 2004 over the Antarctic. These measurements are for the springtime total ozone over Halley Bay, Antarctica. (*Source:* British Antarctic Survey, April 2005.)

followed by

$$ClO + O \rightarrow Cl + O_2.$$

What started as a free atom of chlorine again becomes a free atom of chlorine, leaving it available to repeat the process over and over again, destroying ozone each time it is repeated. It is estimated that one chlorine atom can destroy 100,000 ozone molecules. In addition to the reactions described above, similar but more complex reactions are particularly important for the ozone hole region.

It was in fact the detection of ClO over the Antarctic in amounts about 100 times more than expected that led to man-made CFCs being tagged as the main cause of the ozone depletion. The complete process is more complicated than indicated in the two chemical reactions. For example, the element bromine is thought also to play a role in these reactions, leading to about 25% of the ozone destruction.

There are many compounds of chlorine that come from various sources on the earth's surface. Ocean spray, swimming pools, bleach, and volcanoes can all provide chlorine compounds in substantial quantity. All of the chlorine compounds from these sources, however, are readily dissolved in water and never reach the stratosphere.

It is known that extensive clouds form over the Antarctic in the winter. These clouds include aerosols containing nitric acid as well as water vapor and ice crystals. Normally only a small fraction of the chlorine released from the CFCs in the stratosphere is available to destroy ozone rapidly. The rest is in molecules such as HCl that undergo only slow chemical reactions in the stratosphere. During the Antarctic winter, chemical reactions on the surfaces of the ice crystals convert the chlorine in HCl and other molecules to a compound called chlorine monoxide, or ClO. The catalytic reaction then destroys ozone much more rapidly than in the rest of the stratosphere. Sunlight is required for the reaction, so not much ozone is destroyed in the dark Antarctic winter. Instead, the ozone hole appears in October after sunlight has returned.

This story of ozone depletion illustrates how human activities can have unintended harmful consequences for the environment. Although CFCs were manufactured primarily for use in air conditioners and refrigerators, they ended up being responsible for serious increases in the level of ultraviolet radiation on the earth's surface and contribute to global warming as greenhouse gases. To do this, the CFCs must rise up into the stratosphere after they are released near the earth's surface. Certainly these possibilities never occurred to the engineers who specified their use as refrigerants. As we saw in Chapter 9, ozone is a known air pollutant that plays an essential role in the Los Angeles type smog. At one time it was thought that the ozone created near the earth's surface, and which is so harmful there, would rise up to the stratosphere and make up for any deficiencies caused by CFCs, thus playing a beneficial role in the end. It was soon realized, however, that because ozone is so reactive it could never rise above the troposphere.

The global problem of ozone depletion can be mitigated only by limiting the use of CFCs. Some danger to the ozone layer from CFCs was already known in the 1970s and some less important uses of CFCs such as a propellent in spray cans were restricted. After discovery of the ozone hole in 1985, much stronger measures were put in place. Major chemical companies developed substitutes for CFCs including hydrofluorocarbons (HFCs) and hydrochlorofluorocarbons (HCFCs). For example, HFC-134 (CH_2FCF_3) became the standard refrigerant for automobile air conditioners in the United States in 1994. The newer refrigerants are considerably more costly than the Freons they replace and they cannot be used in older systems which were designed for the Freons.

In 1987 a world conference was held in Montreal to reach an international agreement to reduce the use of CFCs. The Montreal Protocol, which was signed by most nations, mandated a 50% reduction in CFC production by the year 1998. Since that time, because of the deepening of the ozone hole, the protocol has been modified to phase out CFC production on a more rapid schedule. Some developing countries, such as China and India, did not sign the original protocol because they were concerned about the economic implications for their countries. Global stratospheric ozone has until recently been decreasing by about 2.5% per decade, but the rate of decrease now seems to be diminishing. It appears that CFC control measures may be having a beneficial effect. At the 2.5%/decade rate of ozone thinning, were it to continue, the EPA has predicted 200,000 additional deaths from skin cancer in the next 50 years in the United States alone. Since the CFCs are also greenhouse gases, there is an additional incentive for phasing them out.

In early 1992 it was found that the ozone hole phenomenon was spreading to more densely populated areas in the Northern Hemisphere at an alarming rate. Because of this, the protocol was strengthened in 1992 (the Copenhagen Amendments) to call for the stopping of all CFC production in industrialized countries by the end of 1995. This call for an end to production also included methyl chloroform and carbon tetrachloride. By 1997, world production of CFCs had dropped from a high of about 1.2 million tonnes in 1988 to essentially nothing. Projected trends for atmospheric concentrations of chlorine and bromine are shown in Figure 10.2 for three different scenarios: No protocol, adoption of the 1987 Montreal Protocol, and enforcement of the 1992 Copenhagen Amendments. Without the Montreal Protocol and its amendments, the stratospheric concentrations of chlorine and bromine could be expected to triple by the year 2050.

While the major ozone holes are found in the Antarctic stratosphere and to some extent in the Arctic stratosphere, measurable reductions have also occurred across the United States and elsewhere. From 1979 to 2002 the average ozone concentration over the United States was reduced by about 8% from 325 to 300 DU. There is a considerable seasonal variation with increased levels in the winter and spring and lower levels in the summer and fall. Actual measurements of the UV-B ultraviolet radiation have shown increases that are at their greatest at high northern and southern latitudes. UV-B has a wavelength of 280 to 315 nanometers and higher energy per photon than UV-A which has a wavelength of 315 to 380 nanometers. The UV-B radiation is capable of greater biological damage than is UV-A.

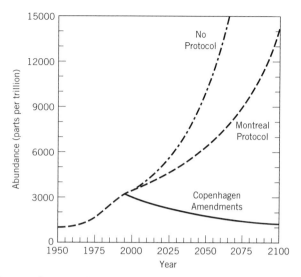

Figure 10.2 Expected stratospheric abundances of chlorine and bromine assuming no Protocol, the Montreal Protocol, and the Montreal Protocol with the Copenhagen Amendments. (*Source:* Scientific Assessment of Ozone Depletion: 1994. World Meteorological Organization Global Ozone Research and Monitoring Project—Report No. 37, Executive Summary, February 1995.)

The substitutes for the CFCs, namely, the HFCs (hydrofluorocarbons) and HCFCs (hydrochlorofluorocarbons), are both significant greenhouse gases with 1300 times and 1700 times, respectively, the warming effects of CO_2 as measured tonne for tonne, but they are less important contributors to greenhouse warming than the CFCs they replace. The latest round of treaty negotiations which took place in Beijing in 1999 imposed a freeze on the production of HCFCs by 2030 for developed countries and by 2040 for developing countries.

While production of CFCs has stopped, their use continues mainly in refrigeration and air-conditioning systems. It is illegal to vent CFCs to the atmosphere, so mechanics servicing an automobile air conditioner, for example, must capture and recycle the refrigerant. The replacement of CFC-12 in newly manufactured motor vehicles with HFC-134 began in 1992 and was completed in 1994. On older model cars, modifications to the air conditioner are needed if HFC-134 is to be substituted for a Freon refrigerant.

As can be seen in Figure 10.1, the amount of ozone in the atmosphere over the Antarctic remains at a level much reduced from what it was several decades ago in spite of the measures taken in the Montreal Protocol and the various amendments. This is because the CFCs, the HFCs, and the HCFCs already in the atmosphere do not go away rapidly even if no more are being added. Their residence times are very long, estimated at 50 to 100 years. This can be seen in Figure 10.2 where, even with the Copenhagen Amendments, the abundance of these atmospheric contaminants is expected to drop only by half in 100 years.

While the problem of ozone depletion in the upper atmosphere is not completely solved, the adherence of most nations to the Montreal Protocol and its subsequent amendments is a shining example of nations getting together to deal with a worldwide environmental problem in an encouraging fashion.

10.3 The Greenhouse Effect and World Climate Changes

The term *greenhouse effect* refers to the idea that incoming solar radiation readily penetrates the glass coverings of an ordinary greenhouse, but the outgoing infrared radiation from the interior does not. The result is that the solar energy is trapped, which leads to heating of the greenhouse. (The detailed description of this process is more complex; it is discussed in Chapter 4.) A similar phenomenon takes place in the earth's atmosphere. Certain gases in the atmosphere, now called the greenhouse gases, act to warm the atmosphere below them. The effect is related to the fact that the sun's radiation temperature is very high compared to the earth's surface temperature. The incoming solar energy easily penetrates the atmosphere, but the thermal radiation from the earth's surface does not. The wavelength of the radiated thermal energy from the earth's surface is in the infrared region (about 4 to 20 microns). Certain gases absorb electromagnetic radiation very effectively in this region of wavelength, and if they are present in the atmosphere, they will trap the infrared radiation moving upward from the earth's surface. This contributes to a warming of the atmosphere.

The greenhouse gases now of greatest concern include carbon dioxide (CO_2), methane (CH_4), nitrous oxide (N_2O), chlorofluorocarbons (CFCs), and tropospheric ozone (O_3). Carbon dioxide now accounts for about 57% of the greenhouse gas effect, methane about 17%, and CFCs about 5%. There is growing evidence that various human activities are leading to an increase of the concentration of these gases in the atmosphere, and hence to the possibility of an increase of the temperature of the atmosphere. This may result in significant changes of climate on a global scale.

Carbon dioxide is not usually classified as an air pollutant, as it is naturally reasonably abundant in the atmosphere (now about 0.038%, or 380 ppm). Any burning of carbon-based fuels leads to the formation of CO_2. Although the chemical reactions in the combustion of the various fossil fuels are not always simple, the general process may be summarized as

$$C + O_2 \rightarrow CO_2 + \text{energy}.$$

Whether the fuel being burned is wood, coal, petroleum, or natural gas, carbon dioxide will result. The natural process of photosynthesis depends on atmospheric CO_2, and when organic matter decays, CO_2 is released to the atmosphere. The problem we are now facing is that atmospheric CO_2 is increasing at about 0.4% annually, and human activities, especially the ever-increasing consumption of fossil fuels and the reduction of forests, are the likely causes of the increase.

Example 10.1

If atmospheric CO_2 increases at a rate of 0.4% annually, in how many years will the present concentration of CO_2 (380 ppm) be doubled to 760 ppm?

Solution

The doubling time (see Appendix) is

$$t_D \text{ (years)} = \frac{69.3}{\% \text{ growth per year}}$$

$$= \frac{69.3}{0.4\%/\text{yr}} = \textbf{173 years}$$

Methane, at 1.7 ppm, is far less abundant in the atmosphere than CO_2, but it is, molecule for molecule, 20 to 30 times more effective at absorbing infrared radiation than is CO_2, and it is increasing at about 1% per year. The main sources of atmospheric CH_4 are thought to be anaerobic bacteria in rice fields, the digestive tracts of cattle, release from handling fossil fuels, termite colonies, sewage treatment plants, and landfills.

Nitrous oxide is present in the atmosphere only to about 0.3 ppm. Its concentration is increasing at 0.2% annually. The main sources are the use of nitrogen fertilizers and other agricultural practices.

As discussed in Section 10.2, the CFCs, mostly Freon-11 and Freon-12, have dominated the problem of ozone depletion. They are also greenhouse gases, but account for only about 5% of the total effect. Since the production of CFCs has essentially ceased, their atmospheric concentration is no longer growing, but their lifetime in the atmosphere is rather long (50 to 100 years). However, the concentration of the new substitutes, which are also greenhouse gases, is increasing. There is a need for even more benign substitutes.

Because of the ongoing depletion of ozone in the stratosphere and the efforts that have gone into studying this depletion, the contribution of ozone to the greenhouse effect is now quite well understood. Ozone now may be making as much as a 15% contribution. Table 10.1 summarizes some details of the various greenhouse gases.

To be somewhat more quantitative about the effects of greenhouse gases on the global climate, the concept of climate forcing in terms of watts/m^2 has been introduced. This is additional heating or cooling of the atmosphere due to some perturbation on the system. For example, at their present concentrations, carbon dioxide is thought to add about 1.4 watts/m^2, methane adds 0.6 watt/m^2, chlorofluorocarbons 0.3 watt/m^2, nitrogen oxides 0.3 watt/m^2, and ozone 0.4 watt/m^2. In addition to these greenhouse gases, it has been realized more recently that aerosols can contribute either negatively or positively. Negative forcings can cause cooling of the atmosphere. Black carbon is now thought to contribute about +0.8 watt/m^2, reflective aerosols −1.3, cloud droplet changes −1.0,

Table 10.1 Characteristics of the Major Greenhouse Gases

Gas	Preindustrial Concentration	Concentration in 2003	Annual Rate of Increase	Lifetime (Years)	Importance to Greenhouse Effect
Carbon dioxide (CO_2)	~280 ppmv	380 ppmv	0.4%	50–200	57%
Methane (CH_4)	~700 ppbv	1,720 ppbv	0.6%	12	17%
Nitrous oxide (N_2O)	~275 ppbv	312 ppbv	0.25%	120	5%
CFC-11	0	255 pptv	0	50	5% for the four compounds listed
CFC-12	0	554 pptv	0	102	
HCFC-22 (a CFC substitute)	0	158 pptv	5%	12	
CF_4	40 pptv	80 pptv	2%	50,000	
Ozone[a] (O_3) (tropospheric)	25 ppbv	40 ppbv			

[a]Because of the complexities outlined in the text, the ozone characteristics and importance are not entirely delineated here.
Source: The Intergovernmental Panel on Climate Change (2001), updated to 2005 by T.J. Blasing and Sonja Jones, Oak Ridge National Laboratory.

land cover changes −0.2, and sun brightness increases over the past 150 years +0.4 watt/m².

According to James Hansen, director of the NASA Goddard Institute for Space Studies, the sum of the forcings added since 1850 is +1.6 ± 1.0 watts/m². While there is an appreciable uncertainty, there is external evidence that this total is approximately correct. It is perhaps somewhat surprising that these relatively minor forcings can cause major effects on the world climate. The system was in a delicate balance and these small perturbations have already caused measurable warming, with retreating glaciers and Arctic ice thinning among other observed effects.

Another greenhouse gas, in fact the most important, is water vapor. Were it not for the naturally occurring water vapor in the atmosphere, the earth would be many degrees colder than it is now. The amount of water vapor in the atmosphere is relatively stable, and not directly subject to man's influence, so it is not normally considered as a greenhouse gas of concern. It may happen, however, that if there is sufficient global warming, the warmer air will take up more water vapor, leading to a positive feedback effect. More warming means more water vapor, more water vapor means more warming, and so on.

The overall question of the effect of man-made CO_2 and other greenhouse gases on the earth's temperature and climate is a complex one involving some processes that are not completely understood. On the other hand, some aspects of the problem are well established. It is abundantly clear that we are injecting large amounts of CO_2 into the atmosphere. Since the amount of coal and other fuels burned each year is approximately known, reasonable estimates can be made of the amount of CO_2 released. This amount has increased steadily from about 0.5 billion tonnes/yr in 1860 to about 5 billion tonnes/yr now. It can be expected that the concentration of atmospheric CO_2 will continue to increase as more fossil fuels are burned, possibly until the total recoverable fossil fuels are consumed. There are about 2×10^{12} tonnes of carbon in the total recoverable resource of fossil fuels.

One aspect of the problem no longer open to vague speculation is the amount of CO_2 present in the atmosphere. Since 1958 accurate measurements of the atmospheric CO_2 concentration have been carried out at an observatory on Mauna Loa, Hawaii. These measurements are shown in Figure 10.3. There is an obvious general upward trend, amounting to an increase of about 1.6 ppm per year, or an annual increase of about 0.4%. The interesting seasonal oscillation in the data is caused by the removal of CO_2 by photosynthesis during the growing season in the Northern Hemisphere. A similar general increase of CO_2 has also been measured at the South Pole and other locations.

Figure 10.4 shows CO_2 concentrations in the atmosphere over the past 1000 years. The recent Mauna Loa measurements are shown as the dashed line. The earlier data are taken from ice cores in Antarctica. There can be little question that beginning with the Industrial Revolution in about 1850 the CO_2 concentra-

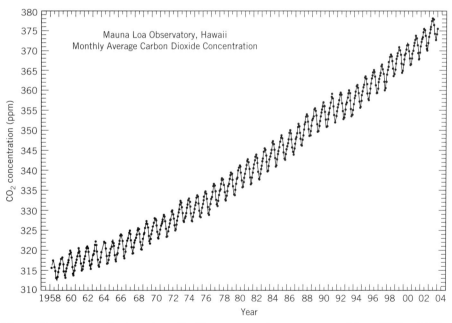

Figure 10.3 Atmospheric carbon dioxide concentrations as measured from 1958 to 2003 at the Mauna Loa Observatory on the island of Hawaii. The data are monthly averages in parts per million by volume. (*Source:* C. D. Keeling and T. P. Whorf, Scripps Institution of Oceanography; data provided by the Carbon Dioxide Analysis Center, Oak Ridge National Laboratory, and U.S. National Oceanic and Atmospheric Administration.)

336 Chapter 10 Global Effects

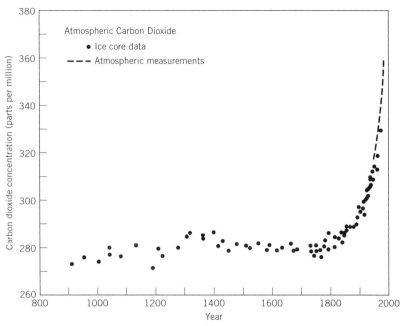

Figure 10.4 Atmospheric carbon dioxide concentrations over the past 1000 years from ice core records and from Hawaii. The ice core measurements were taken in Antarctica. The dashed curve is based on the Mauna Loa measurements. The increase in the atmospheric carbon dioxide concentration since the onset of industrialization is evident. (*Source:* Adapted from "Common Questions about Climate Change," United Nations Environmental Programme, World Meteorological Organization.)

tion in the atmosphere started to rise dramatically. It appears that only about 50% of the fossil fuel emissions are retained in the atmosphere. According to the Intergovernmental Panel on Climate Change (IPCC, 2001) the present carbon dioxide concentration has not been exceeded during the past 420,000 years, and likely not during the past 20 million years. The current rate of increase is unprecedented during at least the past 20,000 years.

In addition to the 700 billion tonnes of carbon in the form of carbon dioxide in the atmosphere, there are extensive pools in the biota, the organic matter in the soils, and in the oceans. The biota hold about the same amount of carbon as the atmosphere, about 700×10^9 tonnes, and the organic matter in the soils holds 1000 to 3000×10^9 tonnes. It is further known that photosynthesis removes about 110×10^9 tonnes of carbon from the atmosphere each year, and the amount of CO_2 added to the atmosphere by plant respiration and the decay of organic matter is about the same. The deep layers of the ocean hold a vast amount of carbon in the form of dissolved CO_2 (about 40×10^{15} tonnes); the surface layers (the top 100 m), where carbon dioxide mixes with the atmosphere, hold very much less, about 600×10^9 tonnes. Although mixing can occur between the deep and surface layers of the ocean, this is believed to happen over a very long time span. The surface layers of the ocean appear to be sinks for about 3×10^9 tonnes of CO_2 per year.

One basic question seems to be how much of the observed increase in atmospheric CO_2 could be due to a reduction in biomass. There has been a notable reduction of the tropical rain forests, which are excellent pools of CO_2, and arguments have been made that this could cause part of the increased atmospheric CO_2. More recent calculations and models, however, indicate that the net biotic pool of CO_2 has probably not changed enough to account for the carbon dioxide increase. Although forests have been cut down, there has also been extensive agriculture established by irrigation of land that was previously not very productive. The combustion of fossil fuels contributes 5 to 6 billion tons per year of carbon as CO_2 and, with larger uncertainties, deforestation contributes 1 to 2 billion tons per year. Table 10.2 shows the most recent information on the global budget for CO_2 along with the uncertainties.

In summary, the average anthropogenic emissions for 1989–1998 are 7.9 ± 1.0 billion tons per year, and of this, 3.3 ± 0.2 billion tons are stored in the atmosphere. There is little doubt that we are adding to the atmospheric burden of CO_2, year by year, mainly by burning fossil fuels, but also by reducing the world's forests.

The next main question relates to the effect on the world's climate of these carbon dioxide increases. The temperature of the earth has changed over geologic time. During the periods of glaciation, it was about 5°C lower than now. The last period of glaciation was 10,000 years ago. Since then there have been fluctuations of a degree or more, the period of 1430 to 1850 A.D. being one of

Table 10.2 Average Annual Global Budget of CO_2 for 1989–1998

CO_2 Sources	Flux (Gt C/yr)[a]
Fossil fuel combustion and industrial processes	6.3 ± 0.6
Net from changes in tropical land use	1.6 ± 0.8
Total anthropogenic emissions	7.9 ± 1.0
Partitioning Among Reservoirs	
Storage in the atmosphere	3.3 ± 0.2
Ocean uptake	2.3 ± 0.8
Terrestrial uptake	2.3 ± 1.3

[a]Gigatons of carbon per year.
Source: Intergovernmental Panel on Climate Change, May 2000.

particularly low temperatures in Europe. The mean annual global temperature from 1861 to 1994 is shown in Figure 10.5. Although there are large fluctuations from year to year, it seems evident that there has been at least a 0.5°C increase in the average temperature over the past 133 years.

There are at least two effects that tend to lower the earth's temperature. One of these is natural. Volcanoes spew forth a very fine ash that can stay aloft for several years and form a layer of sun-blocking haze. A second type of effect is predominantly man-made. Industrialized countries contribute a variety of aerosols to the atmosphere from smokestacks, cars, tilled fields, and burning forests. Sulfur dioxide, for example, can become sulfate aerosols that form a haze that tends to reflect solar radiation. There is concern that these temperature-lowering effects have masked the temperature increases caused by greenhouse gases and that we will eventually be faced with greater greenhouse increases than are indicated by the recent temperature records. In 1992 it was noted that the global temperature took a significant drop. Critics of the global warming hypothesis were quick to point out that this happened in spite of the increased CO_2 concentration. It was soon realized, however, that the sulfur dioxide and ash from the 1991 Mt. Pinatubo volcanic eruption in the Phillipines had caused the drop, and that the CO_2 was playing its normal role.

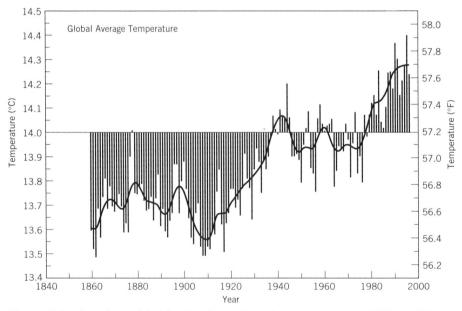

Figure 10.5 Combined global land surface, air, and sea temperatures 1861 to 1994. The solid curve represents smoothing of the annual values shown by the vertical bars. (*Source:* Adapted from *Climate Change 1995—The Science of Climate Change*, Intergovernmental Panel on Climate Change, Cambridge University Press; and "Common Questions about Climate Change," United Nations Environmental Programme, World Meteorological Organization.)

The Greenhouse Effect and World Climate Changes 339

When the term global warming is used, it refers to many related phenomena taking place at the same time:

1. The earth's temperature is increasing. The decade of the 1990s was the warmest since the mid 1800s. The year 1998 was the hottest on record, followed by 2002, 2003, 2001, and 1987.
2. Most of the world's glaciers are melting. When Glacier National Park was established in 1910, it had about 150 glaciers. There are now fewer than 30, and those that remain have shrunk in area by two-thirds (see Figure 10.6). Eighty percent of the snows of Kilimanjaro have melted. The Quelccaya ice cap in Peru, the largest in the tropics, is contracting more than 600 feet a year. There are many other examples.
3. The Arctic sea ice is being reduced both in area and in thickness. The area is shrinking by 9% per decade, and the thickness by 15 to 40% over the past 30 years.
4. The level of the oceans is now rising more rapidly than in the past. The main cause is thermal expansion of the seawater and, to a lesser extent, the melt-

Figure 10.6 Two photographs of the same glacier in Glacier National Park, one recent, one from 1910. (*Source*: top ©Kiser and bottom ©Dan Fagre)

ing of the Antarctic ice. The global mean sea level increased at an average annual rate of 1 to 2 millimeters during the 20th century. Such sea level increases now threaten many island nations such as Tuvalu and the Maldives. More than 100 million people now live within 1 meter of mean sea level.

5. Many other global warming effects are now becoming apparent. These include longer growing seasons, thawing of the Alaskan permfrost, coral reef bleaching, earlier plant flowering, earlier bird arrivals, shifting of plant and animal ranges poleward, and more frequent El Niño events.

What can be said about the future world climate? There are at least three steps to answering this question. The first is the prediction of the increase of the greenhouse gases in the atmosphere. Second is an understanding of the effects of the increased concentrations on global temperature, and third is the effect of the increased temperature on the global climate. The average global temperature over the past 160,000 years is correlated rather closely with atmospheric CO_2 concentration. On a finer scale, however, the correlation over the last 100 years is less convincing. Nevertheless, the majority of the concerned scientific community feels that there is a high probability that the 0.5°C temperature increase in the last 100 years is related to the increase in atmospheric CO_2 concentration.

Models of the world's climate (General Circulation Models, or GCM) are regularly run on powerful computers, and scientists vary parameters in these models to ascertain, for example, the effect of a doubling of the atmospheric CO_2 concentrations or that of the equivalent composite greenhouse gases. These models tend to predict a 3.0 to 5.5°C increase in the global temperature for a doubling of the concentration. Other estimates have been in the range of 1.5 to 4.5°C. Such predictions, however, are quite uncertain because of imperfect treatment of various meteorological phenomena.

The Intergovernmental Panel on Climate Change has developed a series of scenarios of future greenhouse gas and aerosol concentrations based on assumptions of populations, economic growth, land use, energy availability, and fuel mix for the period of 1990 to 2100. With these assumptions and with knowledge of atmospheric chemistry, predictions were made of the future climate. The best current estimate is that the mean global surface temperature will go up by 3°C by 2100. With the lowest greenhouse gas concentration and climate sensitivity assumption, the estimated increase dropped to 1°C. The highest estimate of emissions and climate sensitivity gave an increase of almost 6°C. In every case the average rate of warming would probably be greater than any seen in the past 10,000 years.

An important achievement in comparison of models and observations was reported in the 2001 IPCC assessment. Comparison of the earth's temperature observed year by year from 1850 to 2000 to model calculations that included only natural climate forcings showed only limited agreement. Similarly, a model that included only man-made forcings (anthropogenic) showed little agreement. However, a more advanced model that includes both natural and anthropogenic forcings shows excellent agreement with observation. This finding validates the modeling approach and reinforces confidence in this means of analysis.

According to the IPCC models, the corresponding rise of the sea level from 1990 to 2100 for the three cases gave 50 cm for the best estimate, 15 cm for the low estimate, and 95 cm for the high estimate. Even if concentrations of the greenhouse gases were stabilized by 2100, the sea level would continue to rise. There would also be very marked changes in the rainfall and general climate in any particular region, leading to droughts in some areas and floods in others. Important agricultural areas such as the U.S. Midwest could become arid, and many coastal areas could be inundated by the rising ocean waters. There is concern that the world's food supply could be diminished at a time when the population is growing.

What can be done to avoid such extreme changes of global climate? While the answer of lowering the emission of greenhouse gases is easy to state, it is not so easy to accomplish. The main culprit is CO_2, and there can be no satisfactory solution to the problem without lowering CO_2 emissions. The halocarbons (CFCs and their substitutes, the HCFCs and HFCs) are being phased out under the Montreal Protocol and its Amendments. Some methane gas could be captured and put to beneficial use. Emissions of nitrogen oxides and O_3 are part of the atmospheric pollution problem discussed in Chapter 9, and hopefully further efforts to reduce their emission will lessen their greenhouse contributions.

Carbon dioxide emission is basically an energy-related problem as its reduction involves burning less fossil fuels and changing tropical land-use patterns. With respect to CO_2 any improvement in the efficiency of electricity generation is helpful. There is also an appreciable difference in emissions for the various fossil fuels on an equal-energy basis. As can be seen in Table 10.3, the best available natural gas–fired combined-cycle gas turbine emits less than half the amount of CO_2 that coal-fired steam turbines do. With any of the fossil fuels, cogeneration, as an example of improved efficiency, can often reduce CO_2 emissions.

There has been recent discussion about and some effort in relieving the problem of carbon dioxide emission into the atmosphere by capturing and then injecting the carbon dioxide into deep underground formations. This falls under the general technology of carbon dioxide sequestration, meaning removing the carbon dioxide to a place where it will not be a harmful environmental contaminant. The owners of a natural gas operation in Norway have been injecting about a million tons per year, or 3% of the Norwegian carbon dioxide emissions, into a sandstone bed 3000 feet below the ocean floor. They are doing this primarily to avoid paying a national emission tax of $50 per ton of carbon dioxide. Beyond this effort, other sequestration locations under discussion include the ocean floor, coal beds, depleted oil and gas reservoirs, and deep aquifers. Sequestration technology is still not highly developed; it remains uncertain and it is highly doubtful that it offers any real hope for substantially relieving the greenhouse gas problem.

Example 10.2

The average American household uses 5000 kWh of electric energy per year.

(a) Calculate the amount (in tonnes) of CO_2 released to the atmosphere if the electricity is generated by a conventional coal-fired steam turbine. Use data from Table 10.3.

(b) Calculate the CO_2 released as in (a) but using the best available combined-cycle natural gas turbine.

Solution

(a) *Coal*

$$5000 \text{ kWh}_e \times \frac{325 \text{ grams carbon}}{\text{kWh}_e} = 1.63 \times 10^6 \text{ grams carbon}$$

$$= 1.63 \text{ tonnes carbon}$$

To convert tonnes C to tonnes CO_2, use atomic weight of carbon = 12 and molecular weight of CO_2 = 44.

$$\text{Amount of } CO_2 = \frac{44}{12} \times 1.63 \text{ tonnes}$$

$$= \mathbf{5.98 \text{ tonnes } CO_2}$$

(b) *Natural gas turbine*

$$5000 \text{ kWh}_e \times \frac{140 \text{ grams carbon}}{\text{kWh}_e} = 0.70 \times 10^6 \text{ grams carbon}$$

$$= 0.70 \text{ tonne carbon}$$

$$0.70 \text{ tonne carbon} \times \frac{44}{12} = \mathbf{2.56 \text{ tonnes } CO_2}$$

Any switch to natural gas from coal is going to encounter a resource problem sooner or later because the energy content of our remaining natural gas is very much less than that of our remaining coal. This is shown in Table 2.9. Natural gas is also in great demand for heating and cooking and it will probably see greater use in transportation.

Energy conservation measures will be an important part of reducing CO_2 emissions. As discussed in Chapter 7, there are many practical ways to reduce waste of energy without sacrifice.

Any transfer of energy demand from fossil fuels to renewable energy sources can contribute to reducing CO_2 emissions. While some of this is now happening, the pace is far too slow to prevent significant changes of climate. One of the greatest advantages of nuclear power reactors is that they have essentially no emissions into the atmosphere. No carbon dioxide at all is emitted. If generally acceptable resolutions can be found to the problems of cost, reactor safety, radioactive waste disposal, and weapons proliferation, an expanded nuclear energy program could make an important contribution to reducing the emission of greenhouse gases. If biomass is burned as an energy source, in the long term it does not add to the problem of CO_2 emission because about as much CO_2 is taken up by growing the biomass as is released in burning it. Production and use of biofuels become a closed CO_2 cycle.

Table 10.3 Carbon Dioxide Emissions and Efficiencies for Several Combustion Technologies and Fuels for Electricity Generation

Technology (Fuel)	Efficiency (%)	CO_2 Emission[a] (g C/kWh$_e$)
Average conventional steam turbine (coal)	34	325
Best available steam turbine (coal)	39	280
Pressurized fluidized bed combustion (coal)	42	260
Average conventional steam turbine (oil)	38	235
Best available combined-cycle gas turbine (oil)	48	285
Average combined-cycle gas turbine (natural gas)	36	170
Best available combined-cycle gas turbine (natural gas)	45	140

[a]Grams of carbon equivalent per kWh$_e$.
Source: Intergovernmental Panel on Climate Change (1995).

Other technologies can also be called upon to help reduce CO_2 emissions, but the problem is formidable because of our massive dependency on fossil fuels. Neither the developed nor the undeveloped countries are likely to relinquish their fossil fuel use very easily.

On February 16, 2005, the 140 nation Kyoto Protocol went into effect without the participation of the United States. While many countries have endorsed significant reductions in emissions, the United States has been reluctant to do so because of possible economic consequences. Many in the present administration still do not accept the reality of global warming due to greenhouse gases. There is also resistance to not including developing nations such as China in the emission reduction requirements. Ten large per capita emitters of carbon dioxide from fuel combustion are shown in Table 10.4. The European Union agreed to an 8% reduction and Japan 6%. Overall, 38 nations are scheduled to reduce emissions by a bit more than 5% from the 1990 levels. The remaining developing nations are exempt for the time being from any reduction. Those involved with the problem realize that the agreed-on reductions will not stop global warming. However, these reductions will slow the rate at which it is advancing.

There are serious problems with the Kyoto Protocol in that there are no clear monitoring and enforcement provisions. The treaty, however, is an important first step in the widespread realization that the greenhouse gases are a significant threat to the global climate and that remedial action is called for.

Table 10.4 Ten Large Per Capita Emitters of CO_2 from Fuel Combustion in 2002[a]

Country	Per Capital Emissions (tons/year)	Share of World Total Emissions (%)
United States	19.66	23.5
Canada	16.93	2.2
Russia	10.43	6.2
Germany	10.15	3.5
South Korea	9.48	1.9
Japan	9.47	5.0
United Kingdom	8.94	2.2
Italy	7.47	1.8
China	2.55	13.6
India	0.97	4.2

[a] Several less populous countries have higher per capita emissions of CO_2. For example, Qatar is reported to have emitted 59 tons per capita in 2002 (United Nations, 2005).
Source: CO_2 Emissions from Fuel Combustion, 1971–2002, International Energy Agency, 2004.

The IPCC has explored various scenarios involving different levels of CO_2 emissions and their effects on global warming. They find that if the emissions could be stabilized at the 2000 level of 9 billion tons of carbon per year, the atmospheric CO_2 concentration would still continue to rise, reaching 550 ppm by 2125 with a temperature increase of about 1.7°C. Another scenario of having the concentration rise to 550 ppm by 2125, at which level it would stabilize, would require substantial reduction of emissions.

In the United States there appears to be little understanding on the part of the general public of the probable consequences of the greenhouse global warming problem, and hence there is little political support for any painful restrictions on emissions of greenhouse gases. Powerful lobbying groups from both labor unions and industry are actually opposed to limitations on carbon dioxide emissions.

As an example of the kind of steps that might be necessary to reduce the CO_2 emissions to below the 1990 levels by 2010, a tax of $100 per ton of carbon dioxide emitted might be implemented. Such a tax would at least be directed toward the immediate problem and would give economic incentives to technologies that minimize CO_2 emissions.

Example 10.3

Based on a tax rate of $100 per ton of CO_2, calculate the additional cost to the average American household for 5000 kWh$_e$ (a typical year's usage) of electric energy furnished by (a) a conventional coal-fired power plant and (b) the best available combined-cycle natural gas turbine. Use the results of Example 10.2.

Solution

(a) *Coal*

$$5.94 \text{ tonnes } CO_2 \times \frac{1 \text{ ton}}{0.907 \text{ tonne}} \times \frac{\$100}{1 \text{ ton}} = \mathbf{\$650}$$

(b) *Natural gas turbine*

$$2.56 \text{ tonnes } CO_2 \times \frac{1 \text{ ton}}{0.907 \text{ tonne}} \times \frac{\$100}{1 \text{ ton}} = \mathbf{\$282}$$

There is a major difficulty in controlling emissions in the developing countries, particularly China and India. These countries insist that the restrictions on CO_2 emissions should not apply to them for a number of years. They point out that it is the developed countries that now put most of the CO_2 into the atmosphere, and that restricting the developing countries will only retard their industrial growth. They claim that they could never catch up with the industrialized nations if they implement the necessary restrictions.

Key Terms

Greenhouse gases
Dobson units
Chlorinated fluorocarbons (CFCs)
Hydrofluorocarbons (HFCs)
Montreal Protocol
Copenhagen Amendments
Greenhouse effect
Carbon dioxide sequestration

Intergovernmental Panel on Climate Change
Global warming
Anthropogenic emissions
Deforestation
Glaciation
General Circulation Models

Suggested Reading and References

1. Lyman, F. *The Greenhouse Trap.* Boston, MA: Beacon Press, 1990.
2. Fisher, D. E. *Fire and Ice.* New York: Harper and Row, 1990.
3. Gore, A. *Earth in the Balance.* New York: Houghton Mifflin, 1992.
4. *Climate Change 1995—The Science of Climate Change.* Intergovernmental Panel on Climate Change. New York: Cambridge University Press, 1995.
5. *Climate Change 1995—Impacts, Adoptions, and Mitigation of Climate Change: Scientific–Technical Analyses.* New York: Cambridge University Press, 1995.

6. French, H. F. "Learning from the Ozone Experience." *State of the World 1997.* World-watch Institute. New York: W. W. Norton, 1997, p. 151.

7. Weart, S. R. "The Discovery of the Risk of Global Warming." *Physics Today* (January 1997); The Discovery of Global Warming, Harvard University Press, 2003, 228 pages.

8. Hamil, P., and Owen, B. T. "Polar Statospheric Clouds and the Ozone Hole." *Physics Today*, **44,** No. 12 (December 1991), pp. 34–42.

9. Ramanathan, V. "The Greenhouse Theory of Climate Change: A Test by an Inadvertent Global Experiment." *Science*, **240,** No. 4850 (April 1988), pp. 293–299.

10. Balzhiser, Richard E., and Yeager, Kurt E. "Coal Fired Power Plants for the Future." *Scientific American*, **257,** No. 3 (September 1987), pp. 100–107.

11. Alpert, S. B. "Clean Coal Technology and Advanced Coal-Based Power Plants." *Annual Review of Energy and the Environment*, **16** (1991), pp. 1–23.

12. Spencer, D. F. "A Preliminary Assessment of Carbon Dioxide Mitigation Options." *Annual Review of Energy and the Environment*, **16** (1991), pp. 259–273.

13. *Scientific Assessment of Ozone Depletion: 1994.* World Meterological Organization Global Ozone Research and Monitoring Project—Report No. 37, Executive Summary, February 1995.

14. Herzog, Howard, Eliasson, Baldur, and Kaarstad, Olav. "Capturing Greenhouse Gases." *Scientific American*, **282,** No. 2 (February 2000), pp. 72–79.

15. Socolow, Robert, Hotinski, Roberta, Greenblatt, Jeffrey B., and Pacala, Stephen. "Solving the Climate Problem: Technologies Available to Curb CO_2 Emissions." *Environment*, **46,** No. 10 (December 2004), pp. 8–19.

16. Stavins, Robert N. "Forging a More Effective Global Climate Treaty." *Environment*, **46,** No. 10 (December 2004), pp. 23–30.

17. Trenberth, Kevin E. "Stronger Evidence of Human Influences on Climate: The 2001 IPCC Assessment." *Environment* **43** (May 2001), pp. 8–19.

18. Hansen, James. "Defusing the Global Warming Time Bomb." *Scientific American*, **290,** No. 3 (March 2004), pp. 68–77.

19. The Heat Is On. A Series on Global Climate Change. Photographs by Essick, Peter. *National Geographic*, **206,** No. 3 (September 2004).

    Appenzeller, Tim, and Dimick, Dennis. "Introduction, pp. 2–11.

    Glick, Daniel. "Geosigns—The Big Thaw," pp. 12–33.

    Montaigne, Fen. "EcoSigns—No Room to Run," pp. 34–55.

    Morell, Virginia. "TimeSigns—Now What?," pp. 56–75.

20. Socolow, Robert H. "Can We Bury Global Warming?" *Scientific American*, **293,** No. 1 (July 2005), pp. 49–55.

21. Houghton, John; Global Warming: The Complete Briefing, Third Edition, Cambridge University Press, 2004.

Questions and Problems

1. Assume that the total coal proved reserves of the world are eventually burned and that 50% of the resulting CO_2 is added to the atmosphere. Estimate the resulting increase in

concentration of CO_2 (in parts per million by volume or, equivalently, by number of molecules) in the atmosphere of the world. See Table 2.7 for coal reserves, and Section 9.2 for the mass of the atmosphere. Assume that the coal is 100% carbon.

2. If the rate of increase of the CO_2 concentration in the atmosphere is 0.4% per year, in what year would you expect the concentration to have doubled from the preindustrial level in 1860?

3. Propose a program to reduce by 60% the amount of CO_2 released through burning of fossil fuels by the year 2020.

4. (a) Using data from Table 10.3, calculate the tonnes of CO_2 that would be added to the atmosphere if coal was burned in a conventional steam turbine plant to produce all the 12 QBtu of electricity for one year for the United States.

   (b) Repeat the calculation in (a) for all the nation's electricity derived from natural gas burned in average combined-cycle gas turbines.

5. Suggest ways to increase public awareness of the global warming problem.

6. Calculate the approximate total tax revenue if all of the coal burned in the United States was burned under a $100/ton carbon dioxide tax. Do this (a) for one year at the present rate and (b) for the total U.S. proved reserves. Assume 100% carbon for the coal. See Chapter 2 for U.S. coal statistics.

Multiple Choice Questions

1. The so-called greenhouse effect is due at least partially to _____ .
   a. the opacity of ozone to infrared radiation
   b. the transparency of ozone to ultraviolet radiation
   c. the opacity of ozone to ultraviolet radiation
   d. the opacity of CO_2 and H_2O to infrared radiation
   e. the opacity of CO_2 and H_2O to ultraviolet radiation
   f. the transparency of CO_2 and H_2O to infrared radiation
   g. the opacity of CO_2 and ozone to infrared radiation
   h. the opacity of ozone and H_2O to infrared radiation

2. The increase observed in atmospheric levels of carbon dioxide over the past 100 years is most probably due to _____ .
   a. variations in the rate of lunar eclipses
   b. catalytic converters
   c. the burning of fossil fuels
   d. backyard incinerators
   e. fluctuations in ultraviolet intensity
   f. supersonic transport flight in the statosphere
   g. the advent of nuclear reactors
   h. our excessive use of Freon-11

3. The concentration of CO_2 in the atmosphere, as measured at Mauna Loa, increased from about _____ ppm in 1958 to about _____ in 2005.
   a. wildly fluctuated, no trend can be observed
   b. actually decreased due to increase in particulates
   c. 315 to 380
   d. 315 to 630
   e. 330 to 430
   f. 100 to 300

4. The reason that skin cancer can be caused by solar ultraviolet radiation but not by the visible light at the earth's surface is _____ .
   a. The wavelength of UV is longer, hence it is more destructive
   b. There is much more UV than visible light at the earth's surface
   c. Once you have a good tan, visible light can do no further harm
   d. UV photons have more energy and hence can ionize biological molecules more readily

5. It is feared that because of the past release of Freons the _____ in the upper atmosphere will continue to decrease, which will increase the _____ at the earth's surface.
   a. CO, ultraviolet
   b. Ozone, ultraviolet
   c. CO_2, temperature
   d. O_3, infrared

Appendix

A.1 Linear Plots, Semilogarithmic Plots, and Exponential Growth

Many relationships in nature and the plots of their time dependence are *linear*. An example would be displayed by plotting the distance a person is away from a given city, if that person is driving away from the city at a constant speed. If the speed is 50 mph, the plot would look like Figure A.1. In this case there is an obvious relationship between the distance, d, and the time, t. This relationship may be represented algebraically as:

$$d = s \times t,$$

where the speed, s, is multiplied by the time. If the time is measured in hours, and the speed is in miles per hour for, instance 50, mph, this equation would be written as:

$$d \text{ (miles)} = 50 \, \frac{\text{miles}}{\text{hour}} \times t \text{ (hours)}.$$

This equation tells us that the amount of distance added in any one hour is a constant, that is, 50 miles.

Many interesting time dependencies are not linear. They call for another method of plotting their graphs. The *semilogarithmic* graph, as used in Figure A.4, is a useful and convenient means of displaying data that are nonlinear and that may vary over a large range. This type of graph reveals details of both small and large magnitudes on the same graph. A second property of these plots is that when the vertical axis is constructed so that distance along it corresponds to the logarithm of a quantity and the horizontal axis is a linear progression of time, truly exponential trends will be exhibited as perfectly straight lines. It is then easy to make a straight line extrapolation of an exponential trend into the future.

It is common to have quantities that behave with time such that the increments are not constant for each unit of passing time, but in fact grow larger as time progresses. For example, the amount of interest earned annually on a bank deposit is the annual interest times the amount in the account. If no money is withdrawn, the interest earned the second year will be greater than was earned the first year. The amount earned is proportional to the amount present, and

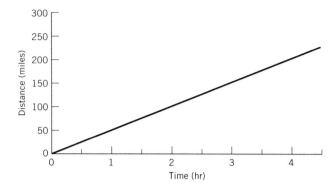

Figure A.1 The distance traveled from some starting point as a function of the time traveled.

there is more present the second year because of the interest added at the end of the first year. The equation that expresses this is

$$\frac{\Delta N}{\Delta t} = (\text{constant}) \times N = \lambda N,$$

where the Greek letter Δ (delta) indicates a change of the variable N, and λ (lambda) (if a positive number) is a growth constant. In words, the equation says that the amount added per unit time is proportional (λ is the proportionality constant) to the amount present. By methods of integral calculus, it can be shown that this equation can be transformed into the more useful form

$$N = N_0 e^{\lambda t},$$

where N_0 is the amount of the variable N at the beginning of the time period under consideration, or at $t = 0$. The letter e is a constant frequently used in calculus; it always has the value of 2.718. A quantity, N, that behaves with time according to this equation is said to grow *exponentially*. A plot of such a variable on a linear graph is shown in Figure A.2.

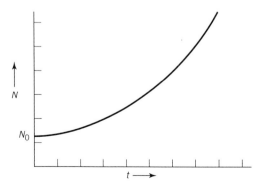

Figure A.2 A growth pattern indicative of exponential growth, or $N = N_0\, e^{\lambda t}$, plotted on a linear graph.

The *logarithm* is the power to which some number must be raised to make it equal to some given number. There are two conventions: one based on the number ten, the other based on the number $e = 2.718$. The first is the common logarithm, abbreviated log, the second is the natural logarithm, abbreviated ln. Because $e^{3.25} = 25.8$, we should say that the natural logarithm of 25.8 is 3.25, or ln 25.8 = 3.25. Similarly $1000 = 10^3$, so log 1000 = 3.

By taking the natural logarithm of both sides of the previous equation, and following the usual rule for the logarithm of a product, that equation can be rewritten in a different way but without changing its meaning:

$$\ln N = \ln N_0 + \lambda t$$

or

$$\ln N - \ln N_0 = \lambda t$$

This last equation shows that ln N is linearly dependent on time. A plot of the equation is shown in Figure A.3. Fortunately, we do not have to get involved with the computation of logarithms either to plot or to understand variables that behave exponentially. If we use a graph that has the vertical axis scaled according to the logarithm, the exponential equation appears as in Figure A.4. It is not necessary to do anything more than plot directly the quantities, N, at their corresponding times, t, and if the quantity is increasing or decreasing exponentially, a straight line will result. Going back to Figure 1.1, it is seen that the total energy consumption line has grown at something close to an exponential rate.

For exponential growth, it is also interesting to consider the *doubling time*, t_D. If we consider a situation in which an exponentially growing quantity doubles, then we could write

$$e^{\lambda t_D} = 2 = e^{0.693}.$$

The last half of this equation merely expresses the fact that ln 2 = 0.693, as can be affirmed by a keystroke on a calculator. Equating the first and third terms in the equation shows that $\lambda t_D = 0.693$. A slight rearrangement gives

$$t_D = \frac{0.693}{\lambda}.$$

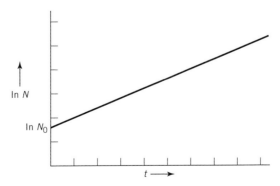

Figure A.3 The equation $\ln N = \ln N_0 + \lambda t$, plotted on a linear graph.

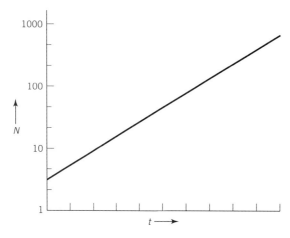

Figure A.4 A semilogarithmic plot of an exponentially increasing variable.

What has been said is that when the exponent of e is equal to 0.693, the quantity N increases by a factor of two. The time period t_D is thus called the doubling time. If the growth constant λ is 10% per year, or 0.10 per year, then:

$$t_D = \frac{0.693}{0.10/\text{yr}} = 6.93 \text{ years.}$$

It is often convenient to use an approximate expression:

$$\text{Doubling time in years} = \frac{70}{\%\ \text{growth per year}},$$

or

$$\%\ \text{growth per year} = \frac{70}{\text{doubling time in years}}.$$

It is not necessary for the times to be given in years; they could just as well be in minutes or any other interval as long as the time units are the same on both sides of the equation.

Population increase is often cited as another example of exponential growth. For example, if at some time the population of a country was 220 million and it increases by 1% per year, for the first year the increase would be 0.01 × 220 million = 2.2 million. The second year it would be 2.22 million, and so forth. It could also be noted that, at this rate, the population would double in 70/1 = 70 years. There is no fundamental reason for the percentage growth rate or the doubling time to be constant for extended periods, as birth and death rates may change for various reasons. Nevertheless, many population groups tend to grow exponentially with an approximately constant value for λ.

The word *exponential* has crept into everyday conversation without exact meaning. It is common to hear of something "increasing exponentially" when in fact it's not, according to the scientific understanding of the term. Exponential does not just mean "very rapidly". For a growth pattern to meet the strict definition of exponential growth, both the doubling time and the rate of increase must be constant over the time period under consideration. These are, of course, equivalent standards. A growth pattern can be tested to see if it is truly exponential by examining its semilogarithmic plot of quantity versus time. If the data fit a straight line, the growth is indeed exponential. For most issues that are dealt with in this book, growth is usually only approximately exponential, but it's a pretty good approximation for many purposes.

A.2 *Fahrenheit, Celsius, and Kelvin Temperature Scales*

In the United States, our everyday temperature scale is the *Fahrenheit* scale. On this scale, water boils at 212 degrees and freezes at 32 degrees. Human body temperature is 98.6, and a normal room temperature is about 70 degrees. Absolute zero is at $-460°F$.

In nearly every other country of the world, the scale in common use is the *Celsius* scale. Sometimes it is known as the *centigrade* scale. Water boils at 100 degrees Celsius and freezes at 0 degrees Celsius. Normal body temperature is 37, and room temperature is about 20 degrees. Absolute zero is at $-273°C$.

The Fahrenheit and Celsius temperature scales serve well in most normal activities, but for many scientific purposes it is necessary for temperatures to be stated in terms of an absolute scale. On such a scale, absolute zero is at zero degrees, a concept that we mention here without going into its precise definition.

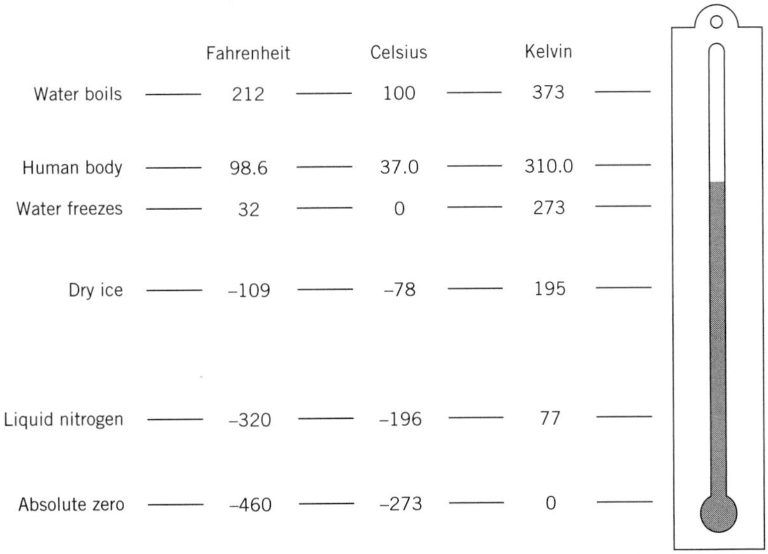

	Fahrenheit	Celsius	Kelvin
Water boils	212	100	373
Human body	98.6	37.0	310.0
Water freezes	32	0	273
Dry ice	−109	−78	195
Liquid nitrogen	−320	−196	77
Absolute zero	−460	−273	0

Figure A.5 The three common temperature scales.

It is related to the idea that at a sufficiently low temperature a gas will exert no pressure on the walls of its container. An absolute temperature scale that has come into common scientific use is the *Kelvin* scale, on which temperatures are expressed in degrees Kelvin (abbreviated K). This scale is similar to the Celsius scale in that there are 100 units between the freezing and boiling points of water. On the Kelvin scale, absolute zero is at 0 K, water freezes at 273 K, and water boils at 373 K at sea level atmospheric pressure. Normal room temperature is just a few degrees below 300 K, and normal human body temperature is about 310 K.

It is a straightforward matter to convert from one temperature scale to another. The relationships can be stated in algebraic terms:

$$C = \frac{5}{9} \times (F - 32)$$

$$F = \frac{9}{5} C + 32$$

$$K = C + 273$$

A comparison of the three temperature scales is shown in Figure A.5.

Answers to Selected End-of-Chapter Problems

1.3 100 foot · pounds; 136 joules
1.4 Heat energy in cart, floor, air, etc., due to friction; Food
1.7 Approximately 12 to 16 tons
1.8 Approximately 5,700 calories or 5.7 Calories. The bicyclist's food energy *input* would have to be very much greater than this.
1.11 3.5 hours
2.4 About 4.5 years
2.7 About 21 dollars per day
2.10 16,800 tons
2.11 About 7.4×10^6 foot · pounds; 0.0016 barrel
2.12 0.1 pound; 0.01 gallon
3.1 20°C; 293 K
3.2 28.5×10^6 Btu
3.4 31% is the Carnot efficiency. This is less than the inventor's claim of 45%.
3.5 36% is the net efficiency for electric heating of water. Direct gas heating has an efficiency of 60%, thereby using less natural gas.
3.10 2.75 ton
3.11 (a) 30,000 Btu
(b) 60%
3.12 $40\% \times 4 = 160\%$ efficiency for the power plant–heat pump combination.
This beats by 60% the 100% efficiency for direct use of the fuel for heating.
3.14 About 50%
3.16 COP = 4
4.9 910 ft^2
4.10 about $5^1/_2$ inches
5.1 426 lb/in^2.
5.2 (a) 7.06×10^{15} joules
(b) 224 MW$_e$
(c) 224,000 people
(d) 98 million dollars

5.3 (a) 265 joules
(b) 251 joules
5.4 78 kW
5.5 (a) 61 W, 484 W, 1636 W
(b) 1, 8, 27
5.6 336×10^6 joules
5.7 38%; 62%
5.8 1.7%
5.10 (a) 7780 m^3/sec
(b) 1,111 m^2, equivalent to a single pipe of 38 meter diameter
5.11 1.3×10^{15} Btu; a little over 1%
6.1 (a) 8.1×10^{13} joules/kg
(b) 5.7×10^{11} joules/kg
6.2 31×10^6 joules/kg
6.3 coal: about 4.6×10^{22} joules; uranium-235: about 0.17×10^{22} joules
6.4 six neutrons
6.8 238 years
6.9 (a) 1,165 tonne/yr of uranium-235
(b) 1.6×10^5 tonnes/yr
(c) 3.1×10^9 tonnes
(d) 28 square miles
6.13 0.005, or 1 chance in 200
6.14 0.31 Curie
6.15 Once every 10,000 years
6.17 0.0008
6.18 5480 bombs
7.1 4930 Btu
7.2 2500 degree days
7.3 18×10^6 Btu
7.5 161×10^6 Btu
7.6 $3220
7.7 $342/yr savings; 2 year payback time suggests it would be a good investment.
7.8 payback time is about 0.8 year
7.9 17%
7.10 3.0×10^6 Btu
7.13 Washer: $1,800, much more than initial appliance cost; dryer; $1,480, much more than

355

Answers to Selected End-of-Chapter Problems

- initial appliance cost
- **7.14** About $7/yr electricity cost
- **7.17** $61 saved
- **8.2** 19 miles per hour
- **9.1** (a) 2940 lb force; (b) 320 lb force
- **9.2** about minus 1 degree Fahrenheit, using an ALR of $-0.65°C/100$ m.
- **9.5** 909 meters
- **9.6** 2.7×10^{25}
- **9.7** 1340 $\mu g/m^3$; 2.7×10^{19} molecules/m^3
- **9.8** 0.128 ppm
- **9.9** (a) 59×10^6 ton/yr coal
  (b) 1.2×10^6 ton/yr SO_2;
  (c) 79,000 tons/yr particulates
  (d) 216×10^6 ton/yr CO_2
- **9.10** 12 $\mu g/m^3$, no
- **9.12** 30 microequivalents per liter
- **10.1** 230 ppm
- **10.2** the year 2033
- **10.6** (a) 350 to 400 billion dollars
  (b) 1.0×10^{14} dollars

INDEX

– A –

Acceleration, definition, *247, 274*
Acceleration force for vehicles, *247*
Accidental deaths, all causes, 273
Acid rain, *301, 308, 315–319*
Adiabatic lapse rate, *290–293*
Aerodynamic drag:
    coefficients, *249*
        force for vehicles, *248, 249, 253, 254*
Aerosols, *304, 308, 310, 311, 329, 338*
Air bags, for automobiles, *275, 276*
Air infiltration of buildings, *219–221, 226*
Air, national quality standards, *296, 297, 312, 313*
Alpha decay, *175–176*
Alpha particle, *176, 197–199*
Alkane, hydrocarbon series, *43*
Alternative sources of energy, Chapter 5
Amorphous silicon, *116*
Anaerobic fermentation, *154–155*
Annual Fuel Utilization Efficiency (AFUE), *223*
Anthracite coal, *50, 51, 310*
Antineutrino, *175*
Appliances:
    home, *44, 227–231*
        energy requirements, *213–229*
Athabasca tar sand deposit, *57*
Atmosphere, of Earth:
    constituents, *288–289*
    pollution of, *287–321*
    pressure, *288*
ATP, *146*
Automobile:
    battery-powered, *256–258*
    flywheel powered, *258–262*
    hybrid, *262–264*
    see also Motor Vehicle
Automobile emissions, *295, 296, 298–307*
Avogadro's number, *296, 297*

– B –

Batteries, storage:
    automotive use, *256–258*
    general, *9, 11*
    energy density, *261*
    various types, *258*
Bay of Fundy tidal basin, *164*
Becquerel, definition, *190*
Beta decay, *175–176*
Beta particles, *175*
Binding energy, nuclear, *196*
Biodiesel, *156*
Biomass:
    as fuel, *149–157*
    production, *147–150*
Bitumen, *56–57*
Bituminous coal, *51, 310*
Boiling water reactor (BWR), *179–181*
Boron control rods, reactor, 179
Breeder reactor, *172, 184–185, 191*
British thermal unit (Btu), definition, *13–14*
Butane, bottled gas, *32, 43*

– C –

Calorie:
    definition, *13*
    food, *13*
Carbohydrates, *32, 146–148*
Carbon, *14, 176*

Carbon dioxide:
  atmospheric concentration, *288, 332–336*
  effects on climate, *4, 100, 332–345*
  sources, *32, 52, 146, 228, 320, 327, 332*
  sequestration, *341*
Carbon monoxide:
  air quality standards, *296, 297*
  concentrations in Denver, *296, 298, 299*
  presence in homes, *220*
  sources, *294–300, 305*
Carbonic acid, *316*
Carboxyhemoglobin, *295*
Carnot, Sadi, *65*
Carnot efficiency, heat engines, *65–67, 69, 76, 80, 105, 106, 141, 177, 252, 270*
Carter, President James, *182, 192*
Catalytic converter, automotive, *306*
Celsius temperature scale, *353–354*
Cesium-137:
  half-life, *176*
  presence in radioactive fall-out, *190*
  radioactive decay, *175–176*
Chernobyl reactor accident, *189–190*
China syndrome, *186*
Chlorofluorocarbons (CFC):
  greenhouse gases, *332–333, 341*
  ozone depletion, *328–331*
Chlorophyll, *146*
Classic smog, *291–308*
Clathrate, see natural gas hydrate, *48*
Claude, Georges, *143*
Clean Air Act, *302, 306, 310, 318, 319, 321*
Clinch River breeder reactor, *185, 192*
Coal:
  consumption, *52–53*
  formation, *50*
  liquefaction, *310*
  reserves, *50–53*
  types, *51*
  world reserves, *52*
Coalbed methane, *48*
Coefficient of performance, heat pumps, *80–81*
Cogeneration of energy, *82–84, 235*
Concentration ratio, solar, *107–108*
Conservation of energy, see Energy, conservation of
Control rods, nuclear reactor, *179–181*
Convection, transfer of heat energy, *103, 104*
Copenhagen amendments, *330–331*
Cost of electric energy, *44, 110, 116, 117–118, 161, 183, 195–196, 228–231*
Critical mass, definition, *174*
Curie, definition, *190*

– D –

Dam failures, *132*
Darrieus rotor, *134, 135*
Decay heat, reactor, *186*
Degree-days:
  definition, *214*
  for various locations, *215*
Delayed neutrons, *179, 189*
Deuterium (deuteron) definition, *196–197*
Deuteron-triton (D-T) reaction, *197, 199–200*
Deuteron-deuteron (D-D) reaction, *197, 199–200*
Diesel engine, *76–78*
Dobson units for ozone concentration, *326*
Double-paned windows, *97, 216–217, 219, 224*
Drag coefficients, aerodynamic, *248–249*

– E –

Einstein, Albert, *9–10, 112, 173, 196*
Electric vehicles, *256–258*
Electricity, cost of, see Cost of electric energy

Electricity, generation of:
    basic process, *24, 67–70*
    by alternative sources, *124–165*
    by direct solar, *111–118*
    by fast breeder reactor, *185*
    by fossil fuels, *2, 20, 24, 46, 52, 68–69, 307, 342*
    by fusion reactor, *202*
    hydroelectric, *125–132*
    by solar thermal, *105–111*
    by steam turbine, *46, 69, 73–74*
    by thermal fission reactor, *172, 174–175, 177–181*
Electrolux process, *119*
Electromagnetic radiation, *11*
Electron volt (eV), *14*
Emergency core cooling system (ECCS), nuclear reactor, *186, 187*
Energy:
    chemical, *9*
    conservation of, *210–239*
    definition, *7*
    density, *256, 260, 261*
    electric, *10–11*
    heat, *9*
    kinetic, *10*
    mass, *9–10*
    potential, *10*
    rotational, *258–259*
    units, *12–14*
    United States, sources in, *2–3*
    various forms of, *9–12*
Energy audit, *223–225*
Energy conservation:
    in homes, *210–235*
    in industry and agriculture, *235–238*
    in transportation, *252–254*
Energy consumption, projections for United States, *211*
Energy efficiency ratio (EER), *229*
Energy requirements, vehicles, *251–254*
Energy storage
    batteries, *256*
    biomass, *147–150*
    flywheels, *259–261*
    hydrogen, *264–267*
Ethane, *43*
Ethanol, *153–154, 208, 271*
Event tree, *186*
Exponential function and growth, *349–353*
External combustion engine, *74*

– F –

Fahrenheit temperature scale, *353–354*
Faraday, Michael, *67–68*
Fast breeder reactors, *184–185*
Fermi, Enrico, *174*
Fireplaces, *221–223, 223–224*
Fission, nuclear reactors, *172–175, 177–183*
Flat-plate collectors, solar, *97–101, 107*
Flywheel, energy storage by:
    automotive, *259–261*
    kinetic energy, *16, 258–262*
Focal point, solar collector, *106, 107*
Food calorie, definition, *13*
Foot-pound, definition, *13–14*
Force:
    automotive requirements, *247–249*
    definition and units, *8*
Fossil fuels:
    in agriculture, *153–154, 156–157*
    see also Coal, Natural gas, Petroleum, Tar sands, Oil shale, Natural gas hydrates
Fractional distillation, *41–43*
Freon-11, Freon-12, *328, 330, 331, 334*
Fresnel lens, *107*
Fuel:
    automotive consumption, *246, 252–253, 254*
    consumption in United States, *2–3, 31*
    from biomass, *149–156*
    efficiency, automotive, *254–255*

Fuel (*Continued*)
    cycle, nuclear reactor, *181–183*
    rods, nuclear reactor, *179–185, 191–195*
Fuel cells:
    use with hydrogen, *266–267*
    use with methanol, *270–271*
Furnaces, for space heating, *44, 221–223*
Fusion, nuclear, *196–198*
Fusion reactors, nuclear, *198–204*

– G –

Gas turbines, *46, 78–79, 83–84*
Gasohol, production, *150, 153*
Gasoline engine, *74–76*
General Circulation Models, *340*
Generation of electricity, *see* Electricity, generation of
Geothermal energy, *157–162*
Geysers field, California, *157–158, 159–160*
Global warming, *53, 100, 329, 332–334, 338–340*
Glucose, *146, 148, 149*
Gram molecular weight, definition, *146*
Gravity, acceleration of, *127, 247*
Green River Formation (oil shale), *54, 55*
Greenhouse effect, *100, 332–344*
Gross Domestic Product, *4, 5, 19*

– H –

Haagen-Smit, A. J., *301*
Half-life, radioactive, *175–177*
Halley Bay, Antarctica, *327–328*
Heat conduction, *213–219*
Heat energy storage, *see* Thermal mass
Heat engines, *65–67, 72–79*
Heat exchangers, *81, 99, 100, 142, 220, 221*

Heat of combustion, *65*
Heat pumps, *79–82*
Heliostat, *107, 108, 109*
Heptane, *43, 64–65*
Hill-climbing force for vehicles, *247–248*
Home heating, standards, *225–226, 228*
Horsepower:
    automotive, *249–251*
    definition, *12*
    per capita in United States, *5, 6*
Hubbert, M. King, *4, 36, 37, 39*
Humidity, *220–221*
Hybrid vehicles, *262–264*
Hydrocarbons:
    atmosphere, sources in, *295, 301–302*
    in photochemical smog, *301–302*
Hydroelectric power
    developed capacity, *130–131*
    potential, *130–131*
    pumped-storage, *163*
Hydrofluorocarbons, *331*
Hydrogen:
    explosive properties, *267–268*
    liquid, as fuel, *265*
    production, *264–265*
    as secondary fuel, *264*
Hydrologic cycle, *126–127*

– I –

Ideal engines, *65–67, 80*
Industrial energy conservation, *235–236*
Infrared, *145, 332*
Insolation, *94–95*
Insulation, thermal, *213–219*
Intergovernmental Panel on Climate Change, *334, 336, 337, 338, 340, 341, 342*
International Thermonuclear Experimental Reactor (ITER), *202–203*

Internal combustion engines, *74*
Isotopes, *174*

– **J** –

Joule, definition, *8, 13*

– **K** –

Kelvin temperature scale, *353–354*
Kerogen, *55–56*
Kerosene, *32, 33, 41, 42, 57*
Kyoto Protocol-1997, *343*

– **L** –

Lasers:
    in fusion reactors, *202–203*
    in uranium isotope separation, *191–192*
Lawson criterion, fusion reaction in plasma, *202*
Lead-Acid battery, *256–258, 261, 264*
Lighting:
    for buildings, *231*
    efficiencies, *231–233*
Lignite, form of coal, *50–51, 310*
Lithium:
    use in fusion reactors, *199, 201*
    resources, *200*
    tritium production, *199*
Load-leveling, electric power plants, *128, 163, 256*
Los Angeles smog, *291, 292, 294, 300, 301–304*
Loss-of-cooling accident (LOCA), nuclear reactor, *186, 187*

– **M** –

Manhattan Project, *174*
Marlstone, *55*
Mass energy, *8–10*
Mass transportation, *279–282*

Mauna Loa, Hawaii, CO_2 measurements, *335*
Mesa Verde, *101*
Methanol:
    as motor fuel, *268–269*
    in fuel cells, *270–271*
Methane:
    production from biomass, *154–155*
    see also Natural Gas
Methane hydrate, *see* natural gas hydrate
Moderators, neutron, *177–178, 189*
Mole, definition, *146*
Molecular weight, *146*
Montreal Protocol, Copenhagen Amendments, *330–331*
Motor vehicle:
    deaths, *270–274*
    energy requirements, *247, 251–252*
    force requirements, *247–249*
    power requirements, *249–251*
    see also Automobiles
Mount Pinatubo volcano, *311, 338*
Municipal solid waste, *151–153*

– **N** –

National Acid Precipitation Assessment Program (NAPAP), *318*
National Ambient Air Quality Standard (NAAQS), *296–299*
Natural gas:
    constituents, *43*
    consumption, *16–19, 22, 47*
    hydrates, *48–49*
    resources, *47–49*
    see also Methane
Neutrino, *197, 198*
Neutron:
    emission in fission, *173–174, 178–179*
    induced fission, *173–174, 178*
    mass, *172*

Newton, definition, *7*
Nitrogen oxides:
    and air quality standards, *297, 300*
    properties, *300–301*
    in photochemical smog, *301–304*
    sources, *295–305*
Nitrous oxide, *332–334*
Nonproliferation treaty, nuclear, *191–192*
Nonrenewable energy, *25, 31, 33*
N-type silicon, *113–114*
Nuclear fission, *173–174*
Nuclear fusion, *196–204*
Nuclear reactors:
    fast breeder, *184–185*
    fusion, *200–204*
    thermal, *177–181*
Nuclear Regulatory Commission (NRC), *187, 188, 195*
Nuclear weapons, *190–192*
NUREG-1150, NRC study, *187–189*

– O –

Ocean thermal energy conversion (OTEC), *138–145*
Octane, *41–43*
Oil, *see* Petroleum
Oil shale:
    formation, *54*
    processing, *54–56*
    resources, *54–55*
Organization of Petroleum Exporting Countries (OPEC), *38, 39, 40*
Oxygenated fuels, *153, 298, 306*
Ozone:
    presence in troposphere, *300–303*
    depletion in stratosphere, *327–332*

– P –

Paraboloidal reflectors, solar, *106–108, 110, 111, 117*

Particulates as pollutants:
    effects, *310–313*
    particle size, *311, 313, 314*
    power plant removal, *313–315*
    sources, *295, 310–312*
    standards, *297, 313*
Passenger efficiency, *245–247, 280*
Passive solar systems, *101–105*
Passive stability, for reactor, *195–196*
Payback time, *218*
Peroxyacyl nitrates (PAN), *302*
Petroleum:
    consumption, *16–20, 22*
    definition, *32*
    enhanced recovery, *37–38*
    formation, *2*
    primary recovery, *36–37*
    refining, *40–43*
    resources, *35–37, 38, 39*
    secondary recovery, *37*
    tertiary recovery, *37–38*
    world production, *39*
pH:
    of surface waters, *316–318*
    scale of acidity, 316
Photochemical reactions, *301, 302, 304*
Photochemical smog, *292, 301–304*
Photoelectric effect, *112*
Photon, *112, 114, 145*
Photosynthesis, *24, 145–148*
Photovoltaic solar cells:
    applications, *116–118*
    cost, *115–118*
    theory, *112–114*
Pinatubo, Mt., volcano, *311, 338*
Plant production, *145–150*
Plasma, fusion reactor, *200–202*
Plug-in hybrid, *363–364*
Plutonium
    alpha decay of, *176*
    in nuclear reactors, *181–183, 184–185*
    in nuclear weapons, *190, 192*
P-n junction, solar cells, *113–114*
Population, United States, *18*

Powder River basin, *310*
Power, definition, *11–12*
Power tower, solar, *108–110*
Power requirements, motor vehicles, *249–251*
Pressurized water reactor (PWR), *179*
Principle of energy conservation, *22–23*
Prompt criticality in reactors, *190*
Propane, bottled gas, *32, 43, 255*
Proton, properties, *172, 174, 198*
Proton-proton cycle in sun and stars, *198*
P-type silicon, solar cells, *113–114*
Proved reserve, *33, 36, 38, 39, 47, 48, 49, 52, 53, 58*
Prudhoe Bay, Alaska, *34*
Public Utilities Regulatory Act (PURPA), *84, 118*
Pumped hydrostorage, *163*

– Q –

Quad (QBtu):
    definition, *16*
    energy equivalents, *16*

– R –

Radiative thermal inversion, *293, 294*
Radioactive waste:
    national inventory, *193*
    reactor inventory, *193–194*
    storage, *192–195*
Radioactivity, *175–177*
Radon, *219*
Rainwater, unpolluted pH, *316*
Rance River, St. Malo, France, tidal power plant, *163*
Rasmussen report (WASH-1400), *187*
Reactors, nuclear:
    BWR, *179–181*
    fast breeder, *181*
    fusion, *200–204*
    plants in United States, *172*
    PWR, *179*

Recycling, for energy conservation, *152–153, 236–237*
Renewable energy, *90–91*
Respiration of plants, *146, 148, 336*
Retorting, *55*
RMBK, Soviet reactors, *189–190*
Rolling resistance for motor vehicles, *248, 252*
R-value:
    definition and units, *215*
    composite layered wall, *216–219*
    recommended values, *219*
    for representative materials, *217*

– S –

Sandia laboratory, *108*
Savery, *72*
Savonius rotor, *134–135*
Scientific notation, *14–15*
Seasonal energy efficiency ratio (SEER), *229*
Seat belts, for automobiles, *275–276*
Semiconductors, *112–115*
Semi-logarithmic plots, *3, 349, 352*
Sequestration, carbon dioxide, *341*
Shale oil, *54–56*
Silicon, use in solar cells, *112–116*
Smog:
    classic, *291, 308*
    photochemical, *292, 301–304*
Sodium (liquid), reactor coolant, *185*
Solar cells, *112–119*
Solar constant, *92–94*
Solar cooling, *119–120*
Solar energy, active system, *97–101*
Solar energy, basic concepts, *92–97*
Solar energy, passive system, *101–105*
Solar furnace (Odeillo, France), *107*
Solar space heating, *97–105, 223–225*
Solar spectrum, *11, 92–93, 99, 147–148*
Solar thermal electricity, *105–111*
Space heating:
    of buildings, *212–226*
    solar, *97–115, 223–225*

Spaceship Earth, *287–288*
Stefan's law, *97–98*
Stoves, heating, *223–225*
Subbituminous, form of coal, *50–51*
Sucrose, *146*
Sulfur oxides:
    in the atmosphere, *307–310*
    air quality standards, *297–310*
    effects, *291–308*
    sources, *46, 51, 295, 307–310*

– T –

Tar sands, *25, 32, 36, 54, 56–57*
Temperature scales, *67, 353–354*
Terminal velocity, *283*
Thermal conduction, *213–219*
Thermal conductivity, *215–218*
Thermal cracking, *41, 43*
Thermal insulation of buildings, *102, 104, 213–219*
Thermal inversion:
    general, *290–294*
    high-pressure subsidence, *292, 293*
    radiative, *293–294*
Thermal mass, *102–104*
Thermal neutrons, *177–179, 181, 185*
Thermoelectricity, *111*
Thermostats, settings for residences, *222–223, 227*
Three Mile Island, reactor accident, *186*
Tidal energy, *25, 26, 90, 91, 157, 162–164*
Tokamak, *200, 201, 202*
Traffic deaths:
    general, *271, 273, 278–279*
    by driver age, *272–273*
    in the United States, *271, 273, 278–279*
    per vehicle mile, *274*
    by type of accident, *274–275*
Traffic safety, *271–279*

Transportation
    fuel consumption in United States, *20, 245–247*
    mass, *279–282*
    passenger, *246, 279*
Tritium:
    formation in ^{6}Li reaction, *199*
    reactions, *198*
    released from nuclear reactors, *185*
Trombe wall, *103, 104*
Troposphere, *288–290, 327*

– U –

Ultraviolet, *91–93, 145, 301, 327–328, 330*
Units, conversion, inside front cover, *7–8, 12–14*
Uranium:
    critical mass, *174*
    general properties and use in reactors, *177–183*
    isotopes, *174*
    resources, *183–185*

– V –

Vehicle:
    energy requirements, *251–254*
    force requirements, *247–249*
    power requirements, *249–251*
    see also Automobile, Motor vehicle

– W –

Wash-1400, (Rasmussen report), *187*
Waste heat recovery, *79, 82–84, 235*
Water heaters:
    and energy conservation, *44, 227–228*
    solar, *99–101*

Watt, definition, *12*
Watt, James, *72*
Wavelength, definition, *11*
Weapons, nuclear, *9, 172–175, 190–192, 197*
Weapons proliferation, *183–185, 190–192*
Wien displacement law, *98–100*
Windpower, *132–138*

Windows, heat loss, *102–104, 214, 216–218, 219, 226*
Work, definition, *7–8*

– Y –

Yellowcake, *183*
Yucca Mountain, Nevada, *183, 192–195*

FUNDAMENTAL CONSTANTS

Gravitational acceleration on earth (g)	9.81 m/sec^2, 32.2 ft/sec^2
Volume of mole of ideal gas at STP	22.4 liters
Avogadro's number	6.023×10^{23}
Atomic mass unit (amu)	1.66×10^{-27} kg
Boltzmann's constant (k)	1.38×10^{-23} J/K
Wien's constant	2.898×10^{-3} m · K
Speed of light (c)	3.00×10^{8} m/sec
Charge of the electron (qe)	1.60×10^{-19} C
Planck's constant (h)	6.63×10^{-34} J · sec
Mass energy of 1 amu	931 MeV
Stefan–Boltzmann constant (σ)	5.67×10^{-8} W/K^4 · m^2
Proton mass	1.673×10^{-27} kg
Neutron mass	1.675×10^{-27} kg
Deuteron mass	3.34×10^{-27} kg
Alpha particle mass	6.64×10^{-27} kg
Electron mass	9.109×10^{-31} kg
Pi (π)	3.14159
Base of natural logarithms (e)	2.71828

USEFUL FORMULAS FROM GEOMETRY

Circle of radius r
   Circumference $2\pi r$
   Area πr^2

Sphere of radius r
   Surface area $4\pi r^2$
   Volume $\frac{4}{3}\pi r^3$